Scientific Imperialism

The growing body of research on interdisciplinarity has encouraged a more in-depth analysis of the relations that hold among academic disciplines. In particular, the incursion of one scientific discipline into another discipline's traditional domain, also known as scientific imperialism, has been a matter of increasing debate.

Following this trend, *Scientific Imperialism* aims to bring together philosophers of science and historians of science interested in the topic of scientific imperialism and, in particular, interested in the conceptual clarification, empirical identification, and normative assessment of the idea of scientific imperialism. Thus, this innovative volume has two main goals. Indeed, the authors first seek to understand interdisciplinary relations emerging from the incursion of one scientific discipline into one or more other disciplines, such as in cases in which the conventions and procedures of one discipline or field are imposed on other fields; or more weakly when a scientific discipline seeks to explain phenomena that are traditionally considered proper of another discipline's domain. Second, the authors explore ways of distinguishing imperialistic from non-imperialistic interactions between disciplines and research fields.

The first sustained study of scientific imperialism, this volume will appeal to postgraduate students and postdoctoral researchers interested in fields such as Science and Technology Studies, Sociology of Science & Technology, Philosophy of Science, and History of Science.

Uskali Mäki is an Academy Professor at the University of Helsinki, Finland and Director of the Academy of Finland Centre of Excellence in the Philosophy of the Social Sciences, Finland.

Adrian Walsh is Professor of Philosophy at the University of New England, Australia.

Manuela Fernández Pinto is an Assistant Professor in the Department of Philosophy at Universidad de los Andes, Colombia.

Routledge Studies in Science, Technology and Society

https://www.routledge.com/Routledge-Studies-in-Science-Technology-and-Society/book-series/SE0054

29 The Fukushima Effect
A New Geopolitical Terrain
Edited by Richard Hindmarsh and Rebecca Priestley

30 Digitizing Identities
Doing Identity in a Networked World
Edited by Irma van der Ploeg and Jason Pridmore

31 The Globalization of American Infrastructure
The Shipping Container and Freight Transportation
Matthew Heins

32 Science, Risk, and Policy
Andrew J. Knight

33 The Ethics of Ordinary Technology
Michel Puech

34 Imagined Futures in Science, Technology and Society
Edited by Gert Verschraegen, Frédéric Vandermoere, Luc Braeckmans and Barbara Segaert

35 Adolescents and Their Facebook Narratives
A Digital Coming of Age
Jill Walsh

36 Scientific Imperialism
Exploring the Boundaries of Interdisciplinarity
Edited by Uskali Mäki, Adrian Walsh and Manuela Fernández Pinto

Scientific Imperialism

Exploring the Boundaries of Interdisciplinarity

Edited by
Uskali Mäki, Adrian Walsh and Manuela Fernández Pinto

Routledge
Taylor & Francis Group
LONDON AND NEW YORK

First published 2018
by Routledge
2 Park Square, Milton Park, Abingdon, Oxon OX14 4RN

and by Routledge
711 Third Avenue, New York, NY 10017

Routledge is an imprint of the Taylor & Francis Group, an informa business

British Library Cataloguing in Publication Data
A catalogue record for this book is available from the British Library

Library of Congress Cataloging in Publication Data
A catalog record for this book has been requested

ISBN: 978-1-138-05934-4 (hbk)
ISBN: 978-1-315-16367-3 (ebk)

Typeset in Bembo
by Taylor & Francis Books

Contents

List of figures

Acknowledgments

The editors would like to thank the Academy of Finland Centre of Excellence in the Philosophy of the Social Sciences at the University of Helsinki for its intellectual, material, and financial support of this project, particularly for hosting the Scientific Imperialism Workshop held at the University of Helsinki in April 2015, from which most of the chapters in this volume stem. Special thanks to TINT's research assistants: Sofia Blanco Sequeiros, Jesse Kuokkanen, Joonas Ottman, Juho Pääkkönen, and Pekka Tolvanen. Also, many thanks to Gabriel Restrepo, research assistant in the Department of Philosophy at Universidad de los Andes, who provided invaluable help with editing and formatting the manuscript. Finally, we are also very grateful to the authors for their contributions and reviews. We feel extremely lucky to have had the opportunity to work with such an outstanding and diverse group of scholars.

Contributors

S.M. Amadae, University of Helsinki, Finland. Research Fellow, Academy of Finland Centre of Excellence in the Philosophy of Social Science. Research Affiliate, Program in Science, Technology, and Society, Massachusetts Institute of Technology.

Sandy Boucher, University of New England, Australia. Lecturer, School of Humanities.

Cléo Chassonnery-Zaïgouche, University of Lausanne, Switzerland. Postdoctoral Researcher, Centre Walras-Pareto.

Stephen M. Downes, University of Utah, United States. Professor, Department of Philosophy.

Manuela Fernández Pinto, Universidad de los Andes, Colombia. Assistant Professor, Department of Philosophy and Center of Applied Ethics. Affiliated Researcher at the Academy of Finland Centre of Excellence in the Philosophy of the Social Sciences.

Roberto Fumagalli, King's College London, United Kingdom. Lecturer, Department of Political Economy. Research Visitor, London School of Economics and Political Science. Visiting Scholar, University of Pennsylvania.

Nicolas Guilhot, New York University, United States. Research Professor, Center for International Research in the Humanities and Social Sciences.

Genco Guralp, Johns Hopkins University, United States. Graduate Student, Department of Philosophy.

Frederique Janssen-Lauret, University of Manchester, United Kingdom. Tenure-Track Research Fellow, Department of Philosophy.

Robert Lepenies, Freie Universität Berlin, Germany. Visiting Researcher, Philosophy Institute. Visiting Researcher, Wissenschaftszentrum Berlin.

Miles MacLeod, University of Twente, Netherlands. Assistant Professor, Department of Philosophy. Affiliated Researcher at the Academy of Finland Centre of Excellence in the Philosophy of the Social Sciences.

Uskali Mäki, University of Helsinki, Finland. Academy Professor, Academy of Finland. Professor of Practical Philosophy. Director of the Academy of Finland Centre of Excellence in the Philosophy of the Social Sciences.

Magdalena Małecka, University of Helsinki, Finland. Postdoctoral Researcher, Academy of Finland Centre of Excellence in the Philosophy of the Social Science.

Alain Marciano, Université de Montpellier, France. Maître de Conférences HDR, School of Economics. LAMETA, Laboratoire Montpellierain d'Economie Théorique et Appliqué.

Patricia Marino, University of Waterloo, Canada. Professor, Philosophy Department.

Steven G. Medema, University of Colorado Denver, United States. Distinguished Professor of Economics.

Kristina Rolin, University of Helsinki, Finland. Fellow at the Helsinki Collegium for Advanced Studies. Affiliated Researcher at the Academy of Finland Centre of Excellence in the Philosophy of the Social Sciences.

Mikko Salmela, University of Helsinki, Finland. Senior Researcher. Academy of Finland Centre of Excellence in the Philosophy of Social Science.

Adrian Walsh, University of New England, Australia. Professor of Philosophy, School of Humanities. Affiliated Researcher at the Academy of Finland Centre of Excellence in the Philosophy of the Social Sciences.

Introduction

Core issues in scientific imperialism

Uskali Mäki, Adrian Walsh and Manuela Fernández Pinto

Scientific imperialism inspires a number of compelling and stimulating questions for philosophers of science, and science studies scholars more generally. Understood as a type of interdisciplinary relation in which one scientific discipline occupies or enters into another discipline's domain, scientific imperialism puts into question at least two central tenets of current scientific practice. First, by uncovering cases in which the expansion of one scientific discipline into another does not necessarily contribute to epistemic success, scientific imperialism challenges the traditional view that the broader the scope of a scientific theory, the better. In other words, it questions whether the scope itself is an epistemic virtue or not. Second, the study of scientific imperialism also explores and contests what have been taken as the indisputable benefits of fomenting interdisciplinary relations within the academy. On the contrary, the study of scientific imperialism calls for a more sophisticated and detailed analysis of interdisciplinary relations, and aims to achieve a more nuanced account of the pros and cons, the benefits and harms, of interdisciplinary interactions. This we take to be desirable. In particular, a focus on cases of scientific imperialism would contribute to a better understanding of the boundaries of interdisciplinarity.

Currently, the growing research and emerging debates over scientific imperialism are placing the topic on the agenda of philosophy of science (see e.g., Mäki 2013; Clarke and Walsh 2009, 2013; Fernández Pinto 2016). Following this trend, the present volume aims at bringing together philosophers of science and historians of science interested in the topic of scientific imperialism and wishing to contribute to its conceptual clarification, empirical identification and examination, as well as its normative assessment.

The volume has two main goals. First, to understand the nature of the interdisciplinary relations that emerge from the incursion of one scientific discipline into one or more other disciplines, such as when the conventions and procedures of one discipline or field are imposed on other fields, or more weakly when a scientific discipline seeks to explain phenomena that are traditionally considered properly part of another discipline's domain. Second, to explore ways of distinguishing imperialistic from non-imperialistic interactions between disciplines and research fields.

Although scientific imperialism is a complex phenomenon, occurring in a broad variety of types and their instances, this volume explores a set of cases of intra-academic interdisciplinary (or inter-field) versions of intellectual imperialism. By privileging cases of intra-academic imperialism, we do not want to suggest that there are not significant cases of extra-academic imperialism, or other cases for that matter, for example, cases where academic disciplines seek to explain phenomena that have been traditionally in the domain of culture, religion, or other non-academic approaches to understanding the world around us. Instead, a choice was made to maintain, particularly with respect to the interdisciplinary dimensions of scientific imperialism, coherence among the papers and to make a focused contribution to its philosophical analysis.

In understanding scientific imperialism as a type of interdisciplinary relation, we are placing the discussion within the broader debate over interdisciplinarity. Studies of interdisciplinarity are broad and varied, and have certainly achieved great popularity in recent decades, particularly in fields such as science education, research management, and scientometrics. The philosophy of interdisciplinarity, however, is still a nascent subfield, and much work is yet to be undertaken on the conceptual analysis and the normative assessment of issues regarding interdisciplinarity. A clear taxonomy of interdisciplinary relations, for example, is still missing.

However, if one considers the variety of interdisciplinary relations as a spectrum ranging from insubstantial interrelations among disciplines, in which disciplines remain intact, to more significant interactions, in which disciplines can be altered in different degrees, all the way to the most radical cases of interdisciplinarity, in which disciplinary boundaries are completely erased, then cases of scientific imperialism would fit somewhere in the middle of the spectrum (Mäki, forthcoming). These are cases where a substantial interaction is taking place between two disciplines, where the disciplines do not remain intact (serious epistemological, and sometimes even institutional, changes are taking place), but where disciplinary boundaries remain, even if moved. Given our interest in intra-academic imperialism, these are also not cases of *trans-disciplinarity*, i.e., when the interaction involves social groups external to the academy and the interchange moves beyond academic settings. In this sense, the volume focuses on a type of cross-disciplinary relation, one in which the interaction between disciplines is asymmetrical (in contrast to an egalitarian collaboration among disciplines).

To use the theoretical frameworks of interdisciplinarity entails the assumption of disciplinarity, or in other words, to talk about interdisciplinary relations presupposes the existence of academic disciplines (Mäki 2016, 331). The institutional organization of knowledge production in scientific disciplines is fairly recent, taking place around 1800, when the use of abstractions (concepts, theories, instruments) began to dominate scientific inquiry, while a growing number of objects of study led the way towards more and more specialization (Weingart 2010, 5–6). Disciplinary differentiation among the social sciences is an even more recent phenomenon occurring at the turn of the 20th century, led by an ideal of social improvement through rigorous empirical inquiry, which brought about the consolidation of new scientific communities (Ross 2003). In addition, if we take into account the rapid expansion of university education in the past half-century, the unprecedented and rather novel character of the present state of academic disciplines becomes even more salient (Turner 2000, 62). In this sense, the study of interdisciplinary relations, including the study of scientific imperialism, is a timely and compelling endeavor.

1 Core issues

The papers presented in this volume touch on a number of different aspects and a variety of case studies regarding both the practices of scientific imperialism and the theories associated with it. However, as it turns out they are also conveniently related to each other, raising similar questions and circling around similar issues. We have identified five core aspects of scientific imperialism as the common threads among the papers, these being: (i) definition; (ii) categorization; (iii) mechanism; (iv) identification; and (v) assessment/evaluation. In what follows, we summarize these core aspects and highlight a series of open questions and puzzles that the papers aim to address in one way or another.

1.1 Definition

Not surprisingly, the issues of how to define "scientific imperialism" and how to distinguish it from other cases of interdisciplinary interaction such as "expansionism," "colonialism," "collaboration," etc., are central to many of the papers in this volume. Here we highlight some of the main debates on this topic.

The central debate concerning the definition of "scientific imperialism" is about whether it should be defined as a moral or political term at the outset or not. Should it be regarded as a positive or a normative term? On the one hand, some philosophers, possibly following John Dupré (1995), want to distinguish between *disciplinary expansion* and *scientific imperialism* where the latter is a pejorative term. On the other hand, others, following Uskali Mäki (2009, 2013), want to define it neutrally and then talk about good or bad imperialism. The chapters in this volume suggest a distinction between a *normative definition* of scientific imperialism (following Dupré), which maintains the strength of the political metaphor and takes scientific imperialism as a pejorative term, and a *descriptive definition* (following Mäki) of scientific imperialism, which is neutral with respect to the appropriateness of the imperialistic move.

A second important definitional matter concerns the use of the political metaphor of imperialism. Dupré (1995), as others before him (e.g., Stigler 1984), relies heavily on the metaphorical aspect of imperialism. Can we shed any light on the question about the proper boundaries between the disciplines through the use of the political metaphor of imperialism? The political connotations remain in the background even under the descriptive definition of scientific imperialism. Accordingly, one goal is to be explicit and systematic about the contents and use of the metaphors, acknowledging that metaphors can also be useful for some purposes.

We should also ask whether it is scientific imperialism, science imperialism, or scientism that is at issue. Even if we keep in mind our interest in intra-academic, interdisciplinary (or inter-field) versions of intellectual imperialism, we can raise the question of what it is that is being colonized: Is one science colonizing another science? Or is science colonizing another area of academic inquiry? Or is Science (with a capital S) colonizing the academy? Another important question concerns the proper unit of analysis. Is the discipline the field of inquiry, a research program, or a theory? The choice of unit should depend on the nature of the particular case at hand and on the particular questions we ask.

1.2 Categorization

Similar to the problem of defining "scientific imperialism," the issue of *typification* or *categorization* of different cases of scientific imperialism is a common theme among the papers of this volume. What kinds or types of scientific imperialism are there? How can we divide the concept of scientific imperialism into relevant categories for analysis? Some of the categorization issues that have emerged are the following.

First, who are the agents conducting the imperialism? While in some cases the imperialism is undertaken by members of the discipline that is expanding, in some other cases members of a discipline import methods from other disciplines. In this sense, we can distinguish between *externally driven imperialism*, i.e., when members of the imperializing discipline are responsible for the expansion, and *internally driven imperialism*, i.e., when members of the target discipline import the foreign theory or methodology. In Chapter 5, Medema illustrates the latter case through an analysis of the law and economics movement.

Second, how deep or concerted is the attempt at imperializing another discipline? How to measure or provide evidence for disciplinary imperialism? How can we tell that the imperializing discipline is really changing scientific practice? Here we can also make a relevant distinction between *domain-only imperialism*, where the expansionist discipline attempts to explain phenomena in the domain of another discipline, without interfering with the practices and conventions of the discipline itself, and *disciplinary imperialism* where the expansionist discipline intrudes in the scientific practices of the discipline and not merely in its explanatory domain (see e.g., Mäki 2009, 2013).

Finally, what happens to the colonized discipline after the expansion? Could there be radical variability in outcomes of expansion? These are questions about how the colonized discipline is treated. Are the views held within the colonized discipline integrated into the intruding or emerging discipline or are they regarded as erroneous or mistaken in some way? Here we can distinguish between a variety of possible outcomes, ranging from total elimination or reduction, to being sidestepped, marginalized, shrunk in size or status, all the way to the acceptance of a pluralism of approaches, wherein both the old and the new approaches coexist.

1.3 Mechanisms

In trying to understand the phenomenon of scientific imperialism, questions regarding the mechanisms leading to its emergence and eventual end also arise. How and why does scientific imperialism originate, and how is it sustained and modified? How and why does it generate these and those consequences and may collapse? It is clear that various epistemic and non-epistemic matters, causes and reasons, are mixed here. Let us distinguish some of these.

On top of playing a role in defining scientific imperialism, this issue also highlights various ways in which imperialism can be implemented: Those that are based on external "intrusion" and those driven by importation of ideas or techniques from outside. So some of the drivers might be regarded as external and some as internal.

There might well be a variety of epistemic triggers or motivations that need to be considered. Cases of scientific imperialism might be triggered by a variety of epistemic aims, such as reductionism, unification, scientific progress, problem-solving, etc. To what extent is scientific imperialism promoted by ideals of scientific progress? By ideals of unification? By reductionism – and if so, of what kind? By expectations of generating more effective problem-solving?

Further, some of the drivers of imperialism might well be non-epistemic, as it were. Cases of scientific imperialism might also be triggered by social, disciplinary, institutional, or other non-epistemic aims. It is worth asking the following questions. Is scientific imperialism epistemically or socially driven? To what extent does it involve social practices or intellectual change? Is it just about the ideas or is it about a substantive change in practices? Perhaps one is not possible without the other, i.e., perhaps all intellectual shifts also involve the change of a social practice?

1.4 Identification

How do we identify cases of scientific imperialism? This is a question that arises persistently, and answering it presupposes questions of definition and categorization to have been answered already. Issues of identification also contribute to one of the main goals of the volume, i.e., to explore ways of distinguishing imperialistic from non-imperialistic interactions between disciplines and research fields.

Delineating clearer ways of identifying cases of scientific imperialism and more particularly of specific types of imperialistic interaction is central to fulfilling the aims of this volume. For it is only when we are capable of signaling that a concrete case of interdisciplinary interaction can be considered to be a case of imperialism, that we are able to apply the conceptual framework of imperialism to illuminate our understanding of that particular interaction.

1.5 Assessment/evaluation

If one adopts the descriptive or positive definition of scientific imperialism – although not only in this case – then further assessment of the imperialism at hand is required. How can we distinguish appropriate from inappropriate imperialistic moves? We can find important suggestions on how to proceed in some of the chapters in this volume. It seems that while some authors endorse a more empirical approach, looking into scientific practice for successful and unsuccessful examples of imperialism, others would like to propose normative constraints to be applied for further assessment.

One important question here concerns the empirical assessment of scientific imperialism. Through an analysis of case studies, authors such as Downes, MacLeod, and Fumagalli aim at empirically assessing whether cases of scientific imperialism are successful or unsuccessful. In other words, they examine which cases of scientific imperialism are considered as successful in the scientific community and which ones are not. There are also questions of normativity. From another perspective, some of the authors in the volume, such as Rolin, Salmela, and Mäki and Fernández Pinto, start their assessment of scientific imperialism from normative criteria, where good and bad cases are distinguished through previously developed constraints.

In addition, we can distinguish between the ethical and political assessment, on the one hand, and the epistemic assessment, on the other hand, of scientific imperialism. We need to consider what arguments can be given for the epistemic disvalue of scientific imperialism. Although there are many moral and political arguments given in the literature, the real puzzle is how to generate plausible arguments for epistemic disvalue.

2 Overview

In order to provide greater coherence, we have divided the volume into three main parts. Chapters in Part I provide a philosophical framing for understanding the phenomenon of scientific imperialism. In Chapter 1, Adrian Walsh and Sandy Boucher address the issue of determining the proper boundaries of scientific disciplines, and the problem of when to consider a disciplinary expansion illegitimate. They identify the threat to conceptions of folk morality and philosophical anthropology as a salient feature of paradigmatic cases of illegitimate expansion (or at least those forms of expansion thought to be illegitimate), while defending a case-by-case analysis of disciplinary expansions in contrast to an overall normative account.

In Chapter 2, Mikko Salmela and Uskali Mäki examine the role of disciplinary emotions in interdisciplinary interaction, particularly in cases of scientific imperialism. Following Mäki's (2013) normative constraints, they argue that a proper justification of disciplinary emotions in cases of scientific imperialism depends on their adaptiveness in promoting epistemically acceptable cross-disciplinary explanatory unification. Thus, some disciplinary emotions, such as the anger and fear that members of a colonized discipline might experience, can be properly justified in cases where the imperializing discipline does not care for explanatory unification or for the clarification of hidden uncertainties. On the contrary, such emotions would be unwarranted, if the imperializing discipline has appropriate ontological, epistemological, axiological, and institutional goals in mind.

The subsequent chapters by Kristina Rolin and Patricia Marino assess the ethical dimensions of scientific imperialism. In Chapter 3, Rolin complements the standard view of scientific imperialism by uncovering a previously ignored moral dimension concerning epistemic access. In particular, she argues that scientific imperialism is morally wrong when it gives rise to a particular kind of epistemic injustice: unfairness in the distribution of credibility, or in other words, when a mismatch occurs between the assigned credibility and the actual expertise of scientists involved in a particular disciplinary interaction. She then applies the concept of epistemic injustice to examine two cases of imperialism, and further identify one as morally dubious. In Chapter 4, Marino discusses the ethical implications of scientific imperialism by focusing on two examples: economic reasoning applied to domains of love and caring; and economic reasoning applied to domains of sexuality. The analysis of these two cases helps illuminate some mechanisms through which even *domain-only* imperialism, i.e., when the expansion does not interfere with disciplinary practices as such, can have negative consequences.

Contributions to Part II of the volume are mainly concerned with the historical origins of the debate regarding "scientific imperialism," going back to representatives of the Chicago School of Economics and their self-identification as "imperialists" (see, famously, Stigler 1984). In Chapter 5, Steven Medema examines the Association of American Law Schools (AALS) and the American Economic Association (AEA) 1969 joint symposium on "Products Liability: Economic Analysis and the Law." He argues that this interaction between lawyers and economists is in fact a case of interdisciplinary boundary crossing, but not a case of imperialism *per se*, given that economists did not impose their methods over the law, but instead, economic methods were considered appropriate and increasingly used by lawyers.

In Chapter 6, Nicolas Guilhot and Alain Marciano examine the interdisciplinary relation between economics and political science after 1945. Taking into account both the internal and the external factors that made it possible for seemingly "economic" methodologies to spread within political science, they argue that the two disciplines were simultaneously transformed as economists and political scientists struggled to develop new approaches for the study of decision-making. Following a similar theoretical thread, S.M. Amadae argues in Chapter 7 that the alleged "economics imperialism" is better described as "rational choice imperialism" and shows – applying Mäki's (2009) category of constraints – that the advances and imperializing power of rational choice are *derivational*, rather than epistemological, conciliatory, or ontological.

In the last chapter of Part II, Cléo Chassonnery-Zaïgouche distinguishes three types of imperialistic relations: intra-disciplinary (or imperialism "within the discipline");, theoretical imperialism; and empirical imperialism, which she applies to three cases. Against the traditional reading of Becker's model of discrimination as a paradigmatic case of economics imperialism, Chassonnery-Zaïgouche argues that this is merely a case of imperialism "within economics." She contrasts Becker's view first with Arrow's statistical discrimination, which she identifies as a humble theoretical imperialism, and then with the more recent empirical measurements of discrimination in economics, which she considers a case of empirical imperialism.

The chapters in Part III of the volume are all concerned with the assessment of new cases of scientific imperialism that arise in current interdisciplinary interactions. In Chapter 9, Miles MacLeod examines the current relationship between molecular biology and modern computational systems biology, and describes this relation as one of *scientific subordination* which is a new type of scientific imperialism in which a discipline co-opts the explanatory and empirical practices of another discipline for its own purposes. MacLeod then evaluates the potential benefits and risks of subordination for both parts.

In Chapter 10, Roberto Fumagalli carefully examines two cases of neuroscience imperialism, one related to the economic modeling of choice and one related to philosophical conceptions of free agency. Based on his analysis of these cases, he challenges neuroscience imperialism, arguing first that the evidential and explanatory relevance of neuroscientific methods are overstated, and second that neuroscience imperialism leads to an untenable reductionism.

Stephen M. Downes in Chapter 11 compares two cases of the application of evolutionary thought to the study of human behavior, i.e., evolutionary psychology and evolutionary linguistics. Against Dupré (1995), Downes argues that evolutionary psychology, insofar as it is not supported by evolutionary biology, should not be considered a case of scientific imperialism at all. In contrast, following Dupré's characterization, evolutionary linguistics seems to be a proper case of imperialism, but, *pace* Dupré, a rather successful one. To conclude, Downes favors an empirically based approach for evaluating appeals to scientific imperialism.

In Chapter 12, Frederique Janssen-Lauret identifies the use of physicalist theories in psychology as a classic case of scientific imperialism, and also as a particularly detrimental one. Showing that first-personal psychology embodies a minor but fruitful increase in expressive strength compared to the impersonal psychology favored by physicalism, Janssen-Lauret argues that this sort of imperialistic interaction undermines the explanatory power of psychology. In particular, she claims that whether or not scientific imperialism is in general harmful, naturalists who reject first philosophy should give up physicalist imperialism.

Magdalena Małecka and Robert Lepenies examine in Chapter 13 the cases of the legal and policy sciences as targets of what they call "behavioral imperialism." With the aim of identifying the conditions under which scientific "trespassing" ought to be considered imperialistic, they propose their account of scientific imperialism and argue that whether a particular interdisciplinary interaction is considered imperialistic or not depends on our understanding of *scientific progress*.

In Chapter 14, Manuela Fernández Pinto attempts a reflexive analysis in which the framework of scientific imperialism is used to understand the relation between philosophy and other disciplines. After examining some cases in naturalized epistemology, in particular social epistemology, as cases of scientific imperialism, she uses Mäki's (2009) theoretical category of constraints to claim that the inclusion of foreign concepts and methodologies in social epistemology should satisfy an epistemological constraint. She advises against the dogmatic commitment of background assumptions, which, she argues, is a particularly detrimental practice in the context of naturalized epistemology.

Finally, the last chapter is a contribution by Genco Guralp, who examines a case in experimental science, i.e., the interdisciplinary interaction between the community of particle physicists and the community of astronomers in the context of the discovery of dark energy. After presenting detailed testimonial evidence of the agents involved, Guralp applies the framework of scientific imperialism as a toolkit to illuminate our understanding of this important episode in the history of science.

References

Clarke, Steve and Adrian Walsh. 2009. "Scientific Imperialism and the Proper Relations between the Sciences." *International Studies in the Philosophy of Science* 23: 195–207.

Clarke, Steve and Adrian Walsh. 2013. "Imperialism, Progress, Developmental Teleology, and Interdisciplinary Unification." *International Studies in the Philosophy of Science* 27: 341–351.

Dupré, John. 1995. "Against Scientific Imperialism." *PSA: Proceedings of the Biennial Meeting of the Philosophy of Science Association* 2: 374–381.

Dupré, John. 2001. *Human Nature and the Limits of Science*. Oxford: Oxford University Press.

Fernández Pinto, Manuela. 2016. "Economics Imperialism in Social Epistemology: A Critical Assessment." *Philosophy of the Social Sciences* 46(5): 443–472.

Fine, Ben. 1999. "The Question of Economics: Is it Colonizing the Social Sciences?" *Economy and Society* 28: 403–425.

Fine, Ben. 2002. "'Economic Imperialism': A View from the Periphery." *Review of Radical Political Economics* 34: 187–201.

Kidd, Ian James. 2013. "Historical Contingency and the Impact of Scientific Imperialism." *International Studies in the Philosophy of Science* 27: 315–324.

Kuorikoski, Jaakko and Aki Lehtinen. 2010. "Economics Imperialism and Solution Concepts in Political Science." *Philosophy of the Social Sciences* 40: 347–374.

Lazear, Edward P. 2000. "Economic Imperialism." *Quarterly Journal of Economics* 115: 99–146.

Mäki, Uskali. 2002. "Explanatory Ecumenism and Economics Imperialism." *Economics and Philosophy* 18: 235–257.

Mäki, Uskali. 2005. "Economic Epistemology: Hopes and Horrors." *Episteme* 1: 211–222.

Mäki, Uskali. 2009. "Economics Imperialism: Concepts and Constraints." *Philosophy of the Social Sciences* 39: 351–380.

Mäki, Uskali. 2013. "Scientific Imperialism: Difficulties in Definition, Identification, and Assessment." *International Studies in the Philosophy of Science* 27: 325–339.

Mäki, Uskali. 2016. "The Philosophy of Interdisciplinarity. What? Why? How?" *European Journal for the Philosophy of Science* 6: 327–342.

Mäki, Uskali. (forthcoming). "Varieties of Interdisciplinarity and of Scientific Progress." Draft presented at the TINT seminar, University of Helsinki, February 27, 2017.

Mäki, Uskali and Caterina Marchionni. 2011. "Is Geographical Economic Imperializing Economic Geography?" *Journal of Economic Geography* 11: 645–665.

Nik-Khah, Edward and Robert Van Horn. 2012. "Inland Empire: Economics Imperialism as an Imperative of Chicago Neoliberalism." *Journal of Economic Methodology* 19(3): 259–282.

Radnitzky, Gerard and Peter Bernholz (eds). 1987. *Economic Imperialism: The Economic Approach Applied Outside the Field of Economics*. New York: Paragon House.

Ross, Dorothy. 2003. "Changing Contours of the Social Science Disciplines." In Theodore M. Porter and Dorothy Ross (eds). *The Cambridge History of Science*. Cambridge: Cambridge University Press, pp. 203–237.

Stigler, George. 1984. "Economics – The Imperial Science?" *Scandinavian Journal of Economics* 86: 301–313.

Turner, Steve. 2000. "What are Disciplines? And How is Interdisciplinarity Different?" In P. Weingart and N. Stehr (eds). *Practising Interdisciplinarity*. Toronto: University of Toronto Press, pp. 46–65.

Udéhn, Lars. 1992. "The Limits of Economic Imperialism." In Ulf Himmelstrand (ed.). *Interfaces in Economics and Social Analysis*. London: Routledge.

Weingart, Peter. 2010. "A Short History of Knowledge Formations." In R. Frodemann, J. Thomson Klein and C. Mitcham (eds). *The Oxford Handbook of Interdisciplinarity*. Oxford: Oxford University Press, pp. 3–14.

Part I

A philosophical framing for scientific imperialism

1 Scientific imperialism, folk morality and the proper boundaries of disciplines

Adrian Walsh and Sandy Boucher

Introduction

Current debates over so-called "scientific imperialism", on one plausible reading, explore significant general issues about the proper boundaries between distinct disciplines. They raise questions about whether some forms of territorial expansion by scientific disciplines into other domains of inquiry are undesirable. Clearly there is a strong normative undercurrent here, as the use of the pejorative term "imperialism" would indicate. However, we face a genuine puzzle here: why should we regard some forms of expansion as illegitimate? Why should any particular boundaries between various disciplines be regarded as sacrosanct? This is especially pressing since presumably not *all* examples of disciplinary expansion are illegitimate.

John Dupré, in a series of influential publications (see especially 1995, 2001), directs his readers' attention to the expansionary ambitions of evolutionary psychologists, and those undertaking the economic analysis of the family which many in the general public find intuitively repugnant. Dupré suggests that what is objectionable in the cases he considers is their imperialistic character, and the fact that they betray a monistic or unificationist approach to science, rather than the pluralistic approach he favours. Against this, we argue firstly that the real concern people have with such examples is not their implications for scientific pluralism: rather, it has its source in the fact that they involve cases where folk conceptions of morality and philosophical anthropology appear to be threatened. (This we label the "disentanglement thesis".) In cases of disciplinary expansion that do not have implications for how humans treat and regard each other, one does not find equivalent levels of opposition to the colonization. We suggest that the concerns over Dupré's central cases (concerns which we regard as legitimate) have been unnecessarily tied to more general questions about pluralism and the boundaries between disciplines, which has ultimately served to muddy the waters. Herein we aim to disentangle the issue of scientific imperialism from the issue of pluralism and proper disciplinary boundaries. Second, we are skeptical of any *a priori* condemnation of any particular case of expansionist science; they must be considered and evaluated on a case-by-case basis.

We should note, before proceeding any further, several points. First, although we share the concerns about the cases that Dupré highlights, we differ over the issue of whether the problem with them is their imperialistic character *as such* and, subsequently, with their connection with pluralism and the proper boundaries between disciplines. Second, our analysis seeks to explain the source of the concern about scientific imperialism; we do not intend either to support or to debunk the critique of scientific imperialism.[1] Third, we are agnostic about the questions of whether we should be scientific pluralists and in this chapter we remain neutral on the topic. We believe there are interesting question in the background about political influences on the development of science itself as well as about the role of science and interpretations of science in everyday political and cultural discourse. However, we do not explore those here.

1 The general features of a pluralist critique of scientific imperialism

There are a number of issues and debates surrounding what has been called scientific imperialism, as the different articles in this volume assert. The primary focus of this chapter is Dupré's discussion employing pluralist notions of the disunity of science and the idea of "proper boundaries", to criticise so-called scientific imperialism. Our arguments are not necessarily intended to apply to *all* critiques of scientifc imperialism that have been, or could be, offered. With this in mind, we shall begin with some brief remarks regarding a number of key assumptions of the critique of scientific imperialism understood as a general criticism of some forms of disciplinary expansion.

A first key assumption of the general idea of scientific imperialism is that of *proper boundaries*. The idea here is that there are proper boundaries between the various disciplines. Here "scientific imperialism", which is a normative term, is contrasted with disciplinary expansion, which is a purely descriptive one.[2] A second assumption is that the violation of these boundaries is a form of imperialism (and in most discussion of scientific imperialism this is regarded as undesirable). There is a third assumption concerning historical contingency. Given that the boundaries between disciplines have shifted over time, any suggestion that the present boundaries (or any particular set of contemporary boundaries) should be maintained must in some sense involve an assumption that the present boundaries more accurately reflect the natural boundaries than past ones. If this is indeed a commitment of those criticising scientific imperialism, then such critics must be committed in a broad sense to the progress of science since defending present boundaries would seem to imply that recent attempts to settle the boundaries of inquiry are superior to those of the past and have been (largely) successful.

It is important here not to confuse or conflate scientific imperialism with *scientism*, the latter which we shall define as the tendency to extend the range of intellectual problems that science and the scientific method should and can solve (at its most extreme it is the claim that science has solved all such problems[3]). There are two important (and closely related) senses in which claims might be thought to be "scientistic". The first – which we take to be the primary sense – involves the suggestion that science has a solution to all possible intellectual problems. The second involves the demand that scientific method be applied to all areas of human intellectual endeavour. (The latter might also be termed methodological naturalism.) Associated with both of these is what Uskali Mäki refers to as the "Kipling attitude", according to which scientists have as their unique burden the obligation to provide these gifts to the rest of humanity.[4] Scientism is distinct from scientific imperialism, as that term is understood herein. Our immediate concern is not with the proper range of science nor with whether science represents the only legitimate method of inquiry – although there are very important issues here – but rather with the relationship between distinct scientific disciplines.[5]

The very idea of scientific imperialism raises a large number of interesting conceptual and methodological issues. First, are there proper boundaries to be discovered between the different fields of intellectual inquiry or should we resist such a suggestion, as Ian Kidd (2013) recommends, on the grounds that the actual boundaries we have are historically contingent social artefacts? Are they just historical artefacts? Second, there is the methodological question of what the proper unit of analysis might be. Is imperialism a matter of the relations between disciplines or fields of inquiry? Might it also be seen as occurring between research programmes and theories? Herein, we shall focus on the colonisation of one discipline by another as the paradigmatic case of scientific imperialism. This seems reasonable since most of the cases discussed in the literature concern disciplinary expansion. However, it is quite plausible for the idea of scientific imperialism to be applied to research programmes, theories and fields of inquiry. Third, one might also ask questions about the *epistemic standing* of the colonised discipline once it has been colonised. What happens to the modes of analysis and theoretical language of the colonised discipline? Are those modes simply discarded and treated as out-dated nonsense or alternatively is it more a matter of their insights being translated into, and redescribed in, a new idiom? There are a range of problems the colonised disciplines have traditionally tried to solve with a certain set of theories or approaches or methods. The imperialists claim the theories or approaches or methods of their own discipline can better solve those problems than the traditional theories, etc., of the colonised discipline. The latter theories are then rejected more or less wholesale. (The difference we have in mind here is roughly equivalent *mutatis mutandis* to the difference between "eliminativism" and "identity theory" in the philosophy of mind.[6]) A final question, which – as the reader will soon discover – is highly relevant to the concerns of this chapter, involves the relationship of any disciplinary expansion to folk views, be they folk views on physics, human nature or morality. Insofar as a disciplinary expansion involves or affects some folk view (or views), it is worth asking whether that influences the merits of any disciplinary expansion. Let us turn now to Dupré's discussion of scientific imperialism.

2 Dupré's scientific imperialism

In his various writings on the topic, Dupré takes the twin ideas that:

i there are *proper* boundaries between the disciplines and;
ii at least some violations of these boundaries are wrong (either normatively or epistemologically);[7]

and connects them to ideas about the disunity of science and epistemological pluralism. He defines scientific imperialism as follows:

i "... the tendency to push a good scientific idea far beyond the domain in which it was originally introduced and often far beyond the domain in which it can provide much illumination" (Dupré 2001, 74).
ii "... the tendency for a successful scientific idea to be applied far beyond its original home, and generally with decreasing success the more its application is expanded" (Dupré 2001, 16).[8]

How does Dupré argue for this controversial claim that we should *not* push ideas beyond their natural domain? His argumentative strategy is as follows. First, he identifies cases where the scientific method of one discipline is expanded and this appears inappropriate for, as Dupré describes it, the disciplinary analysis "no longer provides illumination". Underpinning this claim must be the assumption that there are *proper* boundaries between the sciences. Scientific imperialism thus occurs when disciplines invade other areas beyond their proper domain. Dupré treats the existence of these proper boundaries – if such proper boundaries do in fact exist – as part of a broader story in the philosophy of science concerning the "disunity of science".

Let us begin with the two cases Dupré uses to illustrate his case, these being: (1) *evolutionary psychology*; and (2) *the economic analysis of the family*. We have taken his discussion of these from three separate sources: namely, *The Disunity of Science*; "Against Scientific Imperialism"; and *Human Nature and the Limits of Science*. We are not so much interested in his critical analysis of these disciplines as we are in the question of whether the expansion of these disciplines counts as a form of scientific imperialism that demonstrates the truth of controversial ideas about the disunity of science.

Evolutionary psychology begins, Dupré suggests, with some apparently plausible and even uncontroversial ideas, before slipping into realm of the wrong-headed. First, it is assumed that the best way to understand human behaviour is to understand the structure of human brains. Second, it is assumed that the structure of human brains can be understood, at the most fundamental level, by considering the genes or genetic programme that guide the development of brains. Finally, it is assumed that the explanation of humans having a particular complement of genes is to be sought in reflections on the process of evolution by natural selection (Dupré 2001, 21). He writes:

> In distinguishing themselves from their sociobiological precursors, evolutionary psychologists suggest that where sociobiologists had tended to talk somewhat sloppily about the evolution of a behaviour, say rape by males, they talk instead about the evolution of psychological modules in males for rape. It is a central plank of their project that the mind is to be seen as consisting of a large number of such modules, each designed to address a particular adaptive problem posed by the environment of evolutionary adaptation.
>
> (Dupré 2001, 22)

Dupré is highly critical of this approach. He suggests that evolutionary psychology does not provide genuine insight, but instead offers simplifications and banalities. He is scathing of what he regards as its inability to distinguish between sex and gender (Dupré 2001, 68). Nor does it offer a plausible account of the great differences in behaviour across culture. (Here he argues: if cultures are genetically determined, why then is behaviour so different, given the lack of variety in the genes underpinning them?) Dupré notes also that evolutionary psychology offers no account of why different people develop such different sexual proclivities. Finally, he says that it offers no account of how the complex motivations underlying sexual behaviour interact with the pursuit of many other goals that inform our lives:

> I have looked at just one area of evolutionary psychological speculation, though perhaps the most active one. But most of the difficulties discussed would apply in very similar ways to other areas to which such methods might be directed. Cultural variability and individual variability can be found in most interesting domains of human behaviour.
>
> (Dupré 2001, 69)

Despite such variability being widespread, evolutionary psychology has provided no resources for dealing with problems such as these. Hence, he concludes that in relation to the "illuminations of the real complexities of human nature, the programme may be declared bankrupt" (Dupré 2001, 69).

The second illustrative example of which Dupré makes use is the economic analysis of the family. Here he focuses primarily on Gary Becker's *A Treatise on the Family* (1991) and Tomas Philipson and Richard A. Posner's economic analysis of the epidemiology of the AIDS epidemic (1993), both of which were highly controversial when published. Dupré presents both as examples of inappropriate use of economic styles of thought outside their proper sphere. He criticises these accounts on a number of distinct grounds. First, he argues that they treat altruism, morality and accepted social norms simply as tastes. Second, they are "scientistic" in that they treat themselves as being objective and disinterested reflections on the facts of human behaviour. Third, Dupré claims that adopting the attitudes associated with the economic analysis of home and hearth will result in an undesirable psychological transformation of the characters of those who adopt them. Finally, he suggests that they would become distorted once they get "too far away from home":

> As scientific methodologies move further away from their central areas of application, their abstraction becomes even grosser, and thus relevance to the phenomenon become ever more distant.
>
> (Dupré 1994, 380)

What is significant for our purposes are his claims about scientific imperialism. For Dupré this critique of scientific imperialism is a general claim focused on science and scientific practice as a whole. It is concerned with the boundaries between scientific disciplines. Dupré uses the term to cover both the *practice* of disciplinary expansion and the *desire for unification* that often underpins or drives such expansion.

Dupré's discussion raises important issues in the philosophy of science and involves the defence of some controversial claims, most notably regarding the "disunity of science". According to Dupré, science and scientific methods are fundamentally pluralistic and one should not attempt to reduce them to a single scientific method:[9]

> [B]eyond the attainment of the bland intellectual virtues, there is nothing that it is to be scientific, no common essence to the diverse and variously productive projects generally collected under the heading of science. The contrary assumption is not only mistaken but can be seen to justify bad, and even harmful science.
>
> (Dupré 2001, 71)

Accordingly, Dupré's position is also anti-reductionist: it asserts it is not possible to reduce all knowledge to some single unified set of notions.[10] Thus, whatever progress in science might be for Dupré, it does not involve unification.

All of this is clearly normative or at the very least it is normative in tone. As Ian James Kidd notes, Dupré's account is "accusative". But this raises a puzzle. Why should we regard some forms of disciplinary expansion as illegitimate? What might be the source of any such illegitimacy? From where does the "proper" in the phrase "proper boundaries of science" derive its meaning and force? This was the question that Clarke and Walsh explored in their 2009 article which questioned, *inter alia*, to what extent one could extend the political metaphor of imperialism to understand the harm that might be committed in genuine cases of scientific imperialism (see Clarke and Walsh 2009).

The puzzle is magnified when one considers that in many cases disciplinary expansion has been positive. Think here of the emergence of Newtonian mechanics, and its spectacular success in explaining and unifying a large variety of phenomena in different domains. Given one assumes imperialism to be a normative term, then presumably this was a desirable form of expansion and not imperialistic. For a different example consider the emergence of modern nuclear physics.

Modern nuclear physics provides dating techniques that have helped to solve problems in evolutionary theory, anthropology, archaeology and other historical sciences. Here we have an example of the methods and theories of one science being employed to solve outstanding problems of another science. Yet far from being problematic, this example of expansionist science is celebrated as a great achievement. How might good forms and bad forms of expansion be distinguished?

Some proponents of the critique might draw analogical comparisons with political imperialism in order to demonstrate the moral harms involved with some forms of disciplinary expansion. However, this is somewhat implausible. Is there any reason for thinking that scientific disciplines are like nation-states or other political communities in any relevant respects? Of course, the cynical view might be simply that claims of imperialism reflect the material interests of those groups that are under threat of colonisation. If one discipline loses funding due to its replacement by another then clearly they will have reasons for objecting to the expansion, though, of course, this does not provide genuine grounds for opposition. If the reasons were thus it would be what Alasdair MacIntyre (1984) terms the "external goods" of a practice, as opposed to the internal ones, that are driving resistance. Those external goods would include concerns for funding to conduct research and social status. The concern here would be that if a discipline is colonised then there is a high chance that people from the colonised discipline will eventually lose their employment. Presumably there is more to Dupré's opposition than a concern with such external goods.

Our strategy in what follows involves two main lines of argument. First, we explore the extent to which *the illustrative cases* provide genuine evidential support for a pluralism of the kind that Dupré advocates or for the idea of scientific imperialism. As we shall argue below, the cases to which Dupré appeals involve instances in which the new science attempts to overthrow another scientific discipline that has assumptions closely connected to cherished folk conceptions of human nature or morality; or to make the point conversely, the would-be imperialising science has assumptions that are at odds with widely held folk moral conceptions.[11] It is folk morality, rather than concern with pluralism or scientific imperialism, that is "turning the wheel", as it were. This is not to say that the illustrative cases are not of importance in themselves; our point is simply that they do not provide the grounds Dupré wants them to provide. We are thus running a "*disentanglement thesis*". We distinguish the illustrative cases from the claims about: (i) the proper boundaries of science; (ii) scientific imperialism; and (iii) the disunity of science, none of which is to reject many of the concerns about the illustrative cases themselves.[12] The primary virtue of the disentanglement thesis is that it demonstrates that one can be critical of the illustrative cases without buying into controversial philosophical claims about pluralism, scientific imperialism and the disunity of science; one does not need a sledgehammer, as it were, to crack these walnuts.

Second, we provide a methodological critique of the use of the notion of "proper boundaries" when assessing the legitimacy of new inquiries. We focus here on whether it makes sense to talk of proper boundaries *in se* as a means of explaining why certain expansions might be wrong. Perhaps there are indeed proper boundaries between the disciplines but are we in a position to know what those boundaries might be prior to any such expansion? Thus although we are *agnostic* on whether there are in fact any proper boundaries, we claim that even if there are such boundaries, we are not in a position to know them in a way that is useful for determining how sciences should proceed (these you might refer to as *epistemic* doubts). Instead we endorse what might be called a *piecemeal approach*.

3 Folk morality "turns the wheel" in the two illustrative cases

Dupré's analysis of the two illustrative cases plays both an *explanatory* and a *justificatory* role in his discussion of scientific imperialism. It is explanatory in that it attempts to provide an explanation of why people *might* resist these kinds of invasions. It is justificatory in the sense that it tells us why people *should* resist both the economic analysis of the family and evolutionary psychology. In this section we consider the plausibility of scientific imperialism as an explanation of why these disciplinary expansions cause disconcertion on the part of many.

Let us assume these two cases do involve genuinely disturbing developments – and even if one does not hold that to be true, for the sake of argument, let us assume it to be the case – do they really demonstrate that there are proper boundaries between the sciences and that one engages in an activity that is fundamentally wrong when a science moves beyond its boundaries? Do these cases reveal anything more general about the relationship between the sciences?

We are sceptical that they do. Our alternative explanation is that the resistance arises because the methods, assumptions and conclusions violate or undermine (either directly or by implication) fundamental folk views many people hold on morality and human nature.

Our argument for this concerns the *source* of the *moral heat*, as it were – that is what accounts for the obvious resistance by many to these two particular disciplinary expansions – and partial evidence is to be found in Dupré's writing itself when he critically analyses what he regards as wrong with these cases. Further, the reader should notice that these cases are the most plausible candidates for objecting to a disciplinary expansion on Dupré's pluralistic grounds. The case for objecting becomes less compelling once we examine cases such as the expansion of systems biology into areas once dominated by traditional molecular biology (MacLeod, this volume). Hence if these kinds of cases are not best explained via the pluralist account, then *a fortiori* there are solid grounds for doubting that other cases of expansion will be convincingly explained by anything like the pluralist account.[13]

Let us consider Dupré's work in some closer detail and consider the reasons he furnishes us with for the pernicious nature of these expansions. Evolutionary psychology, Dupré suggests, is epistemologically worthless. However, much of his criticism concerns its account of human nature. In *The Disunity of Science*, Dupré notes the connection of evolutionary theory to concepts of human nature:

> one very understandable basis of the appeal of evolutionary psychology is that if it can give us important insights into human nature, it is natural to conclude that it has implications for how people should live or how societies should be organized.
>
> (Dupré 1993, 71)

He then notes:

> Curiously, perhaps, these implications have been much more insisted upon by critics than by practitioners, and the latter have tended to deny that there are any such implications.
>
> (Dupré 1993, 71)

Although these arguments are appealing, perhaps they rely too heavily on a kind of *ad hominem* form of argumentation since the *mere fact* that it is not the practitioners of the discipline in question who typically press for expansion of its boundaries of use might not in and of itself be grounds for rejecting any such expansion. Note also that this criticism has very little to do with pluralism and the disunity of science. Significantly, there are more systematic grounds for rejecting Dupré's explanation of what is wrong. In the two illustrative cases, one also finds that the methods and content contain assumptions at odds with folk morality. Take, for instance, the view that one can undertake a cost–benefit analysis of the choice whether to have children. For many people this is an inappropriate way to regard one's choice to raise a family. To measure the likely long-term economic benefits of raising a family might be said to be inimical to the attitudes appropriate to family life. Even the patron saint of economic analysis, Adam Smith, believed raw market exchanges to be out of place in the intimate sphere of family and close friendship because of considerations of affection and beneficence.

We can usefully contrast the attitudes here to Dupré's cases of disciplinary expansion with other cases of disciplinary expansion where the disciplinary change has little or no consequences for our moral attitudes or for our conceptions of self. For instance, consider the case of the shift from molecular biology to systems biology (MacLeod, this volume). This particular theoretical shift has little import for folk normativity and, unsurprisingly, is not controversial in the way that our two illustrative cases are. Given the absence of a general concern in cases that do not violate or contravene folk morality, it might reasonably be supposed that it is the implications for folk morality, rather than concerns with the scientific pluralism, that are animating those who find these cases troublesome. Dupré's analysis here involves a misdiagnosis of what is at issue. Further, the cases where the expansion of a discipline has implications for folk morality are exceedingly rare. They are, in fact, the exception rather than the rule. If it is indeed true that the cases to which Dupré adverts are exceptions rather than the rule, then his choice of them is inadmissibly selective.

We agree with Dupré that the kinds of disciplinary practices evident in his illustrative cases are of concern; we simply do not agree they tell us anything about the *proper boundaries* of science. Our view is that it is the violation of folk morality that is driving objections to these particular expansions.

However, this leaves us with a further puzzle – although it is not one we will attempt to solve here. Why might the folk views matter? Who cares if the folk views are violated? One possible, though perhaps not very plausible, answer is that the collapse of folk views be avoided. Some philosophers wish to resist such collapse at all costs, at least for some folk conceptions. Jerry Fodor, for instance, in a related context, argues that: "If common sense intentional psychology were really to collapse that would be, beyond comparison, the greatest intellectual catastrophe in the history of our species" (Fodor 1987, xii). Equally, *mutatis mutandis*, if common moral conceptions of how human beings should relate to one another or related views on philosophical anthropology were to collapse, then this might also be regarded as a catastrophe. But on the other hand, losses of this kind have occurred many times in the past; if economic theory or evolutionary psychology accurately predicts behaviour, why should they be eschewed?

A second justification for those folk normative views (at least in relation to the economic analysis of the family) is to be found in the work of expressivists like Elizabeth Anderson who argues in *Value in Ethics and Economics* (1993) that value is plural and that there are social circumstances in which economic reasoning is inappropriate.[14] The thought here would be that a great deal of the folk moral repulsion at the expansion of economics into the "home and hearth" can be explained (and justified) in terms of the idea that there are some realms where economic reasoning is inappropriate. Expressivists need not suggest that all folk moral views are correct or to be defended, but they can maintain that where economic reasoning colonises disciplines whose basic theoretical assumptions are consistent or congruent with folk morality, and where *those folk mores involve genuinely significant values*, then it should be rejected on expressivist grounds.

Whatever the merits of the preceding alternative justifications of folk moral responses to the disciplinary expansions involved in evolutionary psychology and the economic analysis of the family, our central point is that folk morality underpins the reactions of those who find these expansions repugnant. The examples do not provide grist to the mill of the scientific pluralist. Let us turn now to our second line of response to Dupré's pluralist critique of scientific imperialism.

4 The methodological critique of Dupré's approach

We do not wish to give the impression that we are claiming that the *only* reason one might have for rejecting proposed disciplinary expansion is because it violates, or is at odds with, folk morality. On the contrary, there will be other normative considerations and, further, as we argue in this section, there will be *epistemic* grounds for doing so. In this section we criticise the suggestion that the concept of proper boundaries itself plays any significant role when attempting to determine the legitimacy or otherwise of any example of disciplinary expansion. Instead, we advocate a *piecemeal* approach to evaluating and criticising instances of disciplinary expansion.

A significant problem with Dupré's approach is that there is no plausible general claim to the effect that one discipline encroaching on another's territory, or extending its methodology beyond its normal field of application, is a bad thing, for the simple fact that frequently such disciplinary expansions are worthwhile and successful. Thus, given a particular instance of disciplinary expansion, we doubt there is any *a priori* reason to suppose it will not be successful. In advance of detailed consideration of the particular case it is not possible, *merely* on the basis that it is an example of disciplinary expansion, know whether it ought to be endorsed, or criticised. It is possible to say "it's wrong for *this* theory to be applied in *this* way to *this* problem in *this* field", but it should not be said that "the problem with all these different examples is that they are all instances of this *general* phenomenon or strategy of scientific imperialism". With respect to Dupré's two examples, there is no *a priori* reason why evolutionary models and methods would not be helpful in explaining certain aspects of human behaviour, or why economic models would not be helpful in explaining family patterns.

If Dupré is correct, the approaches he examines are in fact mistaken *because they are instances of* scientific imperialism. So all we would have to know about them in order to evaluate them negatively is that the features they possess determine them to be instances of scientific imperialism, i.e. that they are instances of the application of theories from one domain to another, non-traditional domain. However, there is nothing inherently wrong with this kind of application. So we cannot criticise an instance of "scientific imperialism" just on the basis of knowing that it has such features. On the piecemeal approach that we favour, the reasons why particular instances of disciplinary expansion succeed or fail will vary between cases. With respect to Dupré's cases, for instance, the reasons why the methods and models of economics are not appropriate for explaining the phenomena of family dynamics, or for explaining, and providing the basis for policy recommendations with respect to, HIV-AIDS phenomena, are different from the reasons why the methods and models of evolutionary biology are not appropriate for explaining contemporary patterns of human fertility. Dupré notes that the home turf of economics is real (not metaphorical) markets, and when it is applied to "analogical or metaphorical" markets the results are questionable. *But that is a fact about economics*, not disciplines and disciplinary expansion in general, and does not on the face of it permit any generalisations to other disciplines or to the general phenomenon of expansionist science.

A further problem for Dupré – which is more a question of dialectical strategy than validity of argument – is that it risks begging the question against proponents of the very approaches he is challenging. After all, his opponents are already well aware they are engaged in disciplinary expansion. Arguing that their approach fails because it is an instance of disciplinary expansion is thus unlikely to convince them. For instance, Dupré appears to be assuming *a priori* that evolutionary theory cannot be applied to anything other than the biological domains to which it has traditionally been applied.[15] But this simply begs the question. Evolutionary psychology is an attempt to extend evolutionary explanation into the domain of human behaviour; Dupré argues that it should be rejected because it is attempting to extend evolutionary explanation into the domain of human behaviour. Such a claim is unconvincing.

In sum, the claim is that disciplinary expansions are bad because they are instances of a general philosophical or methodological mistake, namely that all instances of scientific imperialism are overly ambitious and implausible. We suggest, contrariwise, that it is much better to examine and evaluate each example individually.

We have claimed that an approach cannot be criticised merely on the grounds that it involves disciplinary expansion, as some instances of disciplinary expansion are legitimate. However, it is possible that Dupré is using "scientific imperialism" more narrowly, and normatively, to refer to those instances of disciplinary expansion that are illegitimate. If that is the case, however, the task becomes that of distinguishing legitimate from illegitimate instances of expansion, which requires a detailed consideration of the cases; thus this approach collapses into our piecemeal approach. In other words, by "scientific imperialism" Dupré either means (rather improbably) all instances of disciplinary expansion – in which case he is wrong to tar them all with the pejorative brush of "imperialism", as some such instances are legitimate – or he means just the illegitimate ones – in which case all the work is then being done by the (piecemeal) analysis of particular instances of expansion, to determine whether or not they are legitimate, and then the concept of scientific imperialism seems to lose all explanatory value in this context. If one adopts the latter, more plausible interpretation of Dupré's position, then the idea of scientific imperialism itself does not explain which forms of expansion are to be rejected and which are to be accepted.

Dupré's critique can usefully be thought of as consisting of two components:

i The specific analysis of what has gone wrong with each of the cases.
ii The general objection that they are illegitimate because they are examples of scientific imperialism.

We agree with much of the substance of (i), but suggest that this is where the analysis should end. We reject (ii) for the reasons outlined above. Further, we suggest it would be counterproductive in terms of the dialectic strategy to endorse the second thesis, in that those who disagree with (ii) may well then reject (i). *Indeed the baby of (i) is in danger of being lost when the bathwater of (ii) is rightly thrown out.* So we regard ourselves as offering a friendly suggestion to those like Dupré who object to these types of problematic disciplinary expansion. They can be rejected without committing oneself to strong and implausible claims about scientific imperialism. Those concerned about the types of cases

Dupré discusses would do well, we suggest, to employ the more modest piecemeal critical strategy we advocate, as it allows us to object to problematic expansionist science without taking on unnecessary philosophical baggage.

One possible objection to our line of reasoning is that when we look at the specific reasons for failure in each case, we find that there *is* something in common between them: in each case models, methods, etc. that work in the traditional domain do not work in the colonised domain. This is clearly what is being argued when Dupré says "[a]s scientific methodologies move further away from their central areas of application, their abstraction becomes even grosser, and thus relevance to the phenomenon become ever more distant" (Dupré 1994, 380).[16]

Our reply is quite simple. This line of reasoning does not provide plausible grounds for Dupré's pluralist strategy, because sometimes disciplinary expansion is successful and unobjectionable, and sometimes it is unsuccessful. The question is to determine *when* it is legitimate and when it is not, and to decide this we need to look at the specifics of the cases. To know that Dupré's two cases are instances of when it is not legitimate required looking at the specifics of the cases. This supports what we called the piecemeal strategy. Dupré writes that "alien intellectual strategies may import inappropriate and even dangerous assumptions into the colonized domains" (Dupré 1994, 381). Perhaps that is true. It will depend upon the case (and "alien" appears to beg the question – it assumes what is at issue, that the disciplinary expansion in question is illegitimate). This also supports our piecemeal strategy. (Moreover, it supports the suspicion that Dupré is at most offering a warning that in some circumstances disciplinary expansion is a bad idea, and that expansionist science should be treated with caution and sometimes scepticism – which is surely a legitimate point, but it is surely an uncontroversial point, and a considerably weaker claim than a "general critique of scientific imperialism" would lead one to expect.)[17]

One of Dupré's central concerns about expansionist science is that the imperialists typically seek to revolutionise completely the field they are venturing into with simplistic one-size-fits-all approaches, rather than seeking to complement and work cooperatively with traditional approaches in the field:

> A moderate response … might perhaps be to advocate a pluralistic tolerance of scientific imperialism. While acknowledging the complexity of phenomena, and especially human phenomena, should we not welcome complementary approaches from as many different directions as possible? While there is surely something to this thought, I think we should remain skeptical. For one thing, scientific accounts are seldom offered as one tiny piece of a big puzzle: such a modest picture is unlikely to attract graduate students, research grants, Nobel prizes, or invitations to appear on Night Line. For another, the modest picture offers little justification for the displays of sophisticated technique and argument that provide the admired paradigms of application of these approaches, and especially the economic approach.
>
> (Dupré 1994, 380)

There are several things to say about this argument. First, it is not obvious that this kind of arrogant attitude does characterise all or most expansionist science – Dupré is generalising from a small number of cases. Second, even if he is correct that it does characterise expansionist science in general, this would appear to be a contingent matter, rather than something inherent in the very nature of expansionist science. Third, with respect to at least some successful expansionist science, it is not at all obvious that a "we have all the answers" position is unjustified. Was Newton not justified in being ambitious and non-pluralistic with respect to his unification of so many phenomena under his universal laws? Should he have offered his account as one small part of a big puzzle? Again, this line of response supports the piecemeal rather than the general approach to evaluating disciplinary expansion; the fact that sometimes the so-called arrogant attitude is justified, and other times it is not (with a more pluralistic approach being more appropriate) undermines Dupré's claim that the one-size-fits-all approach is never appropriate, and that we can assert in advance of detailed consideration of the case that only a pluralistic approach that is tolerant of a range of approaches, methodologies and models ought to be adopted.

5 Concluding remarks

We have argued that it would be a mistake to regard *mere* inclinations to oppose the disciplinary expansions of evolutionary psychology or Becker's economics as necessarily committing one to philosophical claims about the disunity of science or to the use, as a methodological tool for distinguishing between legitimate and illegitimate disciplinary expansions, of the idea of there being proper boundaries between distinct scientific disciplines. According to our Disentanglement Thesis, one can be opposed to the expansions outlined in the two illustrative cases and yet not endorse pluralism and the disunity of science. Moreover, there are reasons for believing that divergence from the deliverances of folk morality provides a better explanation for whatever intuitive pull these cases have. Turning to the question of policing the boundaries of disciplines, we argue that the idea of proper boundaries itself does not provide an analytic or methodological tool for determining what kinds of expansions are to be avoided. Neither of these arguments, however, provides grounds for rejecting pluralism *tout court*, or even the idea of there being existing proper boundaries between disciplines. On those issues we remain silent. It is *the use to which* Depré *puts those ideas* that we are questioning.

Further, we do not mean to suggest, in pursuing these sceptical lines of argumentation, that those *mere inclinations* associated with concerns about the illustrative cases are trivial or insignificant. Dupré is correct to seek an explanation for both the deeply worrying nature of the cases themselves and the imperialistic impulse more generally. However, the explanations we would proffer might be thought more humdrum – and less philosophically controversial, perhaps – than Dupré's. First, there is the phenomenon of *hubris*. A lack of proper modesty and humility may often lead scientists to believe they have discovered the "key" that will enable them to solve the outstanding problems in diverse areas of inquiry, and a lack of sufficient respect for the traditional approaches and methods of the colonised discipline will also contribute to this.[18] Second, there is the issue of what we might call "ideology". Programmes of disciplinary expansion are sometimes driven by non-scientific, ideological agendas. A number of critics have independently made this point about the cases upon which Dupré focuses: economic explanations of family life and evolutionary psychology have both been subjected to serious criticisms on the grounds that they are ideologically driven. One plausible answer to the question "why should such poor science (a) be done in the first place, and (b) be taken seriously?" is that it serves to support political and ideological interests, movements and ideologies. In this instance the phenomenon of imperialising science might be, in part, an example of the widespread phenomenon of non-scientific forces and interests influencing scientific inquiry. Finally, there is the issue of *exporting success*.[19] Presumably a common motivation for imperialistic scientific behaviour is simply the fact that certain approaches, theories and methods have worked very well in their traditional domain, and thus it is worth seeing whether they may not fare as well in other domains. This is summed up well by Okasha, talking about the general motivation for evolutionary psychology (a motivation he endorses):

> Though solid empirical evidence on the issue is unfortunately in short supply, the idea that the cognitive mind has been fashioned by natural selection must surely merit serious consideration, if only because of the success of the Darwinian mode of explanation in general biology.
>
> (Okasha 2003, 68)

The first two motivations are clearly illegitimate (or at least problematic), but the third, concerning the exporting of success, represents an entirely legitimate consequence of genuine intellectual curiosity. It behoves us to remain vigilant against the dangers of hubris and political ideology, whilst ensuring that opportunities for fruitful expansions are maintained.

Notes

1 Our aim is explanatory, not normative.

2 Notice that some authors in this area, such as Mäki (2003), use the terms "scientific imperialism" in a descriptive way. However, we shall assume that it is fundamentally a normative term.

3 Here we have the work of popular writers like Sam Harris in mind. Alexander Rosenberg makes similar remarks in *The Atheist's Guide to Reality* (2011). Phillip Kitcher sums up the view nicely when he writes:

> Since atheism is thought to be territory already secured, the targets now in view are the Big Questions, questions about morality, purpose and consciousness that puzzle softheaded people who muddle over them. Science brings good news. The answers are now all in. This conviction that science can resolve all

> questions is known as 'scientism' – a label typically used pejoratively (as by Wieseltier), but one Rosenberg seizes as a badge of honor.
>
> (Kitcher, 2012)

4 Mäki writes:

> [t]hese proud and self-congratulatory uses of the term 'imperialism' manifest what may be called the *Kipling principle* in the context of science: the superior discipline's burden to bring scientific enlightenment to other disciplines, or at least to the study of other domains"
>
> (Mäki 2013, 330)

5 Scientific imperialism, understood as the improper colonisation of one discipline by another, should not be confused with the charge that is sometimes levelled at scientists from developed nations who obtain knowledge and techniques from traditional communities in developing nations. To illustrate the point, imagine a tree that has been traditionally cultivated by a community in a developing nation, has leaves which have strong insecticidal powers and which the community have long used to deal with insect pests. Imagine also that a group of scientists working for a large chemical firm discover the tree's power and obtain a plant variety right over the tree and patent the chemical structures found in the leaves without giving any recognition – legal or financial – to the traditional community. This kind of transaction, where there is no due recognition and little if any financial compensation from the scientists involved for the knowledge obtained, has sometimes been referred to as scientific imperialism. Here the term refers to what is regarded as a form of theft perpetrated by scientists from the developed world. It is imperialistic in the sense that the scientists are said to take advantage of the asymmetrical power relations that hold between the developed and the developing world so as to obtain knowledge and resources that they would otherwise not be able to obtain. This is not what is at issue herein. We are simply concerned with the relationship between various scientific disciplines.

6 There are interesting further connections that might be drawn here between scientific imperialism and disciplinary expansion, on the one hand, and reductions in the philosophy of mind from the mental to the physical. In particular, it would be interesting to examine what becomes of the colonised discipline and to what extent it is regarded as mistaken or erroneous, and to what extent it can be incorporated within the colonising discipline.

7 Of course, one might interpret the view more strongly as the claim that any violation is wrong. However, this would be highly implausible – hence, in the interests of fairness, we examine a more plausible opponent.

8 The passage from which this is extracted reads as follows:

> There is another, quite different sense of reductionism that Dennett evidently has in mind, … 'Darwin's dangerous idea [evolution by natural selection]' he writes, 'is reductionism incarnate, promising to unite and explain almost everything in one magnificent vision.' This is an example of something rather different, if related: what I prefer to call scientific imperialism. By this I mean the tendency to push a good scientific idea beyond the domain in which it can provide much illumination.
>
> (Dupré 2001, 16)

9 Dupré is at pains to stress that his view is not relativistic. He does not deny that any scientific account could be true nor does he claim that all scientific theories have equal epistemic credentials and equal value (Dupré 1993, 261–262). He also stresses

that he is not a dualist, reintroducing ghost-like substances into our ontology. See *The Disunity of Science*, where he says: "[A]lthough I want no truck with the non-physical substances just mentioned, a central purpose of the ontological pluralism I have been defending is to imply that there are genuinely causal entities at many different levels of organization" (Dupré 1993, 101).

10 Dupré distinguishes "horizontal" from "vertical" reductionism. The latter involves claims about the "vertical relations between theories" (e.g. whether biology can be reduced to chemistry), while the latter involves claims about "theories of different kinds that deal with objects at the same structural level" (Dupré 1994, 374). He is opposed to both forms of reductionism, but the debate about scientific imperialism is only about horizontal reductionism.

11 Note the difference between this and "scientism", the latter which involves cases where ordinary folk beliefs are replaced by scientific methods and content (often inappropriately). In the discussion mentioned we are concerned with cases where one scientific discipline is more compatible with folk beliefs than another, rather than a case where folk beliefs and systems of belief are replaced by scientific methods and content.

12 We are concerned about the harm in the cases mentioned but endorsing the harm in those cases does not mean one has to commit oneself to the disunity of science.

13 We focus on the most credible cases and if these do not work, then why think that the pluralist account will be plausible in other less compelling cases?

14 For a discussion of Anderson's work on this topic that highlights its political implications, see Walsh (2001).

15 This may be thought a straw man; after all, Dupré (1994, 337) does acknowledge that evolutionary speculations about human behaviour may in principle "provide a legitimate basis for empirical investigations …". However, ultimately his view is that evolutionary accounts of human behaviour fail *because* they are an example of scientific imperialism, in that they transgress the proper boundaries separating disciplines. This is something that can be known in advance of detailed consideration of the nature and merits of this particular expansion.

16 As a generalisation, this would seem to be falsified by the instances of successful disciplinary expansion we have considered.

17 Compare the response to the complaints about "adaptationism" (e.g. Dennett 1995), which suggest the opponents of adaptationism are at most offering unobjectionable warnings about the dangers and problems associated with adaptive hypotheses, while dressing this up as a much stronger and more controversial claim about the deep problems with a general commitment to something they call adaptationism.

18 This is related to a point that Dupré does make about sweeping, unificationist claims having the best chance of attracting funding, getting the attention of the press, selling books, etc.

19 An alternative ground for concern would involve what we shall refer to as "topic loss". The thought here is that there might well be some questions worthy of serious intellectual consideration that cannot be solved straightforwardly by the imperialising science, or at least cannot be solved as well as by the colonised science. Topic loss should be distinguished from "Kuhn loss", which is the claim that within scientific revolutions, alongside gains, there are losses with respect to explanatory power and problem-solving ability. Kuhn loss is fundamentally relativistic with respect to truth. The idea of topic loss, however, is not relativistic with respect to truth, nor are there any claims here about incommensurability of frameworks. Topic loss is simply the claim that certain questions can be removed from a framework of inquiry or systematically overlooked when we adopt a new disciplinary framework at the expense of another.

References

Anderson, Elizabeth. 1993. *Value in Ethics and Economics*. Cambridge, MA: Harvard University Press.

Becker, Gary S. 1991 [1981]. *A Treatise on the Family*, enlarged edn. Cambridge, MA: Harvard University Press.

Clarke, Steve and Adrian Walsh. 2009. "Scientific Imperialism and the Proper Relations between the Sciences". *International Studies in the Philosophy of Science* 23(2): 195–207.

Clarke, Steve and Adrian Walsh. 2013. "Imperialism, Progress, Developmental Teleology and Interdisciplinary Unification". *International Studies in the Philosophy of Science* 27(3): 343–353.

Dennett, Daniel. 1995. *Darwin's Dangerous Idea*. Harmondsworth: Penguin Books.

Dupré, John. 1993. *The Disorder of Things: Metaphysical Foundations of the Disunity of Science*. Cambridge, MA: Harvard University Press.

Dupré, John. 1994. "Against Scientific Imperialism". *Philosophy of Science Association Proceedings* 2: 374–381.

Dupré, John. 2001. *Human Nature and the Limits of Science*. Oxford and New York: Oxford University Press.

Fodor, Jerry. 1987. *Psychosemantics: The Problem of Meaning in the Philosophy of Mind*. Cambridge, MA: MIT Press.

Kidd, Ian James. 2013. "Historical Contingency and the Impact of Scientific Imperialism". *International Studies in the Philosophy of Science* 27(3): 315–324.

Kitcher, Philip. 2012. "Seeing is Unbelieving: Review of A. Rosenberg, *The Atheist's Guide to Reality: Enjoying Life without Illusions*". *The New York Times Sunday Book Review*, March.

MacIntyre, Alasdair. 1984. *After Virtue*. Second edition. Notre Dame, IN: University of Notre Dame Press.

MacLeod, Miles. 2018. "Scientific Subordination, Molecular Biology, and System Biology". In Uskali Mäki, Adrian Walsh and Mauela Fernández Pinto (eds) *Scientific Imperialism: Exploring the Boundaries of Interdisciplinarity*. Abingdon: Routledge.

Mäki, Uskali. 2013. "Scientific Imperialism: Difficulties in Definition, Identification and Assessment". *International Studies in the Philosophy of Science* 27(3): 325–339.

Okasha, Samir. 2003. "Fodor on Cognition, Modularity, and Adaptationism". *Philosophy of Science* 70(1): 68–88.

Philipson, Tomas J. and Richard A. Posner. 1993. *Private Choices and Public Health: The AIDS Epidemic in an Economic Perspective*. Cambridge, MA: Harvard University Press.

Walsh, Adrian. 2001. "Are Market Norms and Intrinsic Valuation Mutually Exclusive?" *Australasian Journal of Philosophy* 79(4): 525–543.

2 Disciplinary emotions in imperialistic interdisciplinarity

Mikko Salmela and Uskali Mäki

Introduction

Emotions are an important aspect of scientific work, yet they are still largely neglected in the research on science. Their role in shaping scientific observation and discovery as well as the justification, revision and rejection of old convictions and new conjectures has nevertheless received some attention in sociology, philosophical epistemology, and philosophy of science, beginning from the early writings of Ludwik Fleck (1935, 1936).[1] He argued that collective emotions and moods are important in scientific activity conducted by "thought collectives" with distinct "thought styles" – shared cognitive frameworks characterized by common research problems, evaluative standards, methods, techniques, and literary styles. Fleck, who believed that "there is no emotionless state or pure rationality as such" (Fleck 1935, 49), suggested that "thought styles" evoke and maintain a collective mood that "produces the readiness for an identically directed perception, evaluation, and use of what is perceived" (Fleck 1936, 101). Recent research supports the view that emotions indeed have these roles in scientific perception and evaluation. Thus, Jack Barbalet (2011) suggests that scientific aesthetics, learned during training in a specific research community, guides scientific perception and gives rise to aesthetic pleasure when a particular research episode corresponds with the emotionally laden values of the research community. Several philosophers have also highlighted the functions of emotions as detectors of salience and in epistemic justification (Elgin 2008; Hookway 2008; Tanesini 2008; de Sousa 1987). Scientists' own descriptions of the research process confirm these roles of emotions in science (e.g. Koppman et al. 2015).

Yet these reflections on emotions in social cognition overlook another important way in which emotions are involved in scientific practice. This is their role in the production and reproduction of scientific disciplines and the associated disciplinary identities of scientists. Indeed, if we understand disciplines along the lines of Fleck's "thought collectives" with distinct "thought styles" as their epistemic cores, the affectivity of thought styles brings emotion into the heart of disciplines.

What is important for our purposes is that *emotions not only shape the internal social lives of disciplines, but also make a significant difference for their external relations.* This is where interdisciplinarity enters the stage. The importance of disciplinary emotions has increased together with the growing stress on interdisciplinarity in research. In interdisciplinary collaboration, scholars from different disciplines are supposed to negotiate the concepts, methods, and theories applied in their joint projects. These situations trigger various emotions such as those that arise if some participants of an interdisciplinary project feel that they are treated unjustly by participants from another discipline or disciplines. Another set of contexts in which disciplinary emotions emerge consists of non-collaborative asymmetric relations characterized by intellectual or institutional expansion across disciplinary boundaries. Versions of such scientific imperialism include the phenomenon in which one discipline seeks to "invade" a domain that is already "occupied" by another discipline with its own concepts, methods, and theories. In these encounters among disciplines, disciplinary emotions motivate both the proponents and opponents of scientific imperialism.

We begin by analyzing the nature of disciplines, which is crucial for understanding the social identification of individual scholars. This background allows us to analyze disciplinary emotions as group-based emotions of a specific kind that are both causally and rationally tied to disciplinary identities. We then proceed to examine the role of different disciplinary emotions in interdisciplinary interaction, and in scientific imperialism in particular, elucidating the situations in which these emotions emerge and the kinds of behavior they motivate. We suggest that the *internal justification* of imperialistic disciplinary emotions as motivational forces derives from the epistemic core of the discipline that contains second-order judgments about the superiority or some other special expediency of the imperialistic discipline in relation to others. However, the *rational justification* of such disciplinary emotions – as well as those that are prompted in the target disciplines of scientific imperialism – depends on their adaptiveness in promoting epistemic progress in scientific inquiry at large.

1 What are disciplines?

There is a wide agreement on the nature of disciplines as normative institutions whose two main aspects, the epistemic and the organizational, always go together (Mäki 2013). Yet there is a third aspect: disciplinary cultures and identities that reflect the ways in which members of a discipline internalize and enact its epistemic posits and social arrangements.

Considering its epistemic aspect, a discipline is a more or less convergent system of concepts, methods, theories, questions, established facts, canonical convictions, standards of evaluation, and styles of argumentation. The "more or less" here is to be taken literally. Some disciplines are epistemically and methodologically more plural and diverse than others, involving different schools and traditions and fashions that coexist (in peace or in conflict), as in the case of sociology and human geography; whereas other disciplines – especially if they are guided by an ideal of unity – seek to suppress diversity, with rival schools perhaps actively trying to undermine the credibility of each other, as was the case, for instance, between analytic and continental philosophy from the early 20th century until recently.

Epistemic rivalries within disciplines often relate to struggles for organizational positions, resources, and power. The organizational aspect of disciplines consists of ingredients such as university departments, endowed chairs and other academic and administrative positions, educational programs and degrees, certifications and affiliations, research centers and institutes, scholarly associations, conferences, and journals. Until recently, the structure of academic departments in many cases used to correspond – roughly – with the structure of disciplines on a one-to-one basis. However, things are changing along with cuts of public funding and increased demands of cost efficiency in higher education. These trends have inspired another trend, that of turning departments into large, multidisciplinary units, especially in many European universities. While this development in administrative structure is often justified in terms of promoting interdisciplinarity (on top of saving in administration costs), it may also serve to shake the disciplinary foundations of interdisciplinarity by putting pressure on the organizational autonomy of disciplines (e.g. Radder 2010).

If the two aspects of a discipline, the epistemic and the organizational, always coexist and entangle with each other, it might be futile to ask which one is primary. In part, this depends on the perspective we take on disciplines. From an internal perspective, the epistemic aspects of a discipline constitute its core even if this core is liable to change. Advances in theories, methods, and concepts drive the development of a discipline and motivate the inception of new organizational forms and forums such as conferences, associations, networks, and journals rather than the other way round. A discipline obviously needs some organizational "bodies" in which its epistemic "mind" can operate its two main functions of research and education. However, the organizational structure can be flexible as long as it allows for sufficient autonomy to the discipline. Evidence from studies on socialization to disciplinary identities supports the primacy of the epistemic aspect of disciplines as well. Disciplinary cultures are maintained and perpetuated by means of identification with disciplinary norms and ideologies that transcend the organizational settings of academic departments (Parry 2007; Henkel 2000).

Exploiting the resources of Raimo Tuomela's (2007) theory of social ontology, we may characterize disciplines as social institutions with an *ethos*: a set of constitutive goals, values, beliefs, standards, norms, practices, and/or traditions that give scholars both motivating and justifying reasons for their disciplinary activities. The constitutive ethos of a discipline consists of a distinct set of epistemic and methodological principles, standards, norms, and values that the members of the discipline seek to satisfy and promote in their activities as group members affiliated with a discipline. Disciplines *host* several organizations such as university departments, research institutes, or academic societies, and their particular members occupy task-roles such as a professor or a president in those organizations. Yet it is sometimes possible to conduct disciplinary research (in some disciplines) even without occupying a task-role in academia by following the relevant disciplinary norms in one's activities. Indeed, most leading European philosophers until the 19th century did not work at universities. This is one reason for not identifying disciplines with their organizational aspects even if *some* organizational structure is necessary for the maintenance and dissemination of the disciplinary ethos and for validating sufficient mastery among the novices.

Another main requirement of social institutions in Tuomela's theory is their *collective acceptance* by group members. All disciplines have some foundational norms and ideas on which those affiliated with the discipline more or less agree. Again, the "more or less" is important here. The degree of agreement varies from discipline to discipline (biology to economics, for example), and from some parts of a discipline to its other parts (say theoretical to methodological aspects). Intra-disciplinary diversity is consistent with collective acceptance and agreement when it is based on acknowledged complementarity, with research fields focusing on their sub-domains, addressing their specific problems and applying their distinctive techniques in harmony with one another. Or diversity can be based on rivalry and disagreement, when alternative perspectives are offered in competition or even in conflict with one another. For example, those agreed-upon norms can be methodological and concern legitimate ways of making and justifying arguments, collecting evidence, forming and testing hypotheses, and so on, even if there are disagreements on substantial theoretical issues. If there is no agreement on deep theoretical convictions or on the appropriate methods of adjudicating disputes between different positions, we can talk about different schools with partly dissimilar ethos within a discipline. Yet different schools typically share roughly the same domain of phenomena, research problems, and recognized classic texts that define those problems, and thereby the ambit of the discipline. The shared ethos of a divided discipline is founded on those shared problems and classic texts rather than on the present standards and theories. Yet, at least typically, there is a more substantial agreement on epistemic and methodological criteria within a discipline.

In addition to epistemic and organizational aspects, disciplines have a third aspect which is often attributed to people working in the discipline. This aspect focuses on the ways in which scholars internalize and enact the epistemic posits and social arrangements of their discipline. *Disciplinary culture* comprises epistemic and methodological commitments of the discipline, but also material practices of disciplinary work and social relations of knowledge production in and through which this work is conducted (Schoenberger 2001). All these aspects of disciplinary culture are reflected in a sense of a *disciplinary identity* in terms of which we understand who we are and what we do in the academic world. Together, disciplinary culture and identity provide an intentional and psychological background for the emergence of what we propose to call *disciplinary emotions*: pride and fear, envy and jealousy; confidence and insecurity; feelings of disciplinary superiority and inferiority; and other similar emotions that are experienced as members of a discipline on specifically disciplinary grounds. Emotions contribute to the process of disciplinary socialization in interaction with senior colleagues and peers in seminars, workshops, and conferences in ways that Collins (1998) has analyzed in detail. Consequently, emotions become rational responses to events that are significant from the perspective of the disciplinary identity. The rationality of emotions must here be understood in a Davidsonian sense that refers to their forming a coherent and holistic pattern with an agent's attitudes and other responses, rather than their being mediated by reason or reasoning.[2]

2 What are disciplinary emotions?

We next analyze disciplinary emotions as emotions that are felt in the social identity of a disciplinary scholar. Disciplinary emotions are emotions felt for reasons that rationally – again, in a Davidsonian sense – relate to the epistemic or organizational aspects of the discipline. The reasons for disciplinary emotions are grounded in the disciplinary ethos, for instance when we evaluate our academic work in research or teaching, or that of our disciplinary colleagues. We feel happy and proud of our (either personal or joint) academic successes; sad and disappointed at our (likewise, personal or joint) academic failures; ashamed if we fail to live up to the core values of our discipline; angry at colleagues who violate those values; and so on. In addition to providing motivating reasons for these emotions, the disciplinary ethos also gives justifying reasons for those emotions within the disciplinary community. However, disciplinary emotions can be felt for organizational reasons as well. Thus, we are shocked if we learn that our university plans to discontinue teaching and research in our discipline, or to reduce its resources substantially, or merge our discipline into a larger academic unit in which its autonomy is seriously jeopardized.

Disciplinary emotions can be either private or shared with others, depending on the context.[3] These emotions are typically adopted and reinforced in social contexts – lectures, seminars, conferences, and debates, both formal and informal – where they are shared with disciplinary colleagues. Collective enthusiasm, admiration, awe, or some other positive shared emotion is experienced when a scholar in a public presentation succeeds in presenting perceptibly original ideas building on previously established research and theorizing. If the presenter fails in this task, in spite of having raised expectations in the audience, the resulting shared emotion is negative: disappointment, dejection, indignation, or annoyance. Examples of private disciplinary emotions include those that we experience as peer reviewers of article manuscripts and research proposals, for instance, or in general reading scholarly works in our field. We are impressed if the manuscript, research proposal, or monograph meets the disciplinary criteria of excellence particularly well, and irritated or disappointed if it fails in light of the same criteria (and especially irritated when it pretends to excel).[4]

Another important distinction in disciplinary emotions concerns their intentional objects. Some disciplinary emotions are felt about *the actions of disciplinary colleagues* that are evaluated on disciplinary grounds. For instance, admiration, awe, and envy of a new theoretical insight or empirical finding in one's field presuppose understanding and appreciation of the relevance and epistemic value of the discovery, and this understanding emerges from the disciplinary ethos. An additional reason for envy in contrast to admiration and awe is the fact that someone else presented the new insight or finding, which may appear to imply one's inferiority to the other researcher. The involvement of this personal reason does not prevent envy from being a disciplinary emotion because the goal of being given credit for success in one's field is part of one's disciplinary identity, even if this goal is shared across all disciplines. Other disciplinary emotions are felt about *other disciplines and/or their representatives*. These emotions focus on the theories and worldviews, research styles and techniques, financial resources and academic prestige of those other disciplines. For instance, it has been argued that economists envied and emulated the mathematical methods of physicists that they perceived to be superior to the methods of their own discipline, whereas other social scientists have later envied economics for similar reasons (Mirowski 1992). Emotions of this kind are important in the context of scientific imperialism.

Finally, not all emotions felt in academia are disciplinary. Disciplinary identities are subcategories of academic identities whose constitutive values, norms, standards, and conventions transcend disciplinary boundaries. Accordingly, disciplinary emotions are a subcategory of academic emotions felt in the social identity of an academic. For instance, some criteria in terms of which the activities of professors are evaluated are the same across disciplines and pertain to their role and academic identity as professors. Consequently, emotions about how individuals meet the demands of the non-disciplinary aspects of their academic identity are not disciplinary even if those emotions are pervasive in academia. For instance, all faculty members may feel angry and irritated about their increased administrative tasks after a poorly planned administrative reform. Regardless of how intense these emotions are, they are not disciplinary insofar as their motivating reasons are not disciplinary. Another important group of non-disciplinary academic emotions are those that relate to power relationships and hierarchies within disciplinary communities, such as processes of inclusion and exclusion from an academic conversation or career (Bloch 2012). Important as these emotions are, they are more or less similar in all disciplines, and therefore generally academic rather than disciplinary. A major shortcoming of previous research on emotions in academia has been its inability to distinguish disciplinary emotions felt in the social identity of a disciplinary scholar from other academic emotions.

3 Disciplinary emotions in imperialist interdisciplinary relations

Interdisciplinary interaction is an important context for disciplinary emotions. We now move on to examine the role of disciplinary emotions in interdisciplinary relations, and in scientific imperialism in particular. Imperialistic contexts are characterized by dimensions such as disciplinary superiority and inferiority, centrality and marginality, upgrading and downgrading, dominance and autonomy, hegemony and subjugation, admiration and scorn, neglect and engagement, intrusion and resistance, expansion and protection, unification and separation, and various asymmetries in cognitive and other resources. We distinguish between disciplinary emotions typically felt by scientific imperialists, such as feelings of pride, superiority, and confidence; and disciplinary emotions typically felt by those in the imperialized fields and disciplines, such as feelings of fear, inferiority, jealousy, envy, humiliation, and anger. We analyze situations in which these emotions emerge and the kinds of behavior they motivate in the context of scientific imperialism.

Status and resource asymmetries among disciplines are a common source of emotional tensions in interdisciplinary collaborations. Natural and life sciences generally enjoy higher status than social sciences and humanities for reasons that seem obvious to some and contestable to others. These asymmetries influence the work of interdisciplinary research teams even if they operate under the rhetoric of mutuality, reciprocity, and exchange. Social scientists and humanists may feel subjugated and weak as they typically need to learn more about the disciplinary perspectives of life scientists than these will learn about the viewpoints of social scientists and humanists (see Callard and Fitzgerald 2016). For example, social science and humanities researchers working in medical faculties have been forced to adjust to the research evaluation criteria of medical research which significantly differ from their own disciplinary standards, with ensuing feelings of frustration, despair, and alienation (Albert et al. 2014). Feelings of subjugation may also emerge from *scientific subordination* in which a discipline or field co-opts the explanatory and empirical practices of another discipline (or its individual researchers) to service its own supposedly more productive theories (MacLeod, this volume). Further still, natural scientists (and economists among the social scientists) enjoy a higher prestige in society as politicians and policymakers tend to seek their advice in practical problems that have technical and measurable dimensions, however richly social these problems otherwise might be. Facing these asymmetries of prestige and power, social scientists and humanists can either respond with anger, trying to achieve more recognition for their substantial efforts and contributions as Rabinow (2009) suggests, or, if they take a more pragmatic approach as Callard and Fitzgerald (2016) recommend, acquiesce in their subjugated position, experiencing occasional "sadomasochistic" pleasures in it. Both choices come with disciplinary emotions of different kinds.

A most consequential context of disciplinary emotions in interdisciplinary interaction is expansionist scientific imperialism. One way for it to begin is when one discipline seeks to explain phenomena that are perceived to belong to the domain occupied by another discipline. "Belonging" and "occupying" are not easy concepts to analyze, but roughly, the idea is that the identity of many disciplines is partially constituted by having certain types of phenomena in their explanatory purview on historical grounds, their having made regular attempts to explain those phenomena, or at least having them programmatically on their explanatory agenda, by including them in the scope of the discipline. Therefore, one discipline's attempt to expand the applicability of its methods, concepts, or theories to explain phenomena in another discipline's domain can be perceived an unwelcome intrusion across disciplinary boundaries, especially if the expanding discipline defines the new domain in its own disciplinary terms, suggesting that the domain "belongs to" its "territory." This kind of activity can be described as a form of scientific imperialism: *imperialism of scope*. In its weakest and apparently most innocent form, imperialism of scope amounts to *domain-only imperialism* in which an imperialist discipline merely seeks to explain phenomena in domains that belong to other disciplines, without any attempt to change the ways of inquiry in these disciplines. In the more intrusive and far-reaching forms of imperialism, those of *style* or of *standing*, the targeted discipline itself – its principles and practices, or its status – is challenged more seriously, by transferring or imposing the research styles and strategies of the imperialistic discipline on other disciplines, and/or by appropriating the academic resources of the imperialized disciplines (Mäki 2009, 2013). One would expect the latter two forms of imperialism to trigger the strongest emotional responses, among scientific imperialists as well as among the practitioners in the imperialized disciplines.

3.1 Disciplinary emotions of the imperialists

Scientific imperialism is associated with feelings of confidence, superiority, and pride in the discipline's theories, methods, and research styles, especially in relation to other disciplines, among its adherents. The feelings of confidence and superiority may have multiple sources. Thus Marion Fourcade and her co-authors (2014) argue that the perceived superiority of economics springs from sources like these: (1) its relative insularity as a discipline and its dominant position within the network of social sciences; (2) a pronounced hierarchy within the discipline, especially in comparison with other social sciences; (3) changing networks of affiliations of economics in consequence of which economics has reoriented towards finance; and (4) the material situation, worldviews, and social influence of economists that have set economists apart from their academic peers. These factors reinforce each other such that the "position of social superiority breeds self-confidence, allowing the discipline to retain its relative epistemic insularity over time and fueling a natural inclination towards a sense of entitlement" (Fourcade et al. 2014, 2).

Focusing on epistemic sources of superiority that belong to the disciplinary ethos, economists believe that their discipline is the most scientific among social sciences by virtue of its mathematical methods that are more reliable and intellectually demanding than the methods of other, less formal, social sciences. With this pride in the superiority of their discipline, economists "are less likely to feel the need to rely on other disciplines or even acknowledge their existence" (Fourcade et al. 2014, 6). Instead, economists are eager to apply their theories and methods to the study of phenomena that have traditionally belonged to the domains of other disciplines. In a famous defense of economics imperialism, Edward Lazear proudly proclaims that "by almost any market test, economics is the premier social science ... [as] the economic toolbox can be used to address a large variety of problems drawn from a wide range of topics" (Lazear 2000, 99).

It is notable that this appeal to the "market test" establishing the supremacy of economics is an exercise of (higher-order) economics imperialism: it applies the notion of the market to the justification of the interdisciplinary spread of the notion of market (together with other economic notions) to traditionally non-economic domains. Rather than invoking principles such as those of evidential support or ontic similarity, it emphasizes the influence of economics on others outside economics: "The effort to extend the field measures its success by inducing others to adopt the economic approach to explore issues that are not part of classical economics" (Lazear 2000, 104). By this criterion, the imperialist expansion of economics to the domains of other disciplines justifies itself! Lazear grants that other scientists such as sociologists, anthropologists, and perhaps psychologists may be better than economists in identifying problems of research. However, this recognition is half-hearted as economics is claimed to be superior to those other disciplines in its capacity to provide concrete solutions to these problems, as "the parsimony of our method and ability to provide specific, well-reasoned answers gives us a major advantage in analysis" (Lazear 2000, 103).

3.2 Disciplinary emotions of the imperialized

The picture is considerably different in the imperialized disciplines. Focusing here on *externally driven* intrusions (rather than internally driven importations),[5] they tend to prompt emotions such as fear, jealousy, envy, humiliation, and anger. Jealousy derives from the fact that disciplines have distinct research domains that on historical grounds or otherwise appear to *belong to* the discipline. Even if jealousy is typically perceived and analyzed as an interpersonal emotion (e.g. Rorty 1980; Taylor 1988; Goldie 2000), it may have organizational forms that are importantly similar to paradigmatic interpersonal cases. Just as interpersonal jealousy involves treating the loved one as a possession, disciplinary jealousy involves possessive inclinations about the discipline's traditional or customary domain. Accordingly, members of a vulnerable discipline may guard the discipline's domain and borders jealously against perceived or anticipated invasions from other disciplines. Jealousy motivates the building of symbolic and institutional walls around the discipline's domain in order to keep intruding scientific imperialists away. An unintended and undesirable effect of such a strategy may be the rejection of innocent, non-imperialistic proposals for interdisciplinary collaboration as well.

The main problem with disciplinary jealousy, as with interpersonal jealousy, lies in the idea of exclusive possession. Disciplinary domains cannot be owned in a literal sense any more than people. Tradition alone cannot give a sufficient justified reason for excluding researchers from other disciplines from attempting to provide alternative solutions to the problems of one discipline with their own methods, theories, and concepts. In an ideal situation, researchers from all sides should have the right to defend their disciplinary approaches in an interdisciplinary debate where the best argument should win. Naturally, there is no guarantee that the discipline whose traditional target phenomena are being studied with the resources of another discipline emerges as the winner from such a challenge. Therefore, disciplinary jealousy – like its interpersonal counterpart – comes with a sense of powerlessness regarding the agent's ability to keep actual or potential rivals away. This awareness generates and maintains fear and insecurity among members of vulnerable disciplines who feel threatened by the advances of imperialist disciplines into their field.

The propensity for disciplinary jealousy may be associated with the permeability of disciplinary boundaries. Becher and Trowler (2001) suggest that disciplines differ from each other in this respect:

> Impermeable boundaries are in general a concomitant of tightly knit, convergent disciplinary communities and an indicator of the stability and coherence in the intellectual fields they inhabit. Permeable boundaries are associated with loosely knit, divergent academic groups and signal a more fragmented, less stable and comparatively open-ended epistemological structure.
> (Becher and Trowler 2001, 60)

In other words, relatively insulated, hierarchical, and methodologically or theoretically unified disciplines, such as economics, are capable of protecting their borders more effectively than more open, less hierarchical, and internally diverse disciplines. We may then surmise how these characteristics are related to dispositions of disciplinary jealousy. On the one hand, the permeability of a disciplinary border may be associated with a higher propensity for disciplinary jealousy because it renders the discipline chronically susceptible to external influence. Yet on the other hand, and more plausibly, methodological and theoretical pluralism within a discipline may be an antidote against this emotion. Therefore, we may conjecture that the highest propensity of disciplinary jealousy is likely to be found in relatively convergent, methodologically unified disciplines such as anthropology whose domain is targeted by other equally convergent disciplines, such as economics, with their own allegedly superior methods.

If disciplinary jealousy is felt in response to perceived or anticipated intrusions into a disciplinary domain, disciplinary envy is more often felt about issues of power and standing – academic and institutional resources and positions. Callard and Fitzgerald (2016), for instance, report the "neuroscience envy" of social scientists and humanists, focusing on power and resource disparities between their disciplines and the neurosciences. It is also possible to envy another discipline's methods and research styles on disciplinary grounds, that is, to regard those of one's own discipline as inferior to the envied discipline. An example is the "physics envy" of economists which according to Mirowski involves "a certain contempt for the history of economics, a tendency toward an uncritical appropriation of a limited range of mathematical formalisms, and constant intrusions by physical scientists seeking to upgrade the scientific status of the discipline" (Mirowski 1992, 61). The self-image of economists as "more scientific" and "more rigorous" than other social scientists derives from this emulation of mathematical physics, which "rests almost entirely upon superficial points of resemblance between physics and economics" (ibid., 67). Insofar as economists have successfully convinced others of their scientific status, their disciplinary envy has brought them a standing and status that members of other social science disciplines now envy in turn for the economists' position as influential and highly paid experts in business and politics.

There is another possible contemporary source of disciplinary envy that also reveals one of the internal tensions in contemporary science policy. The tendency of decreasing public funding of research and higher education in European universities may increase disciplinary envy as different disciplines start perceiving each other as adversaries in zero-sum games. At the same time, it is recognized that addressing today's grand challenges requires interdisciplinary efforts, and research policies are designed to promote close collaborations between disciplines. The left and right hands of science policy may thus be pulling in opposing directions in this respect.

Epistemic injustice is a source of many negative disciplinary emotions that may be triggered in interdisciplinary situations. Miranda Fricker (2007), who introduced this term, did not associate it with interdisciplinary interaction, but used it more generally to characterize the connection between social and epistemic power in the creation and maintenance of social meanings. She maintains that these social processes often work against those in otherwise marginalized positions in society, such as women, and sexual and ethnic minorities. She distinguishes between two forms of epistemic injustice: hermeneutical and testimonial. The former is "the injustice of having some significant area of one's social experience obscured from collective understanding owing to hermeneutical marginalization" (ibid., 158). Fricker argues that hermeneutical injustice is wrong because it results in the victim's loss of confidence in his or her beliefs and their justification or – in the extreme – loss of knowledge. Testimonial injustice has the same consequences even though it operates differently: through prejudice that causes a hearer to give a speaker less credibility as a knower than he or she deserves.

Fricker identifies feelings of *humiliation* and *under-confidence* elicited by being a victim of epistemic injustice. These feelings are intelligible in the face of imbalance of social power between the perpetrator and the victim of epistemic injustice that render the latter incapable of resisting or protesting against his or her unjust treatment. Yet if we consider an individual in a more equal power position, we should not be surprised if she perceives epistemic injustice as *offence* or *insult*, and responds with *anger* instead. Anger is felt in response to perceived offences, slights, and wrongs, as philosophers of emotion from Aristotle onwards have pointed out. Fricker does not mention anger as an emotional response to epistemic injustice, and the reason for this is that in her understanding, the socially and epistemically marginalized groups lack moral and epistemic resources to perceive their treatment as unjust in the first place.[6]

The situation is different in many contexts of scientific imperialism where the members of an imperialized discipline often perceive the imperialists' intrusion into their domain as an offense, sometimes prompting them angrily to resist it. If active resistance will not arise, feelings of humiliation, despair, and under-confidence are among the possible other responses. The Canadian social scientists and humanists who work in medical faculties are a case in point, as testified by their dire feelings about their inability to resist the evaluation of their research practices, productivity, and academic excellence in terms of disciplinary criteria alien to them (Albert et al. 2014). These scholars are victims of testimonial injustice as they are evaluated by medical researchers whom they perceive to lack the expertise to soundly evaluate their work. Yet it seems evident that the feelings prompted by epistemic injustice include feelings of anger besides those of humiliation and despair.

Disciplinary anger can be felt in interdisciplinary interaction when members of a discipline appraise that their discipline or disciplinary colleagues have been treated unfairly or harmed in some other way by members of another discipline or by administrators, funding agencies, or structural conditions unduly favoring the other discipline. The difference between anger and jealousy is that the latter is felt typically when an offensive action is merely anticipated or conceived in imagination, whereas anger presupposes an actually experienced action of another party that the subject(s) of emotion appraises as an offense of some kind. Testimonial injustice is in operation in scientific imperialism insofar as members of an imperialistic discipline seek to undermine the credibility of researchers of another discipline as experts in the relevant domain (Rolin, this volume). Expertise is a matter of competence in a particular domain, and credibility is others' perceived view of another subject's competence. Scientific imperialism in which the expertise available in another discipline is either ignored or regarded as categorically inferior to that of the imperialist discipline amounts to testimonial epistemic injustice.

Two further observations are in order. First, different sets of disciplinary emotions are often associated as the two sides of a skewed imperialist coin. For example, insofar as scientific imperialism is motivated by disciplinary pride, it appears that anger of the target and pride of the proponent function as complementary disciplinary emotions. Second, the disciplinary emotions of both scientific imperialists and their targets are rationally grounded in the disciplinary ethos of the respective disciplines. The disciplinary anger of those who perceive themselves as victims of epistemic injustice derives from the ethos of their discipline that they perceive as being violated by scientific imperialists who claim to have superior methods or theories for studying the relevant phenomena. Likewise, disciplinary envy and anger are grounded in the disciplinary ethos of the imperialized disciplines as these emotions relate to conceptions of research styles and proper domains of disciplines.

Besides first-order epistemic and methodological norms and strategies, the disciplinary ethos may include second-order judgments about the superiority of those norms and strategies in dealing with wider sets of phenomena or problems than those traditionally reserved for the discipline in an academic division of labor. Confidence in the superiority of the discipline's theories and research styles can be accompanied by a *Kipling principle* which states that it is the superior discipline's burden to bring scientific enlightenment and progress to other, less-advanced disciplines (Mäki 2013). Insofar as the imperialists have positive emotions about their first-order epistemic and methodological principles and strategies, these emotions easily attach to their second-order judgments about those first-order standards as well. This affective investment has the potential of reinforcing the capacity of these second-order judgments to help resist or ignore possible evidence against these judgments, thus strengthening scientific imperialists' confidence in their mission.

Economics and evolutionary psychology are examples of disciplines with this kind of self-appointed mission: economics by virtue of its mathematical methods of modeling that economists regard as superior to those of other social sciences with less formal methods, and evolutionary psychology in virtue of relying on its theoretical view that human behaviors can be explained as adaptations formed at the dawn of humankind (Barkow et al. 1992). In contrast to these overtly imperialistic disciplines, some disciplines and theories such as gender research and social constructivism can be covertly imperialistic even if they do not recognize this pattern due to their particularistic and subversive meta-understanding of their own epistemic perspective, which may let these researchers perceive themselves as victims of other kinds of scientific imperialism. Yet those researchers' own imperialistic inclinations manifest in their confidence in expanding their analyses to ever new social and cultural phenomena and in their belief that these analyses are superior to the previous ones.

4 The justification of disciplinary emotions

A major philosophical question about disciplinary emotions in the context of scientific imperialism concerns their rational justification. Disciplinary ethos with its first-order epistemic and methodological norms and second-order judgments about the validity and applicability of those norms provides an internal justification for disciplinary emotions within the disciplinary community. Unfortunately, this intra-disciplinary perspective is not very useful in interdisciplinary contexts. Insofar as a given disciplinary ethos is internally coherent, it seems capable of justifying those same emotions that we have already discussed as examples of disciplinary emotions in imperialistic contexts: anger at epistemic injustice experienced by victims of scientific imperialism, and pride of its proponents in the superiority of their methods and research styles. Jealousy is a borderline case because it is possible to criticize the belief that a discipline has a monopoly over its domain as misguided even if such a belief is part of the disciplinary ethos. Here we point out the possible mistake regarding what can and cannot be determined by a disciplinary community on the basis of an intra-disciplinary rational consensus.

Therefore, the justification of disciplinary emotions is not merely an intra-disciplinary issue, but reaches beyond disciplinary boundaries. Obviously, this makes it an essentially harder problem to tackle. Nevertheless, the way disputes often go between scientific imperialists and their opponents suggests that it is impossible to avoid putting the issues more broadly as relational interdisciplinary issues. Are the imperialists' feelings of pride, superiority, and confidence manifestations of a *collective hubris* (Sullivan and Hollway 2014) and hence categorically inappropriate versions of these feelings, involving aggression and narcissistic derogation or belittling of other disciplines and their members? Or are scholars in the target disciplines overly timid and sensitive in jealously guarding their domain, or in taking offense at sincere attempts to bring epistemic progress in their field? We propose that in answering these tricky questions one needs to consider the ontological, epistemological, axiological, and institutional constraints on scientific imperialism that one of us (Mäki 2009, 2013) has proposed elsewhere. The key idea is that in order to be acceptable, an imperialistic expansion or intrusion must meet these constraints.

The *ontological constraint* requires that the accomplishment of scientific imperialism should be ontological unification of apparently diverse kinds of phenomena using the theoretical and methodological resources of the imperialistic discipline. The successful pursuit of ontological unification is a process of discovery of real unity in the world. It is a discovery of facts pertaining to the extent to which the domains of different disciplines are made of similar components or governed by similar laws or causal mechanisms. Given that it is the task of science to make discoveries about the real world, it should be appropriate to try to unify as far as possible, regardless of whether disciplinary boundaries need to be crossed.

The *epistemological constraint* derives from the acknowledgment of significant epistemic uncertainty in interdisciplinary trespassing. There is no perfect assurance that ontological unification has been achieved when attempted. Epistemic caution and humility is therefore required when accepting and rejecting theories and explanations of a discipline, especially when applied to far-away territories. Epistemologically appropriate imperialistic claims about the cross-disciplinary unity of phenomena must not hide from view the uncertainties involved but are rather obliged to make them visible by stating explicitly the relevant provisos.

The *axiological constraint* gives non-epistemic values a role to play. Judgments about success and failure in unification must be made relative to values that express the human or cultural significance of the unified phenomena. Theories that manage to unify insignificant phenomena while ignoring or concealing significant ones are less justifiable than those that manage to unify significant phenomena. The concern about important human or cultural values that may fail to be expressed and may be suppressed as a result of scientific imperialism (see Clarke and Walsh 2009) is a reason for putting forth the axiological constraint.

The *institutional constraint* is the requirement that the pursuit of cross-disciplinary unification must occur under the guidance of the principles of institutionally virtuous scientific practice. The ideal normative institutions of honorable science should advise for virtues such as respectful humility and against vices like dismissive hegemonic arrogance in interdisciplinary relations. Such institutions prescribe a culture of critical and open-minded conversation sustained by suitable degrees of diversity and tolerance, and they proscribe against dogmatism and intellectual monopolies that are protected by non-argumentative means of exclusion. Scientific imperialism that engages in critical and respectful debate and that spells out and seeks to justify its presuppositional posits is more agreeable than one that marches on just by relying on the academic or non-academic standing of the invasive discipline.

We are now ready to propose that *the justification of disciplinary emotions in the context of scientific imperialism should be evaluated in terms of their adaptiveness in promoting epistemically acceptable cross-disciplinary unification that meets the above four constraints*. We recognize that it may be difficult to evaluate whether a particular attempt of imperialistic expansion satisfies this criterion when the intellectual space is filled with imperialist arguments and their counterarguments. Moreover, it seems obvious that the parties of such interdisciplinary disputes may not be the best judges in this question when their disciplinary emotions run high, influencing their beliefs and evaluative judgments in ways that make their respective emotions appear justified for themselves in the situation. Jealousy in particular is notorious for influencing thinking and imagination in ways that motivate the subject of emotion to act on his or her emotional thoughts. Similarly, disciplinary jealousy may make its subjects dwell on possessive thoughts about what is perceived as a hostile imperialistic invasion of their discipline, thus predisposing them against all (including reasonable) proposals of cross-disciplinary unification or even non-imperialistic collaboration. In a like manner, the pride of scientific imperialists tends to influence their thinking in ways that support their confidence that the necessary constraints of cross-disciplinary unification are being met, whether or not that is actually the case.

Even if difficult, we believe that it is possible to give some guidelines on the basis of the four constraints.[7] *Imperialism of scope* in its domain-only variety seems the most innocent and potentially supportable form of imperialism. No domain of phenomena is the sole property of any discipline, so attempts by neighboring disciplines to explain those phenomena should not be resisted. Accordingly, disciplinary emotions that express this kind of resistance such as anger, jealousy, and fear about such activity should be taken as conditionally warranted at most. Yet the situation is complicated by the fact that the actual justification of these emotions depends on whether imperialism of scope satisfies the ontological, epistemological, axiological, and institutional constraints.

Imperialism of style and standing are even more difficult to assess. *Imperialism of style* may in principle inspire the target discipline to self-examine and improve its methods and standards of inquiry, but it may also lead to an underestimation or even demolition of approaches and styles of inquiry that are valuable in meeting some important epistemic or human demands. Likewise, *imperialism of standing* may be based on misusing weakly justified images of academic prestige so as to enhance the academic or societal status of a discipline at the expense of that of another. Or, in some (lucky but perhaps rare) cases, it may be a way of reallocating institutional resources and power positions in a way that reflects the actual epistemic capacities and performances of the disciplines involved.

Passing judgment on these matters is highly fallible and time-consuming. Therefore, it should take place in circumstances that support criticism and self-criticism, tolerance and pluralism, mutual respect and understanding. Disciplinary emotions should be so tuned that they do not prevent these considerations but rather facilitate them. The problem is that this kind of cognitively complex and future-oriented perspective of deliberation is difficult for emotions to achieve. This is because the evolutionary function of emotions is to motivate spontaneous action in their eliciting situations: fleeing in fear; hiding in shame; attacking in anger; and so on. It is possible to employ the motivational force of emotions to other, more distant purposes, but this requires their reflective regulation (Gross 1998).

Here the challenge is to regulate our disciplinary emotions on the basis of our understanding on the epistemic acceptability of particular instances of scientific imperialism. Our advice is to treat disciplinary emotions as "fire alarms" that detect situations where we have important concerns at stake but where collective reflection within the disciplinary community is required in order to determine whether or not the *prima facie* emotional appraisal of the situation was justified. This is not an easy task, but perhaps still feasible as the example of social and political movements suggests. The joint action in such movements is often motivated by the participants' shared emotions, and even if these emotions fade during collective deliberation on the suitable course of joint action, it is possible to reinvigorate the emotion when an opportunity to act on the emotion emerges (e.g. Salmela and Nagatsu 2016).

Researchers on interdisciplinarity, philosophers of science in particular, have an important role to play here thanks to two advantages: they can aspire to function as neutral experts who do not have stakes in disciplinary disputes between scientists; and they are equipped with the capacity of offering diagnostic analyses of various hard cases to discern presuppositions that may obstruct fruitful interdisciplinary encounters. In principle at least, philosophers are in a position to judge whether constraints on cross-disciplinary unification and collaboration are being observed in actual projects of interdisciplinary transfer or integration. A little more elaboration is needed in the framework we have outlined here to help philosophers succeed in the difficult task. This is a challenge for another study.

5 Final thoughts

In this chapter, we have presented the first theoretical account of disciplinary emotions and their role in interdisciplinary interaction, focusing on scientific imperialism. In many characteristic instances – especially those governed by externally driven intrusions – the situation emerges as asymmetrical. Members of disciplines targeted by scientific imperialism feel inferiority, jealousy, envy, humiliation, anger, and other negative emotions about their position, whereas imperialists enjoy feelings of superiority, confidence, and pride. We suggested that the justification of disciplinary emotions in the context of scientific imperialism should be evaluated in terms of their adaptiveness in promoting epistemically acceptable cross-disciplinary explanatory unification that meets the ontological, epistemological, axiological, and institutional constraints. Finally, we put forward an idea that philosophers of science may be in a relatively good position to act as neutral arbitrators capable of adjudicating in disciplinary conflicts about the acceptability of instances of scientific imperialism – and, consequently, about the justification of disciplinary emotions in those situations.

We close with two final and perhaps more sanguine observations. First, our focus on the externally driven variants of scientific imperialism has resulted in emphasizing negative emotions like fear and anger within the target disciplines. Internally driven variants – acts of importation carried out by practitioners within the target disciplines – are likely to be associated, at least in limited disciplinary circles, with positive feelings of hope and excitement, joy and curiosity, superiority and confidence. Second, interdisciplinary interaction can take mutually rewarding forms as well. There is evidence that successful non-imperialistic interdisciplinary collaboration can produce high levels of emotional energy that imbues the collaborators with confidence, enthusiasm, and strength for scientific work, while reinforcing their bonds of solidarity (Boix Mansilla et al. 2016; Parker and Hackett 2012; Collins 1998). Importantly, these positive emotions are intertwined with cognitive and interactional dimensions of research collaboration. Thus, collective excitement and joy are markers of establishing an intellectual common ground emerging from different disciplinary viewpoints as well as of the formation of shared group identities. Some positive emotions that arise in successful interdisciplinary interaction are not disciplinary. Instead, they are *interdisciplinary* as they reflect an expansion of the researcher's academic identity from a disciplinary perspective to an interdisciplinary one that is experienced as rewarding and empowering (Turner et al. 2015). This finding provides a nice ending to this chapter as it highlights the substantial role of emotions in *all* forms of interdisciplinary interaction.

Acknowledgments

We thank Manuela Fernández Pinto and Adrian Walsh for their insightful comments on an earlier version of this chapter.

Notes

1 Yet see Parker and Hackett's (2012) discussion on Fleck's ideas.
2 Donald Davidson (1982) famously argued that a person's individual beliefs, desires, intentions, and actions as well as many emotions are intelligible only insofar as they fit into a coherent and holistic pattern that also determines the rationality of the

individual constituents of the pattern. The idea is that we, for instance, fear what we believe to be harmful to something we value, and rationality amounts to this kind of coherence among an agent's attitudes and responses. It is possible to apply this view to social groups as well. The idea is that a social group may manifest a shared evaluative perspective in the group members' responsiveness to the world, and these responses are rational by virtue of their fitting into the coherent pattern (Helm 2008; Salmela 2012). For instance, if a group has set for itself the task of painting a house, then the group members should exhibit a coherent pattern of hope, fear, frustration, anticipation, disappointment, joy, relief, and so on in the appropriate conditions and for the right reason that goes back to their goal of painting the house. Some philosophers have argued that the rationality of emotions cannot be evaluated in terms of their coherence with the agent's beliefs and evaluative judgments because emotions are not inferentially related to those attitudes but instead logically similar to sense perceptions (e.g. Döring 2009). However, this argument fails because it is possible to bring normal human emotions in line with beliefs and evaluative judgments, sometimes with time and effort, whereas this is impossible for sense perceptions as the recalcitrant Müller-Lyer visual illusion shows (Salmela 2011).

3 There is a lively theoretical debate among philosophers on what it means for an emotion to be shared or collective. The two main theoretical approaches are *type-identity* and *token-identity* views. According to type-identity views, two or more individuals share an emotion if each of them has an emotion of the same type, such as anger or fear, about the same object or event. According to token-identity views, there is a single emotion with several participants, similarly to there being a single car with several passengers who share a ride. In this chapter, we adopt a type-identity view of emotional sharing. For an argument for this view, see e.g. Salmela (2012).

4 Bloch (2012) points out that other motives besides disciplinary ones may become mixed with a reviewer's anger in peer review:

> Assessor anger can be a morally legitimate emotion, but it can also be an anger that mobilizes for opposition, a personally motivated feeling of anger, and it can be a group emotion that prepares the ground for negative or less respectful assessments under the guise of an academic sense of justice.
>
> (Bloch 2012, 60)

We do not disagree, but we focus here on cases of the first kind in which anger – and other emotions in peer review – are justified by disciplinary criteria of scholarly excellence alone.

5 This distinction is between intrusion into a target discipline or its domain from outside carried out by an expansionist discipline; and importation of items from an expansionist discipline carried out by practitioners active within the target discipline (Mäki 2013). The emotional dynamics is likely to be different in the two types of case. We simplify matters by focusing on the externally driven case.

6 The history of different civil rights movements – black, women, and gay, for instance – shows that it has been both necessary and possible for these both socially and epistemically marginalized social groups to develop an understanding of their treatment as unjust, while still being treated that way by the majority. Moreover, a central aspect of their success has been the transformation of low-energy, reactive emotions like humiliation and shame that are reflections of the oppression into high-energy, active emotions such as pride and anger that reflect an affirmation of the emancipated self.

7 For a more detailed discussion on these criteria and their application to the question of whether geographical economics is imperializing economic geography, see Mäki and Marchionni (2011).

References

Albert, Mathieu, Elise Paradis and Ayelet Kuper. 2014. "Interdisciplinary Promises Versus Practices in Medicine: The Decoupled Experiences of Social Sciences and Humanities Scholars". *Social Science & Medicine* 126: 17–25.

Barbalet, Jack. 2011. "Emotions Beyond Regulation: Backgrounded Emotions in Science and Trust". *Emotion Review* 3: 36–43.

Barkow, Jerome H., Leda Cosmides and John Tooby. 1992. *The Adapted Mind*. Oxford: Oxford University Press.

Becher, Tony and Paul R. Trowler. 2001. *Academic Tribes and Territories*. Buckingham: The Society for Research into Higher Education & Open University Press. Second Edition.

Ben-Ze'ev, Aaron. 2000. *The Subtlety of Emotions*. Cambridge, MA: The MIT Press.

Bloch, Charlotte. 2012. *Passion and Paranoia: Emotions and The Culture of Emotion in Academia*. Aldershot: Ashgate.

Boix Mansilla, Veronica, Michèle Lamont and Kyoko Sato. 2016. "Shared Cognitive-Emotional-Interactional Platforms: Markers and Conditions for Successful Interdisciplinary Collaborations". *Science, Technology & Human Values* 41(4): 571–612.

Callard, Felicity and Des Fitzgerald. 2016. *Rethinking Interdisciplinarity Across the Social Sciences and Neurosciences*. Basingstoke: Palgrave Macmillan.

Clarke, Steve and Adrian Walsh. 2009. "Scientific Imperialism and the Proper Relations Between the Sciences". *International Studies in the Philosophy of Science* 23: 195–207.

Collins, Randall. 1998. *The Sociology of Philosophies*. Cambridge, MA: Harvard University Press.

Davidson, Donald. 1982. "Paradoxes of Irrationality". In Richard Wollheim and Jay Hopkins (eds). *Philosophical Essays on Freud*. Cambridge: Cambridge University Press, pp. 289–305.

De Sousa, Ronald. 1987. *The Rationality of Emotion*. Cambridge, MA: The MIT Press.

Döring, Sabine A. 2009. "The Logic of Emotional Experience: Noninferentiality and The Problem of Conflict Without Contradiction". *Emotion Review* 3: 240–247.

Elgin, Catherine Z. 2008. "Emotion and Understanding". In Georg Brun, Ulvi Doğuoğlu and Dominique Kuenzle (eds). *Epistemology and Emotions*. Aldershot: Ashgate, pp. 33–50.

Fleck, Ludwik. 1979 [1935]. *Genesis and Development of a Scientific Fact*. Translated by Fred Bradley and Thaddeus J. Trenn. Chicago, IL: University of Chicago Press.

Fleck, Ludwik. 1986 [1936]. "The Problem of Epistemology". In Robert S. Cohen and Thomas Schnelle (eds). *Cognition and Fact. Materials on Ludwik Fleck*. Boston Studies in the Philosophy of Science, Vol. 87. Dordrecht: Reidel, pp. 79–112.

Fourcade, Marion, Etienne Ollion and Yann Algan. 2014. "The Superiority of Economics". *Maxpo Discussion Paper* 14/3. Max Planck Sciences Po Center.

Fricker, Miranda. 2007. *Epistemic Injustice: Power and the Ethics of Knowing*. Oxford: Oxford University Press.

Goldie, Peter. 2000. *The Emotions*. Oxford: Oxford University Press.

Gross, James J. 1998. "Antecedent- and Response-Focused Emotion Regulation: Divergent Consequences for Experience, Expression, and Physiology". *Journal of Personality and Social Psychology* 74: 224–237.

Helm, Bennett. 2008. "Plural Agents". *Nous* 42: 17–49.

Henkel, Mary. 2000. *Academic Identities and Policy Change in Higher Education. Higher Education Policy Series 46*. London and Philadelphia, PA: Jessica Kingsley Publishers.

Hookway, Christopher. 2008. "Epistemic Immediacy, Doubt, and Anxiety: On a Role for Affective States in Epistemic Evaluation". In Georg Brun, Ulvi Doğuoğlu and Dominique Kuenzle (eds). *Epistemology and Emotions*. Aldershot: Ashgate, pp. 51–66.

Koppman, Sharon, Cindy L. Cain and Erin Leahey. 2015. "The Joy of Science: Disciplinary Diversity in Emotional Accounts". *Science, Technology, and Human Values* 40: 30–70.

Kuhn, Thomas. 1970. *The Structure of Scientific Revolutions*. Chicago, IL: The University of Chicago Press.

Lazear, Edward. 2000. "Economic Imperialism". *The Quarterly Journal of Economics* 115: 99–146.

Lyons, William. 1980. *The Emotions*. Cambridge: Cambridge University Press.

Mäki, Uskali. 2009. "Economics Imperialism: Concepts and Constraints". *Philosophy of the Social Sciences* 39: 351–380.

Mäki, Uskali. 2013. "Scientific Imperialism: Difficulties in Definition, Identification, and Assessment". *International Studies in the Philosophy of Science* 27: 325–339.

Mäki, Uskali and Caterina Marchionni. 2011. "Is Geographic Economics Imperializing Economic Geography?" *Journal of Economic Geography* 11: 645–665.

Mirowski, Paul. 1992. "Do Economists Suffer from Physics Envy?" *Finnish Economic Papers* 5: 61–68.

Parker, John N. and Edward J. Hackett. 2012. "Hot Spots and Hot Moments in Scientific Collaborations and Social Movements". *American Sociological Review* 77: 21–44.

Parry, Sharon. 2007. *Disciplines and Doctorates. Higher Education Dynamics* 16. Dordrecht: Springer.

Rabinow, Paul. 2009. "Prosperity, Amelioration, Flourishing: From a Logic of Practical Judgment to Reconstruction". *Law and Literature* 21: 301–320.

Radder, Hans. 2010. *The Commodification of Academic Research: Science and the Modern University*. Pittsburgh, PA: University of Pittsburgh Press.

Rorty, Amelie. 1980. "Jealousy, Attention, and Loss". In Amelie Rorty (ed.). *Explaining Emotions*. Berkeley: University of California Press, pp. 465–488.

Salmela, Mikko. 2011. "Can Emotion Be Modelled on Perception?" *Dialectica* 65: 1–29.

Salmela, Mikko. 2012. "Shared Emotions". *Philosophical Explorations* 15: 33–46.

Salmela, Mikko and Michiru Nagatsu. 2016. "Collective Emotions and Joint Action: Beyond Received and Minimalist Approaches". *Journal of Social Ontology* 2: 33–57.

Schoenberger, Erica. 2001. "Interdisciplinarity and Social Power". *Progress in Human Geography* 25: 365–382.

Sullivan, Gavin B. and James Hollway. 2014. "Collective Pride and Collective Hubris in Organizations". In Gavin B. Sullivan (ed.). *Understanding Collective Pride and Group Identity. New Directions in Emotion Theory, Research and Practice*. London and New York: Routledge, pp. 80–92.

Tanesini, Alessandra. 2008. "Virtues, Emotions and Fallibilism". In Georg Brun, Ulvi Doğuoğlu and Dominique Kuenzle (eds). *Epistemology and Emotions*. Aldershot: Ashgate, pp. 67–82.

Taylor, Gabrielle. 1988. "Envy and Jealousy: Emotions and Vices". *Midwest Studies in Philosophy* 13: 11–27.

Tuomela, Raimo. 2007. *Philosophy of Sociality*. New York: Oxford University Press.

Turner, V. Kelly, Karina Benassaiah, Scott Warren and David Iwaniec. 2015. "Essential Tensions in Interdisciplinary Scholarship: Navigating Challenges in Affect, Epistemologies, and Structure in Environment-Society Research Centers". *Higher Education* 70: 649–665.

3 Scientific imperialism and epistemic injustice

Kristina Rolin

Introduction

By scientific imperialism is meant an attempt to extend a scientific theory's explanatory scope across disciplinary boundaries (Mäki 2009, 2013). Such an attempt is, for example, the application of rational choice theory, a theoretical approach developed in economics, to topics traditionally studied in other social sciences, such as voting (Downs 1957), crime and family life (Becker 1968, 1991), and the social organization of science (Kitcher 1990; Zamora Bonilla 2012). Scientific imperialism is welcomed by those scientists and philosophers who value the search for explanatory unification – that is, the ideal of explaining a variety of phenomena with as scarce theoretical resources as possible. While some philosophers question the importance of explanatory unification (e.g., Hacking 1996; Kellert et al. 2006; Wylie 1999), some others think that it increases our understanding (e.g., Kitcher 1981, 1989). The critics of scientific imperialism, such as John Dupré (1996, 2001) and Uskali Mäki (2009, 2013), suggest that while explanatory unification may sometimes increase our understanding, it does not always do so. The reason for this is, they argue, that it can lead scientists to adopt concepts and models that are empirically inadequate in the domain where they are introduced.

My aim is to reassess the case against scientific imperialism. I focus on economics imperialism and a recent controversy over the *normative* basis for claiming that economics imperialism is sometimes an undesirable form of cross-disciplinary interaction (Clarke and Walsh 2009, 2013; Kidd 2013; Kuorikoski and Lehtinen 2010; Mäki 2009, 2013). The standard view on the topic is that economics imperialism should be resisted when it is epistemically or morally harmful (Clarke and Walsh 2009, 2013; Dupré 1996, 2001; Mäki 2009, 2013). The epistemic harms may come in many forms, such as vague concepts or impoverished accounts of human behavior (Dupré 2001, 128–132). Similarly, the moral harms may range from a failure to respect scientists in other disciplines (Mäki 2013, 337), to a failure to account for morally and socially important phenomena in society (Clarke and Walsh 2009, 195). Economics imperialism is morally harmful when imperialists violate the epistemic responsibilities they have towards scientists working in disciplines that are the target for imperialist explorations (Rolin 2015, 414). By epistemic responsibility is meant a duty to justify one's knowledge claims to a particular audience so that the justification appeals to at least some of the standards of argumentation recognized by that audience. Epistemic responsibility can be understood as a moral obligation because epistemically responsible actions contribute to the well-being of other human beings by showing respect to them, especially in their capacity as knowers (ibid., 421).

In this chapter, I argue that we need a conception of *epistemic injustice* in order to analyze a distinctively epistemic kind of unfairness that may be involved in economics imperialism. My main thesis is that economics imperialism may give rise to a particular kind of epistemic injustice: unfairness in the distribution of credibility. A distribution of credibility is unfair when there is a systematic mismatch between credibility and expertise in a particular domain.

In order to explain why the normative basis for the moral appraisal of economics imperialism is in need of further elaboration, in section one, I discuss Mäki's attempt to define the conditions under which economics imperialism is a legitimate form of cross-disciplinary interaction. I argue that Mäki's four criteria are vindicated by much research in contemporary philosophy of science. While some of the four criteria have a moral-political dimension, I argue that they do not capture an important moral-political aspect of economics imperialism: the struggle over credibility. In section two, I introduce Clarke and Walsh's attempt to apply political theory to philosophy of science in order to articulate and defend the normative basis for demarcating between acceptable and unacceptable instances of economics imperialism. I argue that while Clarke and Walsh are right to suggest that there is a moral-political component in the criticism of economics imperialism, their analysis needs to be supplemented by an account of epistemic injustice. In section three, I articulate and defend the principle of a fair distribution of credibility. In section four, I illustrate the moral appraisal of economics imperialism by discussing two examples.

1 Mäki's constraints on economics imperialism

The term "economics imperialism" has not been invented by philosophers. As Mäki (2009) explains, the term has been launched by economists (see e.g., Lazear 2000; Stigler 1984). In his view, it is up to philosophers to find out whether the use of the term is based on a concept that can be explicated, and whether there are criteria for the normative assessment of the phenomenon referred to as "economics imperialism." While the term "imperialism" could in principle be replaced by another term, philosophers are better off by anchoring their conceptual and normative analysis to a vocabulary that is already familiar to many economists. Such a strategy is likely to increase the impact of philosophical work. In this section, I explain Mäki's definition of economics imperialism and the four criteria he proposes for its appraisal. While his criteria are supported by much research in contemporary philosophy of science, they are not sufficient for the evaluation of economics imperialism.

Mäki defines economics imperialism as a "form of economics expansionism where the new types of explanandum phenomena are located in territories that are occupied by disciplines other than economics" (Mäki 2009, 360). While the core phenomenon in economics imperialism is *explanatory* expansionism, it can involve other dimensions. An imperialism of *style* is an attempt to transfer styles or strategies of research from one discipline to another (ibid., 354). An imperialism of *standing* is an attempt by some scientists to appropriate the prestige, power, and resources of another discipline (ibid., 354). In sections three and four, I suggest that an imperialism of standing is extended to include also attempts to diminish the credibility of scientists in another discipline. While prestige may be associated with many things such as wealth and social status, credibility is a specific kind of prestige which is associated with the power to elicit trust in one's knowledge claims. Since credibility does not receive as much attention as it should in Mäki's normative account of economics imperialism, I suggest that his normative account is in need of further development.

Given Mäki's definition, the term "imperialism" is used as a metaphor suggesting that political and economics imperialism have similar features. As in international politics, one scientific discipline can try to invade a "territory" which has been "occupied" by another discipline. To say that "a territory is occupied by a discipline" is to say that certain phenomena have *traditionally* been studied in a particular discipline (Mäki 2009, 360). For example, an attempt to apply rational choice theory in an explanation of family life is an instance of economics imperialism because rational choice theory has traditionally been part of the toolkit of economics and family life has traditionally been studied by psychologists, sociologists, and anthropologists (see also Marino in this volume). Also, an "occupation" does not need to be externally driven; it can be internally driven by "allies" working in the "colonized" discipline (Mäki 2013, 335). Thus, the metaphor "imperialism" appears to be plausible insofar as research traditions and disciplinary identities matter for many scientists and scholars in a somewhat similar way as cultural traditions and national or ethnic identities matter for many people in the world.

This said, it is important to keep in mind that scientific research is not always meant to be bound by traditions and boundaries (Kuhn 1962). New empirical discoveries may lead scientists to break off from a tradition and seek collaboration across disciplinary boundaries (Andersen and Wagenknecht 2013; Thagard 1999). Sometimes discoveries lead to conceptual changes which give impetus to the emergence of novel specialties (Wray 2005, 2011). In order to accommodate such "revolutionary" developments in science, I have decided to adopt Mäki's epistemically and morally neutral usage of the term "economics imperialism." Given Mäki's definition, economics imperialism is neither good nor bad *intrinsically* (Mäki 2009, 374). Whether it is good, bad, or neutral depends on the circumstances where explanatory expansionism takes place (Mäki 2013, 337).

In Mäki's view, an instance of economics imperialism is legitimate if and only if it fulfills four conditions: ontological; epistemological; axiological; and institutional (Mäki 2009, 373–374; Mäki 2013, 338). The *ontological condition* requires that economics imperialism be based on the pursuit of "ontological unification" rather than mere "derivational unification" (Mäki 2009, 363). Whereas derivational unification is a matter of being able to derive empirically successful predictions from a theory, ontological unification requires that there are similarities in the underlying causal factors or the structural features of the world (Mäki 2009, 365; see also Kuorikoski and Lehtinen 2010; Mäki and Marchionni 2011). The ontological condition strikes a chord with all those philosophers who think that the covering law model of scientific explanation is inadequate because it construes explanation merely as a matter of deriving the explanandum from the explanans. A more adequate account of explanation requires that one is capable of giving an account of some of the causal factors responsible for the events or the state of affairs to be explained (Woodward 2014).

What Mäki calls the *epistemological condition* requires that imperialists make explicit the uncertainties involved in their attempts to apply concepts and models to topics traditionally studied in other disciplines (Mäki 2009, 373). The epistemological condition resonates with the claim that uncertainties in evidential reasoning should be made as explicit as possible not only for epistemic reasons but also for moral reasons. As Heather Douglas argues, scientists, like other humans beings, are morally responsible for their actions and the foreseeable consequences of their actions (Douglas 2009, 67). This means that they are morally responsible for the potential harm caused by their making overly strong knowledge claims and downplaying the risk of error (ibid., 87).

The *axiological condition* states that the expansion of explanatory scope should not take place at the expense of explaining morally and socially significant phenomena (Mäki 2013, 337). In order for an instance of economics imperialism to be legitimate, it is not sufficient that a theory aims to explain a *high* number of phenomena; the set of explained phenomena should include also those phenomena that are perceived as morally and socially significant. The axiological condition is in line with the widely endorsed view that the epistemic goal of science is properly understood to be *significant* truth rather than plain truth (Kitcher 1993), and that significance typically reflects the moral and social concerns of the age even when such concerns appear to be internal of science (Anderson 1995; Kitcher 2001).

While the first three conditions are concerned with scientific explanations, the *institutional condition* is concerned with scientists' cross-disciplinary interactions. It requires that imperialists respect certain norms of epistemically ideal communities. As Mäki explains:

> The optimal institutions of good science should advise for respectful humility and against dismissive hegemonic arrogance in the relations between disciplines. They would support critical and open conversational culture as well as sufficient degrees of diversity and tolerance, while proscribing against closed dogmatism and uniform intellectual monopolies protected by non-argumentative means of exclusion.
>
> (Mäki 2013, 337)

The institutional condition is supported by much philosophical work in the social epistemology of scientific knowledge. For example, Helen Longino argues that an epistemically ideal scientific community should satisfy the four conditions of public venues, uptake of criticism, shared standards, and tempered equality of intellectual authority because such conditions facilitate "transformative criticism" (Longino 1990, 76–81; Longino 2002, 129–131). The fourth condition is especially interesting here because it implies that an ideal epistemic community is open to cross-disciplinary interaction. As Longino explains, the fourth condition requires that scientific communities attend to criticism originating from "outsiders" (Longino 2002, 133–134). However, the first three conditions set constraints for outsider participation in scientific debates. The first condition requires that outsiders respect the public nature of scientific debates. The second condition requires that outsiders be responsive to criticism coming from the members of a scientific community. The third condition requires that outside criticism appeal to at least some of the standards of argumentation that are recognized in the target community. So, while cross-disciplinary interaction is a hallmark of epistemically ideal communities, it is constrained by other conditions.

If Longino is right about epistemically ideal communities, Mäki's institutional condition is vindicated. When Longino's four conditions are realized to a high degree, a scientific community approaches what Mäki calls a "critical and open conversational culture" (Mäki 2013, 337). Whereas the condition of uptake is targeted especially against "closed dogmatism" (Mäki 2013, 337), the condition of shared standards and the condition of tempered equality would prohibit "uniform intellectual monopolies protected by non-argumentative means of exclusion" (Mäki 2013, 337).

So far I have argued that Mäki's constraints on economics imperialism are in line with normative conceptions of good science advanced by many other philosophers of science. Thus, I am in a position to endorse each constraint trusting that it is not likely to meet any immediate objections. Yet, I suggest that Mäki's normative account of economics imperialism is in need of further elaboration in order to understand the kind of epistemic injustice that may be involved in economics imperialism: unfairness in the distribution of credibility.

While Mäki's epistemological, axiological, and institutional conditions are concerned with the moral-political dimensions of economics imperialism, the distribution of credibility is yet another moral-political dimension which requires more attention than it has received so far. Mäki's epistemological condition has a moral-political dimension because it is concerned with uncertainties which may have serious moral-political consequences. The axiological condition has a moral-political dimension because it is concerned with the moral and social significance of the phenomena that are explained. The institutional condition has a moral-political dimension because it is concerned with respect towards other scientists. However, while the institutional condition warns against "dismissive hegemonic arrogance," it is not sufficient to relieve us from worries concerning the distribution of credibility.

In my view, a normative account of economics imperialism should take a more explicit stand with respect to attempts to diminish someone's credibility as a scientific expert vis-à-vis a particular domain. That such attempts are involved in economics imperialism should not be surprising. When an imperialist scientist claims that he or she can explain a social phenomenon traditionally studied in another discipline, and that his or her explanation is superior to other explanations, the imperialist is suggesting implicitly that he or she is also a better expert when it comes to understanding the social phenomenon. In other words, when scientists aim for an imperialist takeover, they not only apply the concepts and models of the intruding discipline to a topic traditionally studied in another discipline; the imperialists also lay a claim to credibility in a novel domain (see also Lõhkivi et al. 2012).

In sections three and four, I argue that it is this aspect in economics imperialism which is in need of further analysis. Before moving on to my analysis of the moral-political dimensions of economics imperialism, I discuss yet another attempt to explain what is wrong with economics imperialism.

2 Clarke and Walsh's political theory of economics imperialism

While the metaphor "imperialism" highlights features that scientific disciplines share with nations, the analogy does not work all the way down. Obviously, scientific disciplines do not "govern" their "territories" in the same way as nations do theirs. Despite the limits of the analogy, Clarke and Walsh (2009, 2013) explore whether the political theory of imperialism might be helpful in assessing economics imperialism. In this section, I explain their view and argue that it does not capture an important moral-political aspect of economics imperialism: the dispute over credibility.

Clarke and Walsh (2009, 199) argue that the normative force in the accusation of *illegitimacy* in economics imperialism is parasitic upon theories of political imperialism. As they explain:

> This strikes us as by far the most plausible way of reconstructing what a charge of economics imperialism might entail. Indeed, it is hard to see how else one might reconstruct it, since the predominant use of the term is in political discourse.
>
> (ibid., 199)

Clarke and Walsh identify two normative elements that underpin the critiques of political imperialism. Political imperialism is thought to be morally wrong because: (1) it is a violation of autonomy; or (2) it involves exploitation (ibid., 201). Exploitation, they argue, has two dimensions. On the one hand, it involves an unjust extraction of human and natural resources; and on the other hand, it unfairly prevents the exploited from realizing their potential to develop (ibid., 201). Clarke and Walsh suggest that it is the latter aspect in exploitation that can be found also in illegitimate instances of economics imperialism. Thus, they come up with the view that economics imperialism is morally wrong insofar as: (a) it prevents a discipline from progressing in ways that it might otherwise progress; or (b) it prevents the expression of important social and moral values in the "colonized" discipline (ibid., 195).

Clarke and Walsh's condition (b) emerges from Dupré's (2001) criticism of economics imperialism which suggests that economics imperialism is wrong in the moral sense insofar as it prevents morally justifiable values from guiding scientific inquiry. While Dupré does not object to the use of mathematic models in the social sciences, he suggests that sometimes such models can lead to an impoverished account of human behavior. The reason for this, he argues, is that moral and social values cannot be conceptualized adequately in terms of preferences and utility functions (ibid., 131). Thus, the underlying assumption in condition (b) is that moral and social values can legitimately guide scientific inquiry insofar as they have an impact on what epistemic criteria and desiderata scientists emphasize when they evaluate scientific theories (see e.g., Elliott 2011, 2013; Kitcher 1993; Kuhn 1977; Longino 1995, 2002; Solomon 2001). For example, moral and social values can legitimately inform the conceptualization of the object of inquiry thereby guiding the way the epistemic value of empirical adequacy is understood (Anderson 2004). Also, moral and social values can legitimately inform the way the epistemic value of explanatory power is interpreted. Sometimes a causal explanation is expected to offer the kind of causal information that is needed to change social conditions seen as unjust (Cudd 2005). Given these arguments, condition (b) is supported by other philosophers' analysis of the proper role of moral and social values in science.

Clarke and Walsh's condition (b) can be seen as an improvement on Mäki's axiological condition which states merely that the expansion of explanatory scope should not take place at the expense of understanding morally and socially significant phenomena. It is an improvement because it clarifies the nature of the harm that may be involved in economics imperialism. The problem is not only that some phenomena are not accounted by imperialist explanatory frameworks; more specifically, the problem is that important moral and social values are overlooked.

While I agree with Clarke and Walsh (2009) that it is important to pay attention to the moral and social values that can legitimately guide scientific inquiry, I think that more can be said about the moral-political dimensions of economics imperialism.[1] In order to analyze further the normative basis for the assessment of economics imperialism, I shift the focus from the evaluation of scientific explanations to the evaluation of individual scientists' expertise and credibility. In the next section, I argue that we need an account of epistemic injustice in order to understand the politics of credibility that may be involved in economics imperialism.

3 A fair distribution of credibility

In my view, the principle of a *fair distribution of credibility* is a key to understanding why economics imperialism is sometimes wrong. In this section I explain first what is meant by epistemic injustice. Second, I explain how economics imperialism can give rise to a particular kind of epistemic injustice: unfairness in the distribution of credibility.

According to one definition of epistemic injustice, it is a wrong done to someone specifically in their capacity as a knower (Fricker 2007, 1). As David Coady (2010) explains, epistemic injustice can be either distributive or discriminatory. Distributive epistemic injustice involves an unfair distribution of epistemically valuable goods such as education or information. For example, distributive epistemic injustice takes place when a person is left in a position in which she is ignorant of something she is entitled to know (Coady 2010, 109). Discriminatory epistemic injustice involves an epistemic wrong done to a person because of social identity prejudice against the person. While discriminatory epistemic injustice may come in many forms, one much-discussed form is testimonial injustice which occurs when "prejudice causes a hearer to give a deflated level of credibility to a speaker's word" (Fricker 2007, 1). An example of testimonial injustice is a situation in which a hearer finds a person's testimony suspicious due to the hearer's racist perception of the testifier.

I suggest that the concept of distributive epistemic injustice can be extended to cover not merely an unfair distribution of education and information but also an unfair distribution of credibility. In order to understand how economics imperialism can give rise to this kind of epistemic injustice, it is necessary to analyze the injustice that is inherent in epistemic injustice. As Miranda Fricker argues, "[f]or something to be an injustice, it must be harmful but also wrongful whether because discriminatory or because otherwise unfair" (Fricker 2007, 151). For example, the harm caused by testimonial injustice is both epistemic and moral. It is epistemic because testimonial injustice is a form of maintaining ignorance. As Fricker explains, it can be seen as "an epistemic disadvantage to the individual hearer, and a moment of dysfunction in the overall epistemic practice or system" (ibid., 43). The harm is also moral because to be wronged in one's capacity as a knower is to be wronged in a capacity "essential to human value" (ibid., 44). Fricker argues also that *prejudice* is the "ethical poison" which explains why testimonial injustice is not merely harmful but also wrong (ibid., 22). All well-functioning adult human beings are entitled *qua* human beings to be taken seriously as speakers. Therefore, to deprive someone of their capacity to function as a knower is unfair when it happens merely on the basis of social identity prejudice (ibid., 35). This means also that not all cases of credibility deficit are cases of testimonial injustice. For example, when a credibility deficit results from an innocent error, no injustice is involved in it (ibid., 21).

Given the twofold nature of injustice, an account of epistemic injustice should do two things: (1) give an account of the nature of *harm* that may be involved in economics imperialism; and (2) explain why inflicting such harm on someone is *wrong* or *unfair*. While the second condition is dependent on the first condition, it cannot be reduced to the first condition because in some cases there is a justification for imposing a burden on someone. For example, when a person is found guilty of crime, there is a justification for forcing a particular kind of burden upon the person, namely a punishment. Thus, in order to argue that economics imperialism can sometimes be wrong or unfair, one needs to be able to show that imperialists are responsible for harm that is imposed on someone else without any justification.

I argue that one harm caused by economics imperialism is a credibility deficit it imposes on scientists in another discipline, thereby making it more difficult for these scientists to function as a scientific expert vis-à-vis a particular domain. While an individual imperialist does not always succeed in imposing such a deficit on other scientists, he or she often attempts to do so by claiming that an economic approach is superior to other approaches in the social sciences (see e.g., Cudd 2005; Lazear 2000). When such claims accumulate, they can give rise to prejudice against scientists working in another discipline, thereby undermining their credibility as scientific experts. So, whereas testimonial injustice harms someone's capacity to function as a knower, economics imperialism can harm someone's capacity to function as a scientific expert. Whereas testimonial injustice harms someone on the basis of their social identity, economics imperialism can harm someone on the basis of their perceived disciplinary identity.

My analysis of the harm involved in economics imperialism supplements Mäki's (2009, 2013) diagnosis of wrong-headed economics imperialism. As we have seen in section one, in Mäki's view economics imperialism is objectionable when it leads to bad science both in an epistemic and a moral sense of badness. One could, of course, argue that bad science, as Mäki understands it, is harmful also because it is a waste of socially valuable resources. Moreover, bad science can be harmful for some social groups or for the society at large when it is applied in public policy or in the development of technologies. Yet, my analysis differs from Mäki's analysis because in my view some instances of economics imperialism are harmful not only because they lead to bad science (and its deplorable social consequences), but also because they undermine someone else's capacity to function as a scientific expert.

My analysis of the harm involved in economics imperialism supplements also Clarke and Walsh's (2009, 2013) analysis of wrongful economics imperialism. Whereas Clarke and Walsh suggest that the harm is done primarily to the development of science by suppressing some research topics or justifiable moral and social values, I suggest that damage is also done to scientists working in the "colonized" discipline.

Having identified the harm that may be involved in economics imperialism, the next step is to explain why it is wrong or unfair to inflict such harm on someone else. After all, there seems to be nothing wrong in questioning someone's credibility in a particular domain if the person is not an expert in that domain. In order to understand the unfairness that may be involved in economics imperialism, it is necessary to articulate and defend the principle of a *fair distribution of credibility* (FDC). I propose the following formulation:

> (FDC) *A distribution of credibility is fair just in case each person's credibility vis-à-vis a particular domain matches the expertise the person has in that domain, and no person is granted too much or too little credibility given the person's expertise.*

In order to understand how it is possible to have a mismatch between credibility and expertise, it is necessary to explain why credibility is not the same as expertise. While the term "expertise" can be understood in many ways (see e.g.,

Collins and Evans 2007; Turner 2014), I use it to refer to legitimate epistemic authority a person has in virtue of her relatively high level of knowledge, her ability to deploy knowledge in answering questions, and her ability to generate new knowledge in a particular domain (Goldman 2006, 19–20). Whereas expertise in this sense is a matter of having knowledge and an ability to deploy and generate knowledge in a particular domain, credibility is about how others perceive a person's expertise as well as her honesty. Expertise is not always transparent to others, and hence, there can be a mismatch between credibility and expertise so that someone is an expert without being recognized as such, and someone else is granted credibility without being an expert (see also Fricker 1998, 167).

It is important to notice also that both credibility and expertise are domain-specific. Thus, a *fair* distribution of credibility is not always an *even* distribution of credibility. In many cases a fair distribution of credibility is an *uneven* distribution of credibility in which each person is assigned credibility in the domain where the person has expertise and not in some other domain. This should not be surprising because the very idea of expertise is that some people are more knowledgeable in some domains than others.

In order to defend FDC, it is important to notice that it involves both epistemic and moral-political criteria. Epistemic criteria are needed to determine who is rightfully a scientific expert in a given domain. In FDC the epistemic criteria are decisive because we need to be able to identify experts before we can determine whether a particular distribution of credibility is fair. In the next section, I will take up the challenge of explaining how lay people can determine who is rightfully a scientific expert in a given domain.

The moral-political criterion proposes that fairness is a matter of having one's fair share of credibility as an expert. While all well-functioning adult human beings are entitled *qua human beings* to be taken seriously as speakers, credibility *as an expert* is a special kind of credibility which needs to be deserved. In my analysis, I assume that all scientists deserve some degree of credibility as experts in some domains. Thus, economics imperialism is morally wrong when it changes the distribution of credibility in a way that is not fair. In the case of unfair economics imperialism, the unfairness stems from the claim that someone has gained more or is given less than the amount of credibility she or he is due.

Given the principle of FDC, we are in a position to make moral judgments of economics imperialism. We have a normative basis for claiming that it is unfair to withhold credibility from someone in a particular domain if that person is an expert in that domain. Insofar as economics imperialism diminishes the credibility of scientists in the "colonized" discipline without any justification for it, these scientists suffer from an unfair credibility *deficit*. Besides an unfair credibility deficit, economics imperialism can generate also an unfair credibility *excess* (see also Fricker 2007, 17). This happens when imperialists enjoy more credibility in a particular domain than they deserve on the basis of their expertise. Thus, whereas some scientists can be harmed by economics imperialism, some others can benefit from it. Economics imperialism is a political issue because it has the capacity to generate conflicts of interests among social groups.

To sum up the discussion, I have argued that economics imperialism is wrong when it gives rise to a particular kind of epistemic injustice: unfairness in the distribution of credibility. I have also explained how the epistemic and the moral criteria interact in the assessment of whether a distribution of credibility is fair. Next, I discuss two examples of economics imperialism in order to respond to an objection that may be raised against FDC.

4 Evaluating economics imperialism

Against FDC, someone may argue that it is based on a question-begging assumption, namely, that there is a method of determining who is rightfully a scientific expert in a particular domain independently of the contestation over credibility. Unless such a method is provided, FDC remains an empty phrase because the question of expertise is not less contested than the question of credibility.

In response to the objection, I argue that there is a method of evaluating experts even when there is contestation over credibility. Some philosophers suggest that we are entitled to defer to experts when we are not in a position to evaluate relevant evidence and arguments directly (Hardwig 2006). I suggest that we adopt a reverse strategy when we are faced with a disagreement concerning experts. The reverse strategy aims to evaluate experts by assessing relevant evidence and arguments *indirectly*. Even when we are not in a position to evaluate relevant evidence and arguments directly, we can evaluate them indirectly by focusing on the dialectical performance of the expert (Goldman 2006, 22–24). Assessing the putative expert's dialectical performance includes such things as assessing the extent to which she engages alternative perspectives and responds to criticism. While this method is hardly an adequate answer to the question of how a lay person can decide between disagreeing expert testimonies (see e.g., Almassi 2012; Anderson 2011), it can be used to probe the depth of the expertise economics imperialists claim to have in a novel domain.

I argue also that the reverse strategy is made possible by scientific standards that are sufficiently general so that they provide scientists (and others) with a shared ground for evaluating the dialectical performance of experts across disciplinary boundaries. For example, the four criteria Mäki (2013) advances for acceptable economics imperialism are general enough to be acknowledged by most scientists across disciplinary boundaries. Thus, it would be a mistake to assume that otherwise different disciplinary communities subscribe to radically different and non-overlapping sets of scientific standards. A more likely scenario is that the sets are at least partially overlapping. As Kuhn (1977) argues, empirical adequacy and consistency are recognized widely as epistemic values in science. It follows that the assumption underlying FDC, namely that there is a method of determining who is a scientific expert in a particular domain independently of the contestation over credibility, is not question-begging. Given that distinct disciplinary communities are likely to share some scientific standards, there is no mystery in understanding how the dialectical performance of experts can be evaluated across disciplinary boundaries.

In order to demonstrate how the indirect evaluation of arguments can be used to assess experts, I discuss two examples, Edward Lazear's "Economic Imperialism" (2000) and Ann Cudd's *Analyzing Oppression* (2006). Whereas Lazear is an economist who attempts to export economic modeling to business school disciplines such as management and organization studies, Cudd is a philosopher who attempts to import economic modeling to feminist studies. Both Lazear and Cudd claim that an economic approach is superior to some other approaches in the social sciences, thereby aiming to undermine the credibility of scientists in the other disciplines. Given FDC, the distinction between exporting and importing is not morally relevant. The morally relevant question is whether the two imperialists succeed in justifying a deflated level of credibility for scientists in the other disciplines, and whether they succeed in demonstrating their own expertise in the domain where an economic approach is introduced.

Let me assess Lazear's imperialism first. His main thesis is that economics is superior to other social sciences (Lazear 2000, 99). By making such a claim, he suggests not only that economics is capable of explaining phenomena traditionally studied in other social sciences; more significantly, he suggests that economists deserve more credibility than other social scientists in a number of domains, such as consumer behavior, demography, discrimination, family life, religion, the theory of the firm, the theory of human capital, human resource management, accounting, business strategy, organizational behavior, political economy, and the organization of healthcare. As Lazear explains, "[b]y almost any market test, economics is the premier social science. The field attracts the most students, enjoys the attention of policy-makers and journalists, and gains notice, both positive and negative, from other scientists" (ibid., 99). In Lazear's view, the success of economics is due to its rigor and generality. Economics applies a rigorous language which "allows complicated concepts to be written in relatively simple, abstract terms" (ibid., 99). While he grants that complexity may add to the richness of description in the social sciences, he thinks that it prevents social scientists from seeing what is essential (ibid., 99–100).

One does not need to be familiar with the technical details of modeling in economics in order to identify three weaknesses in Lazear's argument. To begin with, passing the "market test" is hardly a hallmark of good science. For example, until recently the theory of anthropogenic climate change has not been very successful when it has been put to the "market test." Yet, there has been a relatively wide consensus in the relevant scientific community that the theory is supported by a range of empirical data. As Elizabeth Anderson (2011) argues, such factors as irresponsible mass media, segregation of social networks, and a particular kind of cultural cognition can explain why an empirically successful theory is not well received by the public.

Another weakness lies in the claim that economics is superior to other social sciences due to its rigor and generality. While other social scientists may have different ideas of what it takes to aim for rigor and generality, Lazear assumes that rigor and generality are best realized when human behavior is represented by means of mathematical models, and this in turn requires that economists are ready to make highly idealized assumptions of human beings. Lazear's argument is based on the assumption that rigor and generality, as economists understand them, are overriding theoretical virtues in the social sciences (see also Sappinen 2003). However, this assumption is false. As Kuhn (1977) argues, simplicity and broad scope are not more important theoretical virtues than, say, empirical adequacy and consistency with other scientific theories. Lazear's argument fails to be convincing because he does not appeal to widely shared scientific standards.

Finally, the most serious problem lies in the gaps in Lazear's argument. Despite his claim to know that economics is superior to other social sciences, he does not engage in any scientific research beyond his own area of expertise: economics. Thus, he does not give us any reason to believe that he is an expert in the domain where an economic approach is claimed to be superior to other approaches. All we are left with is an overconfident claim to credibility without evidence of relevant expertise. Given FDC, it would be unfair to allocate credibility to Lazear merely on the basis of his 2000 article.

Let me turn to my second example, Cudd's (2005, 2006) analysis of oppression. Unlike Lazear, Cudd does not claim that a rational choice theory of oppression is superior to all other theories of oppression. Her view is more complex than this. Whereas some accounts of oppression are rejected (such as Hegel's recognition theory, psychoanalytic theory, and the social dominance theory found in evolutionary psychology), some others are integrated into her analysis of oppression (such as non-economic accounts of stereotypes and group formation).

Another difference between Lazear's and Cudd's imperialism is that unlike Lazear, Cudd engages competing theories of oppression in an epistemically responsible way. For example, she rejects Hegel's recognition theory of oppression because it does not measure up to a certain ontological criterion. As Cudd explains, it is difficult to see how the historical development of the society in itself can act as a causal force independently of individuals as Hegel suggests (Cudd 2005, 40). Also, Cudd rejects a psychoanalytic theory of oppression on the grounds that it is not responsive to empirical evidence (ibid., 41), and she objects to the social dominance theory not only because it is based on scarce empirical evidence but also because it fails to meet an important moral criterion (ibid., 42). According to Cudd, the moral criterion proposes that theories of human social phenomena must treat human individuals as morally primary (ibid., 36). In her view, the social dominance theory violates the moral criterion because it equates oppression with dominance and social hierarchy, thereby collapsing a normative notion into a descriptive one (ibid., 44). As a result, the social dominance theory does not offer any tools to distinguish between legitimate and illegitimate forms of dominance and social hierarchy. Also, the social dominance theory attempts to explain oppression as an outcome of behavioral patterns which are thought to be a function of evolutionary processes. Given the hypothesis that some human beings are evolutionarily programmed to dominate, it is hard to see how oppressive institutions and practices could be changed (ibid., 44).

In order to assess Cudd's arguments indirectly, we need to ask whether she appeals to widely shared scientific standards in the social sciences. I argue that whereas the criterion of empirical adequacy is widely shared, the ontological and the moral criteria are in need of further justification. Cudd defends the ontological criterion by arguing that it is needed to counter explanations that appeal to mysterious collective forces (Cudd 2005, 40). Insofar as social collectives figure in explanations, one should be able to understand their constitution and causal powers in terms of individual behavior (ibid., 28). Cudd also argues that the moral criterion is needed because the concept of oppression is morally value-laden (ibid., 20). In her view, theories of morally value-laden social phenomena should be evaluated not only on the basis of whether they are responsive to empirical evidence but also on the basis of whether they fulfill the moral criterion. When the moral criterion is applied to theories of oppression, it is perfectly legitimate to expect that the theory not only explains how oppression is maintained but also gives advice for ending it (ibid., 32).

To assess Cudd's imperialism, we also need to ask whether the economic approach is actually doing explanatory work in her account of oppression. Cudd (2006) argues that rational choice theory is capable of explaining how the oppressed can be co-opted into participating in their oppression. By the paradoxical term "oppression by choice" she means those economic forces that cause inequality through the choices and decisions of the members of the oppressed social group (ibid., 146). The paradox is that the allegedly oppressive social situation seems to be an outcome of the choices and decisions the oppressed have made themselves, and consequently, the social situation does not seem to count as oppression at all. The paradox dissolves as soon as one shows that the choices and decisions are actually an outcome of oppressive institutions and practices, and therefore, they are not aptly described as free choices. Cudd argues that particular choices and decisions are an outcome of institutions and practices insofar as any "rational" agent (rational in the sense of utility maximization) would have made the same choices and decisions under the same circumstances. Thus, rational choice theory can explain how oppressive institutions and practices put individuals into the kind of circumstances where it is "rational" for them to participate in the maintenance of the same institutions and practices. Instead of blaming the victims or claiming that their choices necessarily undermine the very notion of oppression (one cannot be oppressed if one has chosen one's situation freely), rational choice theory can explain how such paradoxical choices come about (ibid., 152–153).

Given these arguments, we have evidence suggesting that Cudd's claim to credibility in the domain where oppression is theorized is supported by relevant expertise. While her disciplinary background is in philosophy of economics, she has made an effort to understand alternative accounts of oppression in other disciplines. The credibility loss generated in some areas of philosophy, psychoanalytic theory, and evolutionary psychology is not without justification. Thus, no epistemic injustice is involved in her imperialism.

To wrap up the lessons from the two examples, the principle of FDC can actually be used to evaluate economics imperialism. Even when there is contestation over credibility, we can evaluate imperialists on the basis of their dialectical performance. Evaluating dialectical performance involves such things as assessing the extent to which the putative expert engages alternative perspectives and responds to criticism. In the case of Lazear's economics imperialism, there is a mismatch between the claim to credibility and the demonstrated expertise. The demonstration of expertise fails to be convincing because he does not engage alternative perspectives on the many topics he covers in his article on economics imperialism. In the case of Cudd's economics imperialism, there is no such mismatch. Expertise is demonstrated by engaging alternative perspectives in an epistemically responsible way. Thus, to grant credibility to Lazear (2000) in the domains he covers in his article would be unfair, whereas it is not unfair to grant credibility to Cudd (2006) in the domain of oppression.

5 Conclusion

Economics imperialism is not wrong in and of itself. It is wrong when it gives rise to a particular kind of epistemic injustice: unfairness in the distribution of credibility. The unfairness in the distribution of credibility is due to a mismatch between credibility and expertise in a particular domain. When the principle of FDC, a fair distribution of credibility, is applied to the cases of economics imperialism, the epistemic and the moral-political criteria go hand in hand. The epistemic criteria are needed to assess expertise in a particular domain and the moral criterion is needed to assess the fairness of an attempt to change the distribution of credibility among scientists. Economics imperialism can be fair when, for instance, an imperialist engages relevant scientific literature across disciplinary boundaries, as Cudd does in her analysis of oppression. An imperialist is morally blameworthy when she or he makes an excessively confident claim to credibility without relevant expertise, or when she or he attempts to undermine the otherwise well-deserved credibility of other scientists. Whereas a credibility excess may be unfair, the most serious harm that may be involved in economics imperialism is a credibility deficit that is unfairly imposed on some scientists merely on the basis of their perceived disciplinary identity, and not because of their lack of expertise.

Acknowledgments

I wish to thank Aki Lehtinen, Patricia Marino, Uskali Mäki, Manuela Fernández Pinto, Paul Roth, Adrian Walsh, Brad Wray, and anonymous reviewers for their comments on earlier versions of the manuscript. I am grateful also to the audience at the FEMMSS conference in Waterloo in 2014, the ENPOSS conference in Madrid in 2014, and the Workshop on Scientific Imperialism in Helsinki in 2015. Funding for this project has been provided by the Academy of Finland Centre of Excellence in the Philosophy of the Social Sciences.

Note

1 For a criticism of condition (a), see Kidd (2013).

References

Almassi, Ben. 2012. "Climate Change, Epistemic Trust, and Expert Trustworthiness". *Ethics & the Environment* 17(2): 29–49.

Andersen, Hanne and Susann Wagenknecht. 2013. "Epistemic Dependence in Interdisciplinary Groups". *Synthese* 190(11): 1881–1898.

Anderson, Elizabeth. 1995. "Knowledge, Human Interests, and Objectivity in Feminist Epistemology". *Philosophical Topics* 23: 7–58.

Anderson, Elizabeth. 2004. "Uses of Value Judgments in Science: A General Argument, with Lessons from a Case Study of Feminist Research on Divorce". *Hypatia* 19(1): 1–24.

Anderson, Elizabeth. 2011. "Democracy, Public Policy, and Lay Assessment of Scientific Testimony". *Episteme* 8(2): 144–164.

Becker, Gary. 1968. "Crime and Punishment: An Economic Approach". *The Journal of Political Economy* 76: 169–217.

Becker, Gary. 1991. *A Treatise on the Family*. Enlarged edition. Cambridge, MA: Harvard University Press.

Clarke, Steve and Adrian Walsh. 2009. "Scientific Imperialism and the Proper Relations between the Sciences". *International Studies in the Philosophy of Science* 23(2): 195–207.

Clarke, Steve and Adrian Walsh. 2013. "Imperialism, Progress, Developmental Teleology, and Interdisciplinary Unification". *International Studies in the Philosophy of Science* 27(3): 341–351.

Coady, David. 2010. "Two Concepts of Epistemic Injustice". *Episteme* 7(2): 101–113.

Collins, Harry and Robert Evans. 2007. *Rethinking Expertise*. Chicago, IL and London: The University of Chicago Press.

Cudd, Ann. 2005. "How to Explain Oppression: Criteria of Adequacy for Normative Explanatory Theories". *Philosophy of the Social Sciences* 35(1): 20–49.

Cudd, Ann. 2006. *Analyzing Oppression*. Oxford and New York: Oxford University Press.

Douglas, Heather. 2009. *Science, Policy, and the Value-free Ideal*. Pittsburgh, PA: University of Pittsburgh Press.

Downs, Anthony. 1957. *An Economic Theory of Democracy*. New York: Harper and Row.

Dupré, John. 1996. "Against Scientific Imperialism". In M. Forbes, D. Hull and R.M. Burian (eds). *PSA 1994: Proceedings of the 1994 Biennial Meeting of the Philosophy of Science Association*, vol. 2. East Lansing, MI: Philos Sci Association, pp. 374–381.

Dupré, John. 2001. *Human Nature and the Limits of Science*. Oxford and New York: Oxford University Press.

Elliott, Kevin. 2011. "Direct and Indirect Roles for Values in Science". *Philosophy of Science* 78(2): 303–324.

Elliott, Kevin. 2013. "Douglas on Values: From Indirect Roles to Multiple Goals". *Studies in History and Philosophy of Science* 44: 375–383.

Fricker, Miranda. 1998. "Rational Authority and Social Power: Towards a Truly Social Epistemology". *Proceedings of the Aristotelian Society* 98(1): 156–177.

Fricker, Miranda. 2007. *Epistemic Injustice: Power & the Ethics of Knowing*. Oxford and New York: Oxford University Press.

Goldman, Alvin. 2006. "Experts: Which Ones Should You Trust?" In Evan Selinger and Robert P. Crease (eds). *The Philosophy of Expertise*. New York: Columbia University Press, pp. 14–38.

Hacking, Ian. 1996. "The Disunities of the Sciences". In Peter Galison and David J. Stump (eds). *The Disunity of Science: Boundaries, Contexts and Power*. Stanford, CA: Stanford University Press, pp. 37–74.

Hardwig, John. 2006. "Epistemic Dependence". In Evan Selinger and Robert P. Crease (eds). *The Philosophy of Expertise*. New York: Columbia University Press, pp. 328–341.

Kellert, Stephen, Helen Longino and Kenneth Waters (eds). 2006. *Scientific Pluralism*. In Minnesota Studies in the Philos SciXIX. Minneapolis: University of Minnesota Press.

Kidd, Ian James. 2013. "Historical Contingency and the Impact of Scientific Imperialism". *International Studies in the Philosophy of Science* 27(3): 315–324.

Kitcher, Philip. 1981. "Explanatory Unification". *Philosophy of Science* 48: 507–531.

Kitcher, Philip. 1989. "Explanatory Unification and the Causal Structure of the World". In Philip Kitcher and Wesley Salmon (eds). *Scientific Explanation*. Minneapolis: University of Minnesota Press, pp. 410–505.

Kitcher, Philip. 1990. "The Division of Cognitive Labor". *The Journal of Philosophy* 87(1): 5–22.

Kitcher, Philip. 1993. *The Advancement of Science: Science without Legend, Objectivity without Illusions*. New York and Oxford: Oxford University Press.

Kitcher, Philip. 2001. *Science, Truth, and Democracy*. New York and Oxford: Oxford University Press.

Kuhn, Thomas. 1962. *The Structure of Scientific Revolutions*. Chicago, IL: The University of Chicago Press.

Kuhn, Thomas. 1977. "Objectivity, Value Judgment, and Theory Choice". In *The Essential Tension: Selected Studies in Scientific Tradition and Change*. Chicago, IL: The University of Chicago Press, pp. 320–339.

Kuorikoski, Jaakko and Aki Lehtinen. 2010. "Economics Imperialism and Solution Concepts in Political Science". *Philosophy of the Social Sciences* 40: 347–374.

Lazear, Edward. 2000. "Economic Imperialism". *Quarterly Journal of Economics* 115: 99–146.

Lõhkivi, Endla, Katrin Velbaum and Jaana Eigi. 2012. "Epistemic Injustice in Research Evaluation: A Cultural Analysis of the Humanities and Physics in Estonia". *Studia Philosophica Estonica* 5(2): 108–132.

Longino, Helen. 1990. *Science as Social Knowledge*. Princeton, NJ: Princeton University Press.

Longino, Helen. 1995. "Gender, Politics, and the Theoretical Virtues". *Synthese* 104(3): 383–397.

Longino, Helen. 2002. *The Fate of Knowledge*. Princeton, NJ: Princeton University Press.

Mäki, Uskali. 2001. "Explanatory Unification: Double and Doubtful". *Philosophy of the Social Sciences* 31(4): 488–506.

Mäki, Uskali. 2009. "Economics Imperialism: Concept and Constraints". *Philosophy of the Social Sciences* 39: 351–380.

Mäki, Uskali. 2013. "Scientific Imperialism: Difficulties in Definition, Identification, and Assessment". *International Studies in the Philosophy of Science* 27(3): 325–339.

Mäki, Uskali and Caterina Marchionni. 2011. "Is Geographical Economics Imperializing Economic Geography?" *Journal of Economic Geography* 11: 645–665.

Rolin, Kristina. 2015. "Economics Imperialism and Epistemic Cosmopolitanism". *International Studies in the Philosophy of Science* 29(4): 413–429.

Sappinen, Jorma. 2003. *Stretching the Scope of Economics: On the Nature and Impact of Gary Becker's Economics Imperialism*. Lappeenranta: Lappeenranta University of Technology.

Solomon, Miriam. 2001. *Social Empiricism*. Cambridge, MA: MIT Press.

Stigler, George. 1984. "Economics – The Imperial Science?" *Scandinavian Journal of Economics* 86(3): 301–313.

Thagard, Paul. 1999. *How Scientists Explain Disease*. Princeton, NJ: Princeton University Press.

Turner, Stephen. 2014. *The Politics of Expertise*. New York and London: Routledge.

Woodward, James. 2014. "Scientific Explanation". In Edward N. Zalta (ed.). *The Stanford Encyclopedia of Philosophy* (Winter 2014 Edition) http://plato.stanford.edu/archives/win2014/entries/scientific-explanation/.

Wray, K. Brad. 2005. "Rethinking Scientific Specialization". *Social Studies of Science* 35(1): 151–164.

Wray, K. Brad. 2011. *Kuhn's Evolutionary Social Epistemology*. Cambridge: Cambridge University Press.

Wylie, Alison. 1999. "Rethinking Unity as a 'Working Hypothesis' for Philosophy of Science: How Archaeologists Exploit the Disunities of Science". *Perspectives on Science* 7: 293–317.

Zamora Bonilla, Jesus. 2012. "The Economics of Scientific Knowledge". In Uskali Mäki, Dov M. Gabbay, Paul Thagard and John Woods (eds). *Handbook of the Philosophy of Science. Volume 13: Philosophy of Economics*. Amsterdam and Oxford: Elsevier, pp. 823–862.

4 Ethical implications of scientific imperialism

Two examples from economics

Patricia Marino

Introduction

This chapter discusses the ethical implications of scientific imperialism by focusing first on two examples and then exploring more general considerations. In a characterization that sparked much recent discussion, John Dupré says that scientific imperialism involves "the tendency to push a good scientific idea far beyond the domain in which it was originally introduced and often far beyond the domain in which it can provide much illumination" (Dupré 2001, 74). Drawing on examples from economics and evolutionary psychology, Dupré goes on to suggest that imperialism can result in science that is misguided or otherwise problematic. In a series of papers, Steve Clarke, Adrian Walsh, and Uskali Mäki have agreed that this basic expression of the idea is vague and underspecified, especially with respect to normative considerations, but have expressed a diverse set of ideas concerning how best to clarify it (see e.g. Clarke and Walsh 2009, 2013; Mäki 2013).

I draw my examples from economics. Economics imperialism has been seen as the paradigmatic example of scientific imperialism, often prompting negative reactions such as that of Dupré. However, as Mäki has emphasized, the evaluation of economics imperialism can be complex, since there are many relevant factors (Mäki 2009). For this reason I begin here with specific cases, exploring some ethical considerations arising in two specific examples. By "ethical implications," I mean practical or epistemic consequences following from the adoption of imperialistic ideas that have ethically significant – and often potentially problematic – aspects. First, in section one, I discuss the example of economic reasoning applied to domains of love and caring, and then, in section two, the example of economic reasoning applied to domains of sexuality. In section three, I discuss various ways of characterizing economics, explicate the sense in which the examples fit Dupré's original formulation, and explain what I take to be the theoretical connection between the mere expansion into a new domain and the ethical implications that result.

In section four I discuss what I think we can learn from the examples, focusing on three ideas. The first point concerns assessment and values. Here, it is sometimes thought that when it comes to economics imperialism, ethical considerations have to do specifically with issues concerning value pluralism and the possibility that economic values are incompatible with other values. I argue, though, that there is a way to articulate the ethical disvalue of some economics imperialism without appeal to value pluralism. The second point concerns assessment and epistemic disvalue. It might seem that expansion into a new domain can be evaluated as positive when it represents an appropriate form of unification, leading, for example, to successful explanations of a range of phenomena; however, my examples show the potential difficulty of determining what should, and should not, count as a successful explanation. This casts doubt on the possibility of a straightforward criterion. Finally, there is the question of how the proper evaluation of a case of imperialism is related to issues concerning domains and discipline. It has been suggested that domain-only imperialism – where one discipline takes up topics typically studied by another but does not interfere with the practices of that discipline – is mere "expansion" and thus might not carry the normative weight that imperialism does. Though there are many factors to consider in such cases, I argue here that my examples help illuminate some mechanisms through which even domain-only imperialism can have negative consequences.

1 Economics applied to love and caring

The first example is that of applying economic forms of reasoning to domains of love, family life, and caring relationships, and I shall focus first on the work of Gary Becker – to whom Mäki has referred as an "arch imperialist" (Mäki 2009, 371). The central difficulty of using economic approaches in caring contexts is obvious. Standard economic approaches model agents as having self-interested preferences, while in love and caring relationships people are typically seen as other-regarding or altruistic. Indeed, it is often seen as characteristic of love that one must have attitudes that are genuinely other-regarding, that is, concerned ultimately with the good of another (see e.g. Frankfurt 1999; Cave 2003; Stocker 1976).

The difficulty can be finessed in various ways. Becker uses the concept of "altruism": in his basic model, a person is "altruistic" when their utility function depends on that of another. For example, if A's utility is dependent on that of B, so that A's utility goes up when B's does, A is considered to be altruistic toward that person. In Becker's model, when the altruist does something to benefit the target, both are made better off: the target as the recipient, and the altruist as receiving "psychic income" from the exchange (Becker 1991, 299–300). In Becker's standard model of family life, there is only one altruist per family – this assumption is made partly to allow for the application of more familiar and elegant mathematics (England 1989, 16).

Becker's ideas have prompted a great deal of discussion and criticism, much of which focuses on the question of whether families should be theorized as units or as themselves sites of conflict and negotiation (see e.g. Sen 1990). However, my focus here is a bit different, and has to do with ethical implications of modeling caring attitudes economically by interpreting preferences as possibly other-regarding. Before we get to those implications, it is worth noting that Daniel Hausman's recent work on preferences also understands economic agents as possibly other-regarding. In his book *Preference, Value, Choice, and Welfare*, Hausman argues that our understanding of economic preferences should be modified: we should understand them as "total subjective comparative evaluations" (Hausman 2011, 7). So, for example, if A has a preference for enough food, and a preference for B to have enough food, these preferences would be on a par with one another.

To begin to see some implications of allowing other-regarding preferences in economic contexts, let us look first at the "double-counting problem" in the contexts of bargaining and negotiation. In the context of explaining why we might restrict preferences to self-interest in such contexts, the contractarian David Gauthier writes:

> [Supposing that] I, considering us equally fond of cake, prefer that each of us get half, not only to your having a larger share but also to my having it, and [that] you prefer more cake for yourself to less, whatever I get, then it seems implausible to suppose that a rational and fair division gives you three-quarters of the cake and me one-quarter.
>
> (Gauthier 1988, 214; see Morris 1991)

Intuitively, the idea here is that if two people come together, and one person's preferences concern only their own well-being, and another person's preferences concern both their own well-being and that of the first person, then taking all preferences into account means that the first person's preferences get counted twice. So efficient outcomes of preference satisfaction would result in the former gaining over the latter – at least with respect to exchangeable goods. The more other-regarding a person is, the worse off they will be. This, Gauthier says, seems an irrational and unfair outcome.

A simple example might involve caring work in the home. Suppose A and B live in a domestic partnership and A is more other-regarding than B – perhaps by nature, or perhaps through socialization and upbringing. If A and B bargain or maximize preference satisfaction to determine who does the caring work around the home – cooking, cleaning, caring for the other's well-being – and they allow both self-regarding and other-regarding preferences, then a result that is efficient, maximizes preference satisfaction, and is likely to arise through negotiation will be one in which A does more of the caring labor than B. Though the social factors and causes are complex, it is worth noting that one standard explanation of the gender wage gap – that women earn substantively less money than men – is that women do more caring labor in the home; this is often attributed to caring preferences arising at least in part through socialization. Another example might be found in commercial surrogacy: as Elizabeth Anderson explains, some surrogates are partly motivated by altruism; if they are offered only what "the norms of commerce demand in return," she argues, they are open to exploitation. Again, the more other-regarding are, unfairly, made worse off (Anderson 1990, 81).

The implication with ethical significance here is that deploying economic models to certain situations in certain ways results in our taking as efficient and appropriate outcomes that are intuitively unfair – that is, outcomes in which other-regarding people end up worse off. Of course, the connection between the use of the model and the ethically loaded conclusion is contingent and rests on certain specifics and assumptions. First, the ethically significant implications arise only when the models are used in a certain way – in connection with ideas about bargaining, efficiency, preference satisfaction, and well-being. The models need not be used that way: Hausman, for example, argues for alternative views of well-being that go far beyond preference satisfaction. However, the relevant use of the models is itself one deeply associated with economic reasoning: it encodes the basic idea that if an outcome is efficient and satisfies personal preferences then that is a reason to value that outcome.

Second, the claim that some outcomes are "unfair" because they mistreat those with other-regarding preferences follows only when we embrace certain conceptualizations of concepts like fairness, and these could be challenged. Indeed, Becker's idea of psychic income at least suggests an interpretation in which the inequality that Gauthier finds "unfair" would not be unfair at all. If we count the altruist as gaining through the satisfaction of other-regarding preferences in just the same way they gain through self-interested ones, then there would be no intuitive unfairness. It is worth noting, however, that this way of conceptualizing things leads to its own evaluative conclusions, some of which might be contrary to widely shared ethical judgments. For example, if the satisfaction of caring preferences is evaluatively on a par with the satisfaction of other preferences, then it would seem to follow that the "gender-wage gap" is not a gap and not a problem, even if the asymmetry is a result of patriarchal social norms.

The preceding considerations are conceptual and abstract, so to see how there may be practical ethically significant implications in this example, let us consider a few specific ways in which the economic model could be used in practice. For example, consider first what Becker's model would recommend for the policymaker who wants to increase overall efficiency. As we have seen, if we take seriously Becker's idea that the altruistic gain in "psychic income" is a gain relevantly similar to material gains, then we will see as efficient results that are intuitively unequal, and the wage gap is not a gap at all. A policy strategy that aims for efficient outcomes would then be insensitive to any wage gap. This is an ethical implication because it means evaluating policies without regard to whether they decrease or increase the wage gap, a gap often thought to be an unjust manifestation of sexism.

Second, we might think about what the model recommends as rational for the person who believes the model is apt and has the goal of increasing their own preference satisfaction. One possibility is that a less altruistic person will rationally exploit the altruism of others by trying to increase it at no cost to themselves. For example, a non-altruist who wants to do less housework might try to instill a social value of altruism in others. Indeed, though it is hard to say how the causal relations function, this has a certain coherence with the way patriarchal norms have historically propped up a social norm that women should be nurturing. Once caring relationships are understood in the context of bargaining and negotiation, the less altruistic can use these tools to exploit the more altruistic. There is some evidence of this happening in the surrogacy debate. In a study of Indian surrogate mothers, Amrita Pande points out how the surrogates are encouraged to see themselves as altruistic, as engaging in a helping and caring activity. Among other things, this may make them easier to control and exploit. One counselor and hostel matron tells her surrogates, "don't treat it like a business … Don't be greedy" (Pande 2014, 978). Again, any conclusions about exploitation rest on assumptions about how that concept should be understood, but the idea here is that when an individual deploys a Becker-style model to figure out what is rational for them to do, the consequences can be intuitively exploitative.

Third, we might think about what the model recommends for the person who believes the model is apt and wants to use the model to further their own interests in light of the factors mentioned in the first and second points. In her work arguing that economic models can serve feminist ends, Ann Cudd recommends withholding caring labor when others fail to do their fair share. Sensitive to the "double-counting problem," Cudd explains how we can distinguish "self-interested" from "selfish" and avoid the difficulty. Rather than taking other-regarding people as having preferences concerning the preferences of others, we can take them as having preferences concerning the well-being of others: no preferences are then "double-counted." Cudd argues that women who are doing an unfair share of caring labor can use rational choice theory and game theory strategically to withhold caring labor so others will do their fair share. She writes:

> If we understand caregiving behavior as, in Gauthier's terms, constrained maximization and noncaregiving behavior as straightforward maximization, then game theory tells us when it is rational to be a caregiver and when it is rational to increase the penalties against non-caregivers in order to increase their numbers in the environment. Game theory suggests that one's disposition is a choice to be caring or to be selfish, and thus that the relative number of caregivers and noncaregivers might be affected by how the collective of the caregivers treats the noncaregivers.
>
> (Cudd 2001, 412–413)

Interestingly, Cudd talks about a disposition as a choice to be caring or selfish, suggesting that a person subject to inequality as the result of other-regardingness might try to cultivate in themselves a less caring, more selfish disposition. While it is possible that others will respond by cultivating in themselves more caring dispositions, other results seem possible as well. The total amount of caring labor required is not fixed: perhaps people will respond by having fewer children, so that less overall caring labor is required. Also, as we have seen, the person who wants to increase their own selfish preference satisfaction can also use the model to develop strategy: if the targets of this proposal want to continue doing less caring labor, why would they not cultivate in themselves a less caring disposition as well? In this case, we would see a kind of arms race of selfishness. Using the economic model in this way may well help feminist ends, but may also lead to the ethically significant outcome that it is rational for each person to cultivate a less caring disposition in themselves.

2 Economics applied to sexuality

The second example concerns rationality, preference attribution, and sexuality. For a long time it has been standard practice in economic reasoning to assume that people are rational – in the sense of acting to maximize their utility or preference satisfaction. Preferences are typically understood through "revealed preference" theory, so that what a person does is taken to reveal what their preferences were. As long as a person's preferences fit the formal criteria of rational choice theory – such as transitivity – those preferences are considered rational.

Over the last few decades this picture has come under sustained attack from psychologists, neuroscientists and others who have pointed out that it is false to the facts: people are not always rational; in fact, they have identifiable patterns of irrationality and biases. Behavioral economists point to examples like "endowment effects" – where we come to value goods more when we already possess them. In a classic paper on the subject, Christine Jolls, Cass Sunstein, and Richard Thaler set out a range of ways in which people fail to act according to the standard theory: they have "bounded [epistemic] rationality" – that is, they do not always reason well; they have "bounded willpower" – that is, they sometimes act contrarily to what they know is best for themselves; and they have "bounded self-interest" – that is, they sometimes behave according to norms of morality, fairness, and altruism (Jolls et al. 1998a).

Against this, however, defenders of the classical approach have pointed out that it is often possible to interpret a person's behavior so that it is rational, by ascribing certain preferences to them. For example, in response to the claim that a person is irrational to eat lobster contentedly if he does not see it when it is alive, but not if he is asked to pick it out of a lobster tank, Richard Posner writes: "an alternative interpretation is that this person simply has different preferences for two different goods: One is a lobster seen only after being cooked, and the other is a lobster seen before, in his living state, as well as after" (Posner 1998, 1553). Posner says the behavioralists' approach "[fails] to distinguish between impediments to clear instrumental reasoning and preferences that we enlightened observers may think silly" (ibid., 1553). With respect to the claim that the classical model cannot help us understand self-destructive preferences like drug addiction and unsafe sex because it requires that people "choose what they prefer," Posner says: "Not so: in fact, the standard conception will assume that people do not want to become addicted or to contract AIDS … [then explore] the conditions under which the costs of such behavior, steep as they are, are nevertheless outweighed by perceived benefits" (ibid., 1557). In response to reasoning like this, behavioralists argue that the rational preferences attributed by the standard approach will be artificial and ad hoc – as if the possibility of finding an explanation that makes a behavior rational shows that this explanation is the right one, when it obviously does no such thing (Jolls et al. 1998b, 1595).

The basic epistemological problem here is that when a person does an unexpected thing, we do not know whether the person acted anomalously in satisfying expected preferences or whether they acted rationally in satisfying anomalous preferences. In his 2012 book *A Course in Behavioral Economics*, Erik Angner discusses the epistemological problem this way:

> We have seen that some of the empirical results invoked by behavioral economists might be possible to accommodate within the standard framework, either by redescribing the choices available to the agent or by admitting additional arguments into the utility function. And it is important not to attribute irrationality to people when their behavior is better described as consistent with standard theory. That said, neoclassical economists frequently bend over backwards to accommodate empirical results in a manner that is both artificial and ad hoc. It is often simpler and more plausible to infer that people are sometimes in violation of standard theory.
>
> (Angner 2012, 210)

This passage suggests that when it is possible to describe someone's behavior as rational or irrational by attributing to them different sets of preferences, and we are trying to decide which description best fits the case, among the factors we ought to consider are overall plausibility and whether the attribution of one set of preferences seems "ad hoc." Scientifically, these are good norms. However, as I shall argue, in some domains – like that of sexuality – it is difficult to judge when one set of preference attributions is more artificial and ad hoc than another. In those cases, because we must appeal to judgments about whether a preference attribution is ad hoc or artificial in making our evaluation, we must rely on intuition or background judgments about how anomalous a preference attribution is – that is, on judgments about what it makes sense to prefer.

I think there are many fascinating examples of the relevant difficulty, but I shall focus here on sexuality. In a much-discussed book originally published in 1975, Kristin Luker presents and analyzes the results of an in-depth study she performed interviewing women about their decisions to abort their pregnancies (Luker 1975). For the purposes of the study, she selected women who knew and understood the risks of getting pregnant through unprotected sex and also knew how to use contraception.

On the face of it these seem like paradigm cases of irrational behavior. However, classifying the behavior this way would require making assumptions about what agents prefer. If we reason analogously to the lobster example – and in accordance with what Posner says about unsafe sex – we get the following analysis. We assume the women do not want to get pregnant, and then we attribute to them preferences for "perceived benefits" that outweigh this preference and make their behavior rational.

In her own analysis, Luker argues that the women are rational, not by assumption but on the basis of preferences she ascribes to them from her interviews. Luker offers several considerations for how we might understand the costs and benefits involved. She points out that for the women in the study there were many costs associated with using contraception. For some women, using contraception meant admitting to themselves that they were sexually active, which conflicted with powerful social norms. Some women had internalized religious norms against using contraception. Some women, in relatively stable relationships, resisted discussing contraception with their partners because of possible friction over the implied suggestion of couple-dom. Some women assumed contraception was a man's responsibility. Some women resented the pill because it seemed to give their partners worry-free access to their bodies, so their partners could use them for sex and without having to take them into consideration in a deeper or more long-term way; stopping the pill felt like a way of restoring equality to the relationship.

However, even with all of this information, personal and painstakingly gathered, it is unclear whether the preferences are rational. Elizabeth Anderson points out that Luker's subjects often acted in what seem like paradigmatically irrational ways. They ignored the fact that they could get pregnant, they engaged in wishful thinking, they were passive when they could have acted, they judged risks incorrectly, and they procrastinated about talking things over with their partners. Anderson adds: "Luker mistakenly supposes that every motivation against doing something is capable of counting as a reason against doing it [...] but some motives repel action without supplying reasons – that is, considerations that can survive conscious scrutiny" (Anderson 2001, 385).

Anderson summarizes this way:

> While it is true that, for any observed behavior, one can construct a rational choice model that generates it, it doesn't follow that the model identifies the actual causes of that behavior. [Furthermore], what counts as a good explanation of a phenomenon depends on what aspects of that behavior one wants to understand.
>
> (Anderson 2001, 389)

Even with information directly from the people in question, we are still not sure when they acted rationally and when they did not.

Now, recall the proposal that when we must make a decision from among different ways of ascribing preferences we ought to consider which approach is overall more or less artificial and ad hoc. I claim that in this kind of context, such a strategy may introduce certain risks. A theorist trying to decide which of two preference ascriptions is more plausible and less ad hoc overall is likely to appeal to their own judgment and life experience about what it makes sense to want. Even in the relatively simple case of the lobster preferences, how are we to decide? Is it more plausible and less ad hoc to attribute irrationality to a person, or to attribute to them a set of condition-specific preferences? To some, the latter approach – in which we say the person preferred lobster if they had not seen it alive and something else to eat if they had – might seem to introduce a complexity. To a person who does not often experience changes of preference from similar factors, the preference ascription may seem ad hoc. However, to a person who often experiences such changes in preferences, it may seem plausible and true to life.

The danger is that judgments about rationality will involve projections of the attitudes of the person making the judgment. In domains like sex, people's preferences vary widely and there is deep disagreement over which sexual preferences make sense and which seem like outliers. This means that when it comes to sexuality, we are especially in danger of attributing to others the preferences we think are intuitively likely. For a person with one sexual history and context, a decision not to use contraception to prevent unwanted pregnancy might seem obviously irrational. For a person with a different history and context, the decision not to use contraception to prevent unwanted pregnancy might seem familiar and intuitive, a result of the very pressures Luker points to.

The ethical implications of the epistemological problem are complex, but there are several I find particularly significant. First, policymakers may rely on judgments about rationality in thinking about appropriate interventions. Thaler and Sunstein have used their behavioral theory as a basis for their "Nudge" proposal: that we should exploit persistent mistaken biases to make people do what is really in their interest and to increase efficient outcomes (Thaler and Sunstein 2008). This is an explicit attempt to influence behavior in favor of rationality and against irrationality. Conversely, Posner and his followers have argued that we have reason to think that consensual exchanges generally make everyone involved better off, since no one would rationally choose to make the exchange unless they preferred the outcome. However, this implicitly rests on an assumption that behavior generally involves a rational reflection of a person's preferences. Both approaches are problematic because they rest on ungrounded assumptions about what people prefer.

With respect to nudging, notice that if we try to nudge people toward behavior that is seemingly more rational when we have attributed that rationality based on misjudgment about a person's preferences, we are nudging them in the wrong direction. Suppose Luker is right when she says the women in question were choosing rationally because of complex preferences. To nudge them toward more rational behavior would be to nudge them toward unprotected sex, even when children are unwanted. This seems a peculiar conclusion.

The peculiarity of this conclusion helps to illustrate, I believe, that if nudges sometimes do seem appropriate, this is because of background beliefs about norms, values, and what it makes sense to want, and not because of rationality and irrationality. From one normative perspective – that abortion is bad but contraception is not – nudges might include reminding people of the possibility of pregnancy or making contraception more visible and available. There are, of course, people who believe that abortion is bad and contraception is also bad, and from this perspective, appropriate nudges would push people who do not want children not to have sex at all. The point here is that when it comes to influencing people's behavior, judgments about rationality are never sufficient for justification: there are always ethical or socio-political value judgments in play. The ethically significant implication, then, is that when we try to use economic models intended to nudge people toward rationality, we do so on the basis of implicit and thus unexamined pre-theoretical judgments about what it makes sense to want.

Ethical implications also arise from taking up Posner's classical strategy. The classical strategy – assuming that people do not want the adverse consequences then exploring the conditions under which the perceived benefits outweigh the costs – excludes the possibility that Luker's subjects were not better off for having chosen as they did. We would, contrary to Anderson's analysis, have to conclude that those subjects were increasing their own well-being. A policy based on assumptions that behavior can be explained as the rational satisfaction of preferences leads us to evaluate choices as good for the choosers, even when they may not be.

Posner's broader theorizing about sexuality in his book *Sex and Reason* provides another interesting example of how assumptions about preferences in economic theorizing can have ethically significant implications (Posner 1994). Posner's book is an attempt to explain patterns of sexual activity (and legal regulation of that activity) through economic analysis. Posner takes a broadly sociobiological account in which sexual preferences are biological, men have a strong drive for sex and women do not, and the trade-offs men make to satisfy their preferences as best they can drive much of the explanation of sexual behavior. Among other things, it follows that the diversity of sexual behavior which is often thought to be evidence for the socially constructed and variable nature of sexuality is attributed instead to variation in circumstances: distinguishing between "real" male homosexuality and "opportunistic" male homosexuality, Posner argues that seeming "diversity" in behavior can be explained through situational trade-offs that opportunistic men make.

Here, necessary background about what people prefer is provided by the sociobiology. Though Posner argues that the biological aspect of his account is inessential to it, I agree with Gillian Hadfield when she claims that this is not so. Hadfield writes:

> If sex drives are not biological but rather are the product of social factors such as deeply ingrained sexual mores or the nature of the erotic lessons that a person learns during the development of a sexual self, if "sex drive" itself is socially constructed, then it is no longer possible to take sexual preferences as given and explain sexual practices and morality as a function of substitution among options of varying costs in order to satisfy these fixed preferences.
> (Hadfield 1992, 490)

That is, the mode of analysis requires making assumptions about the preferences in order to derive explanatory conclusions, and seeing them as biological and fixed is a way to do that. As several commentators have argued, this way of proceeding introduces several factors with normative and ethical dimensions.

The sociobiological account in which men's sexual pursuit of women is fundamental and paramount has come under criticism in part for its sexist assumptions and methodology (see Hadfield 1992). The reification of "real" versus opportunistic male homosexuality is incompatible with Foucauldian ideas about social construction and also with the idea that recognizing the fluidity of desire and orientations might be essential to political progress (see e.g. Rich 1980; Eskridge 1992, 374). Finally, whole categories of people are erased from view: as Hadfield says, lesbians are essentially invisible in Posner's theory, since female desire is not seen as relevant to the analysis. Hadfield concludes: "Posner's biology sets the stage for an economic theory in which men are economic actors and women are economic objects; in which men are rational and women are, essentially, rationed" (Hadfield 1992, 491).

Ethical implications of applying economic analysis to decisions about sexual activity, then, can include the following. Because of epistemological problems, we must make assumptions about preferences and rationality. Partly because of the widely varying and socially complex nature of sexual preferences, it is easy to get these wrong: theorists risk importing their own judgments about what things it makes sense to want, or relying on inappropriate generalizations: e.g., that men are sexual aggressors and women are sexual providers. As in the nudge and contraception example, the use of the theory in policy can obscure the operation of values and normative beliefs in judgments about which nudges are good ones. Some generalizations erase from view people and phenomena we think are important, such as female sexual agency, and by extension, same-sex activity between women.

3 Economics and its domain

Let me now consider some issues related to the definition and domain of economics and how our examples bear on these (see Backhouse and Medema 2009, on definitions of economics discussion). First, as I explain below, I believe that the examples nicely illustrate Dupré's rough characterization of an idea being "pushed" beyond a certain "domain."

In the early phases of its development, economics was often understood as the study of a certain domain or certain restricted aspect of human behavior. Focusing on economics as a branch of the science needed to lead and legislate, Adam Smith concerned himself with the wealth of various nations – thus a certain domain. Writing in the late 19th century, Mill described economics as "concerned with [persons] solely as [beings] who [desire] to possess wealth," and defined it as

> the science which traces the laws of such of the phenomena of society as arise from the combined operations of mankind for the production of wealth, in so far as those phenomena are not modified by the pursuit of any other object.
>
> (Mill 2000, 99)

With respect to assumptions and domain, Mill says that economics proceeds by investigating

> laws [...] under the supposition that man is a being who is determined, by the necessity of his nature, to prefer a greater portion of wealth to a smaller in all cases, without any other exception than that constituted by the two counter-motives already specified. Not that any political economist was ever so absurd as to suppose that mankind are really thus constituted, but because this is the mode in which science must necessarily proceed.
>
> (ibid., 98)

Even Robbins' famous definition of economics in the 1930s as "the science which studies human behavior as a relationship between ends and scarce means which have alternative uses" – which is scope-expanding in one sense – characterizes economics with reference to a certain domain (Robbins 1932, 15).

Though modern definitions of economics are famously varied, taking any of these early definitions we can see how my examples concern economic forms of reasoning being applied beyond the domains mentioned by these early theorists. Love and sex are not only different from wealth, but also resist description as "scarce resources" that can be used for other ends. Love and caring are not fixed in amount and their degree of scarcity just depends on us and our feelings for one another; in addition, love is often thought to be essentially particular and thus ill-suited to serving a variety of alternative ends. Sex is not a commodity to be exchanged, and is also not a scarce resource; in fact, it is often thought that mutuality, in which sexual desire and pleasure are mutually interdependent and increasing together, is a kind of ideal for sexuality.

Furthermore, the examples illustrate some mechanisms through which expansion beyond a domain can be related to implications with ethically significant aspects. One of these mechanisms has to do with the way that simplifying and universalizing assumptions that seem innocuous when applied in one domain can lead to problems when applied in another. When Mill says that economics concerns itself with only one aspect of human behavior, falsely assuming for the purpose of progress that people are subject only to the limited range of motives concerned there, because "this is the mode in which science must necessarily proceed," he is aptly prefiguring contemporary thoughts about the importance of idealizing assumptions in science. The assumption that people are rational, self-interested maximizers of preference satisfaction can be seen in this light: a universalizing and generalizing assumption about people that is not always true but that works well for some purposes in some domains. Looking at wealth and its acquisition, or more broadly at human behavior concerning ends and scarce means, it may be functional and innocuous to assume that people are self-interested rational maximizers of their own preferences. However, in domains like sex and love the assumption is an odd fit: here, people are often other-interested and potentially irrational, the goods are not scarce resources that can be used for other ends, and questions about givenness and diversity of preferences are hotly debated.

4 Implications for imperialism

In this section I consider three ways the discussion so far bears on debates about scientific imperialism. The first concerns the expression of values. As part of the evaluative assessment of imperialism, Clarke and Walsh mention the possibility that imperialism may cause us to "fail to appreciate the irreducibly pluralistic nature of human values" (Clarke and Walsh 2009, 203). They cite the work of Margaret Radin, Elizabeth Anderson, and others as showing how the expansion of economic thinking leads us to value things in the wrong kind of way: family relationships, for example, represent values that are not reducible to, and not commensurate with, market values.

Of course, in some ways my analysis echoes these concerns, but in my opinion, the examples I have discussed importantly extend Clarke and Walsh's expression of the relevant idea. In both examples, I have aimed to draw out ethically significant considerations associated with the application of economic reasoning: that some models of other-regardingness lead to unfairness or a decrease in caring, that some models of sexuality risk incorrect judgments about preferences in ways that lead to problematic conclusions. While I believe it is apt in a general way to characterize these conclusions as cases in which the "expression" of certain values is prevented, I think the mechanism goes beyond considerations related to "irreducible plurality of values" and does not rely on theses about plurality or incommensurability. Judgments about which ethical implications are problematic, and how they are so, depend on what ethical judgments we come to the table with, but for anyone who thinks those ethical implications are problematic, the relevant concerns will carry over – whether they believe the problems have to do with plural values and incommensurability or not. For example, a person might think the gender wage gap unfair, and might find it objectionable to theorize away the existence of women with same-sex desires, without any particular commitment about the values supporting their beliefs or whether those values are plural or incommensurable.

Next let me turn to epistemic disvalue. With respect to the epistemological issues, Clarke and Walsh note that Dupré's thought about ideas being "unilluminating" when used in the target discipline raises questions. If uses of ideas in new ways were obviously unilluminating, they would be unlikely to catch on or be seen as successful – so what is the concern? In a related vein, Mäki points out that expansion into new domains can sometimes exemplify a salutary attempt at unification. Since scientific unification can sometimes be a form of progress, we should try pursuing unifying projects subject to certain constraints – trying, for example, to find ontologically unifying explanations as long as those explanations are successful (Mäki 2013, 336).

With respect to the question of "unilluminating" new approaches that are nonetheless seen as successful, Clarke and Walsh propose that researchers may make a "subtle error": they may "fail to appreciate that such methodologies bring with them a way of approaching their subject matter that sheds light on some aspects of the subject of study but obscures others" (Clarke and Walsh

2009, 202). In response to Mäki's idea about unification, they note that the question of what constitutes a good or successful explanation can be complex:

> An economics imperialist, such as Becker (1991), may believe that he has successfully discerned and then explained a range of social phenomena, when in fact he has been able to explain these phenomena, to his own satisfaction, only by abstracting away from social reality and rendering himself oblivious to aspects of human behaviour that are not well explained by economics.
> (Clarke and Walsh 2013, 348)

I take this to mean: since what counts as a good or successful explanation can be contextually sensitive, the proposed constraint would not function to rule out the proper examples.

I believe my two examples bring to light several ways in which the relevant sorts of complexities about explanation can arise. First, in the case of love and caring, from the point of view of Becker's style of analysis it comes out as axiomatic that when a person acts altruistically toward another, it is because they preferred to do so: in a sense, then, the theory allows us to "explain" altruistic behavior. However, from other perspectives, the causes of altruistic preferences are complex, and can include cultural and social factors, factors typically left out of the economic approach. Judgments of whether the new, more unified explanations are successful may reflect practitioners' judgments over the significance of these aspects of human behavior. From a theorist steeped in an economic approach, the new explanations may seem more successful. From a theorist steeped in a sociological one, they may seem less successful.

Another example of complexity in evaluating explanations has to do with the gender wage gap. If women do more caring labor because they are more likely to have altruistic preferences, in one sense this "explains" the wage gap: people are simply doing what they prefer to do. Again, this leaves out potentially relevant factors: if women face reduced ranges of options from which to choose, this bears importantly on the wage gap, but insofar as the theory takes preferences as basic, this level of analysis becomes irrelevant. Again, whether we see a more unified explanation as successful may depend on pre-theoretic commitments over whether these other factors are important to consider or relatively trivial.

In the case of rationality decisions about sexual activity, we have seen how both the classical approach and the behavioralist one can "explain" the decision to have unprotected sex when children are not wanted: the classical theorist by attributing rationality and asking which preferences would cause the person to act this way, the behavioralist by subsuming the behavior under one of a range of exceptions like "bounded willpower." However, as we have seen, knowing why people do what they do is complex and it is possible that neither theory is getting the right explanation. How would we determine whether the new, more unified explanation is an improvement?

These are all examples where what seems like a good explanation is discipline-specific: an explanation might seem like a good one from within the imperializing discipline, but be importantly lacking when viewed from colonized ones. Among other things, in these kinds of cases assuming the superiority of the imperializing explanation may be a form of epistemic injustice to those in the colonized discipline (see Rolin, this volume). These factors complicate attempts to evaluate attempted unification, and this shows how the "subtle error" mentioned by Clarke and Walsh is a real possibility.

The final issue has to do with what Mäki calls "domain-only" and "disciplinary" imperialism. While disciplinary imperialism involves intrusion into other disciplinary cultures and practices, "domain-only" imperialism involves explaining phenomena in target disciplines while "ignoring those disciplines themselves" (Mäki 2013, 334–335). While acknowledging that different cases raise different issues, Mäki characterizes domain-only imperialism as generally the "weakest" and "most innocent" imperialism of scope. If what I have said so far here is on the right track, though, it seems to me that these ethical implications – and the epistemic disvalue mentioned above – arise primarily from expansion into new domains, and that any wrongs arising from disciplinary issues would be additional to these.

5 Conclusion

In this chapter I have used the specific examples of economics applied to love and sex to show how economics imperialism can have ethically significant implications, and I have used the results to remark on issues concerning assessment and epistemic disvalue. With respect to assessment, I have suggested that negative aspects of economics imperialism can go beyond disciplinary matters and may concern large-scale sociocultural implications. With respect to epistemic disvalue, I have argued that knowing what counts as a successful explanation can be a contextually sensitive and difficult matter. In the broadest terms, I have aimed to show how the use of assumptions appropriate for more traditional economics domains leads to problems when applied in other domains, such as love and sex, where they do not fit well.

Acknowledgments

I am grateful to Doreen Fraser, Manuela Fernández Pinto, Kristina Rolin, and Adrian Walsh for helpful comments on earlier drafts. Thanks also to all the conference participants for their insightful questions and remarks.

References

Anderson, Elizabeth. 1990. "Is Women's Labor a Commodity?" *Philosophy and Public Affairs* 19: 71–92.

Anderson, Elizabeth. 2001. "Should Feminists Reject Rational Choice Theory?" In *A Mind of One's Own: Feminist Essays on Reason and Objectivity*, Second edition. Boulder, CO: Westview Press, pp. 369–397.

Angner, Erik. 2012. *A Course in Behavioral Economics*. Basingstoke: Palgrave Macmillan.

Backhouse, Roger and Steven Medema. 2009. "Retrospectives: On the Definition of Economics". *Journal of Economic Perspectives* 23: 221–234.

Becker, Gary. 1991 [1981]. *A Treatise on the Family*. Boston, MA: Harvard University Press.

Cave, Eric. 2003. "Marital Pluralism: Making Marriage Safer for Love". *Journal of Social Philosophy* 34: 331–347.

Clarke, Steve and Adrian Walsh. 2009. "Scientific Imperialism and the Proper Relations between the Sciences". *International Studies in the Philosophy of Science* 23: 195–207.

Clarke, Steve and Adrian Walsh. 2013. "Imperialism, Progress, Developmental Teleology, and Interdisciplinary Unification". *International Studies in the Philosophy of Science* 27: 341–351.

Cudd, Ann. 2001. "Rational Choice Theory and the Lessons of Feminism". In Louise Antony and Charlotte Witt (eds). *A Mind of One's Own: Feminist Essays on Reason and Objectivity*. Boulder, CO: Westview Press, pp. 398–417.

Dupré, John. 2001. *Human Nature and the Limits of Science*. Oxford: Clarendon Press.

England, Paula. 1989. "A Feminist Critique of Rational-Choice Theories: Implications for Sociology". *The American Sociologist* 20: 14–28.

Eskridge, William. 1992. "A Social Constructionist Critique of Posner's *Sex and Reason*: Steps Toward a Gaylegal Agenda". *Yale Law Journal* 102: 333–386.

Frankfurt, Harry. 1999. "Autonomy, Necessity, and Love". In *Necessity, Volition, and Love*. Cambridge: Cambridge University Press, pp. 129–141.

Gauthier, David. 1988. "Morality, Rational Choice, and Semantic Representation". *Social Philosophy and Policy* 5: 173–221.

Hadfield, Gillian. 1992. "Flirting with Science: Richard Posner on the Bioeconomics of Sexual Man". *Harvard Law Review* 106: 479–503.

Hausman, Daniel. 2011. *Preference, Value, Choice, and Welfare*. Cambridge: Cambridge University Press.

Jolls, Christine, Cass Sunstein and Richard Thaler. 1998a. "A Behavioral Approach to Law and Economics". *Stanford Law Review* 50: 1471–1550.

Jolls, Christine, Cass Sunstein and Richard Thaler. 1998b. "Theories and Tropes: A Reply to Posner and Kelman". *Stanford Law Review* 50: 1593–1608.

Luker, Kristin. 1975. *Taking Chances: Abortion and the Decision Not to Contracept*. Berkeley and Los Angeles: University of California Press.

Mäki, Uskali. 2009. "Economics Imperialism: Concept and Constraints". *Philosophy of the Social Sciences* 39: 351–380.

Mäki, Uskali. 2013. "Scientific Imperialism: Difficulties in Definition, Identification, and Assessment". *International Studies in the Philosophy of Science* 27: 325–339.

Mill, John Stuart. 2000. "On the Definition of Political Economy and the Method of Investigation Proper to It". In *Essays on Some Unsettled Questions of Political Economy*, Second edition. Kitchener: Batoche Books, pp. 86–114.

Morris, Christopher. 1991. "Moral Standing and Rational Choice Contractarianism". In Peter Vallentyne (ed.). *Contractarianism and Rational Choice: Essays on David Gauthier's Morals by Agreement*. Cambridge: Cambridge University Press, pp. 76–95.

Pande, Amrita. 2014. "Commercial Surrogacy in India: Manufacturing a Perfect Mother-Worker". *Signs* 40: 969–992.

Posner, Richard. 1994. *Sex and Reason*. Boston, MA: Harvard University Press.

Posner, Richard. 1998. "Rational Choice, Behavioral Economics, and the Law". *Stanford Law Review* 50: 1551–1575.

Rich, Adrienne. 1980. "Compulsory Heterosexuality and Lesbian Existence". *Signs* 5: 631–660.

Robbins, Lionel. 1932. *An Essay on the Nature and Significance of Economic Science.* London: Macmillan.

Sen, Amartya. 1990. "Gender and Cooperative Conflicts". In Irene Tinker (ed.). *Persistent Inequalities: Women and World Development.* New York: Oxford University Press, pp. 123–149.

Stocker, Michael. 1976. "The Schizophrenia of Modern Ethical Theories". *The Journal of Philosophy* 73: 453–466.

Thaler, Richard and Cass Sunstein. 2008. *Nudge: Improving Decisions about Health, Wealth, and Happiness.* New Haven, CT: Yale University Press.

Part II

Historical origins of scientific imperialism debate

5 Scientific imperialism or merely boundary crossing?

Economists, lawyers, and the Coase theorem at the dawn of the economic analysis of law

Steven G. Medema[1]

Introduction

Economics features prominently in discussions of "scientific imperialism," and with good reason. Two of the most prominent examples of it, economic analysis of law and the economic approach to politics (the latter often referred to as public choice analysis), have long and transformative histories, and there are few corners of the humanities and social sciences untouched by economic analysis. Critics of the imperialistic turn in general (e.g., Dupré 1994, 2001) and of economics imperialism in particular (e.g., Fine 1999, 2002; Fine and Milonakis 2009) conjure up visions of colonizing economists attempting to impose their visions of scientific methodology and of a society organized on market principles on fields of inquiry ill-served by either, all the while ignoring the fact that one discipline cannot "impose" its methods on another and thus that someone must be buying what the economists are selling. Economist cheerleaders (Stigler 1984; Lazear 2000) only feed the fires, and even less overtly messianic scholars, such as Gary Becker (1976), emphasize that the economic approach to human behavior has almost boundless application. It is difficult to deny that economics offers an easy target for critics of scientific imperialism as well as wonderful case studies in its origins and development.

To this point, however, relatively little effort has been devoted to historicizing the development of economics imperialism and to relate this history to more recent work in the philosophy of science on the nature of scientific imperialism.[2] This is surprising for a number of reasons, not least of which is that we would expect that those discussing scientific imperialism would devote significant attention to how these so-called imperialistic movements developed – that is, ground their philosophy of science in historical context. This would seem to be a necessary step on the path to the development of accurate philosophical categories and the application of those categories or labels in a way that accurately describes or reflects the history of science. This chapter attempts to move the discussion forward through the examination of an illuminating early moment in the history of the economic analysis of law.

In 1966, the American Economic Association (AEA) and the Association of American Law Schools (AALS) established an ad hoc Joint Committee of professors to explore the prospects for increased interaction between lawyers and economists. The committee was the brainchild of Henry Manne, a George Washington University law professor who had been trained at the University of Chicago and was actively engaged in bringing economic analysis to bear on corporations law. The story of the Joint Committee's origins, trajectory, and eventual demise is itself of historical interest – in no small part because of its relationship to the conquest aspect of "scientific imperialism" – but that story will not be pursued here.[3] Instead, this chapter focuses on the one major activity undertaken by this committee: a 1969 conference centering on the question of whether products liability law could be usefully informed by the application of economic analysis to it.

This conference took place when the economic analysis of law was very much in its infancy. Guido Calabresi's development of an economic analysis of accident law was attracting a good deal of attention,[4] and Ronald Coase's application of economic reasoning to legal issues in "The Problem of Social Cost" (1960) was also getting a small amount of play in the legal literature – in no small part due to Calabresi's use of it.[5] Becker (1968) had only recently published his economic analysis of criminal law, and Richard Posner's wide-ranging treatise, *Economic Analysis of Law* (1973), was still several years in the future. The products liability conference brought together a group of economists and legal scholars, from a variety of perspectives or schools of thought, to assess the utility of an economic approach to the subject and, by extension, the possibilities of a more broad-based application of economic analysis to legal thinking.

From the perspective of the present, the economic analysis of law is considered one of the most – if not *the* most – successful examples of "economics imperialism," where "successful" is defined to mean having developed a significant long-term place in the literature. Legal scholarship is now replete with economic analysis, and it is the rare law student who does not encounter the economic approach to the subject on multiple fronts during her legal education. "Law and Economics" is a recognized field within the discipline of economics,[6] with economists regularly engaged in analyzing the impact of legal rules via both theoretical modeling exercises and the application of econometric techniques. However, the modern economic analysis of law was barely a blip on the radar in the late 1960s,[7] and the workshop organized by the AALS-AEA Joint Committee thus provides an interesting, albeit small-scale case study in how those at the forefront of suggesting possibilities of the application of economics to the law went about the task of arguing their case, and of how legal scholars and other economists reacted to this effort.

As we shall see, this conference can be regarded as something of a microcosm of the early reactions to that which goes by the name, "economics imperialism." Economists were by no means of like mind on the utility of such efforts, nor were legal scholars. At least as important, though, is the fact that this episode, and other data from the period, show that the development of the economic analysis of law is far more nuanced than the "economists as colonizers" imperialism stories suggest. Legal scholars were importing every bit as much as economists were exporting, and they were making use of economic analysis in traditionally "non-economic" areas of law well before economists began their own forays.[8] While Mäki (2013, 335), for one, would label this an instance, at least in part, of "internally" driven imperialism, substantial questions remain as to whether the term "imperialism" works as an appropriate label for this slice of intellectual history.[9]

1 Roland McKean and the marketing of economics to lawyers

Products liability was selected as the topic for the Committee's initial effort because it was, as Henry Manne and Gerald Meier put it in their "Foreword" to the published proceedings of the conference, "[a] legal area which had not been the subject of extensive economic analysis" (Manne and Meier 1970, 1). This statement is certainly accurate if, by "economic analysis," one means "analysis undertaken by an economist." The fact is, however, that the second half of the 1960s saw a not insignificant number of articles – authored by legal scholars – that discussed issues of products liability from the perspective of economics. The application of economics here was motivated in part by the related analysis of accidents and accident law by Guido Calabresi (1961, 1965a), and by Walter Blum and Harry Kalven (1964), in the first half of the decade. However, for those concerned about the trend toward strict manufacturer liability for harm or, worse, that a social security-type system should be set up to manage accident compensation claims, this literature would have been of little comfort. Those whose economics disposed them to favor a market process which let the chips fall where they may, so to speak, would have problems with the literature extant.

The conference itself had an almost missionary rationale to it – a perception that meshes nicely with traditional imperialist narratives save for the crucial fact that this entire effort was instigated by a legal scholar. As Manne and Meier noted in their "Foreword" to the symposium, the hope of the organizers and the Joint Committee was that "economists might become more aware of current intellectual problems faced by lawyers and that lawyers might become more aware of the contributions that the discipline of economics could make to solving these problems" (Manne and Meier 1970, 1). To accomplish this, the Committee asked an economist to prepare a lengthy monograph on how economic reasoning could inform the understanding of issues in products liability law, with the particular goal that the paper provide "a critique of the implicit and explicit economic content of the more important legal literature" in this area (Manne and Meier 1970, 1).

Given that this was a *joint committee* of the AEA and the AALS, and that the committee was the brainchild of a legal scholar (albeit one with strong economics inclinations), it is significant that this initial effort – the monograph writing – was not envisioned as an *interdisciplinary collaboration* between economists and lawyers. While the monograph author was free to consult with whomever he wished, the Committee was explicit in its view that "a collaborative effort was not indicated" because "the desired critical flavor might thereby be lost" (Joint Committee of the American Economic Association and the Association of American Law Schools 1967, 717). The interdisciplinary nature of the venture would commence only after the monograph had been completed, through appraisals by the four selected commentators – two economists and two legal scholars – and general discussion among the invited conference participants. The goal, clearly, was to show how economic analysis could enhance, and potentially supplant traditional methods of, legal reasoning.[10]

Finding an author for the products liability monograph proved to be difficult, as Manne's entreaties were turned down by several individuals at the top of his list. The Committee finally settled on University of Virginia economist Roland McKean. McKean had been educated at the University of Chicago, worked as a research economist at the RAND Corporation from 1951–1963, and held a professorship in economics at the University of California, Los Angeles (UCLA) before moving on to the University of Virginia in 1968. His earliest research focused on monetary theory and policy, but his operations research work at RAND quickly gave rise to a research agenda focused on the efficiency of alternative allocation systems, with defense economics and public economics being the primary areas at which this work was targeted. McKean had done no work in the economic analysis of law prior to penning his essay on products liability, but his attempts to bring economic principles to bear on governmental budgeting processes may provide some explanation for his willingness to probe the implications of applying economic analysis to the law. Having no legal background, it is perhaps no surprise that McKean's acknowledgments included thanks to several law school professors at Harvard and one at UCLA for background information and suggestions related to his paper.

That McKean prepared his monograph with one eye toward the law and another toward economics is apparent from the start, as he pointed out in his introduction that "[i]n choosing among alternative products liability arrangements, one must consider the effects of these arrangements (1) on fairness or equity, and (2) on the behavior of various persons affected and hence on resource allocation" (McKean 1970, 3). In spite of law's long-standing emphasis on a conception of justice defined against a background of fairness, McKean argued that "*[m]ost thoughtful persons* will also be concerned with the ways in which behavior and hence resource allocation are affected" by the laws, or lack thereof, relevant to the situation (ibid., 4, emphasis added).

McKean's stated goal was to survey the products liability literature with a view to ascertaining "the economic implications of products liability law to determine whether these implications are being discussed critically and carefully" (McKean 1970, 4). One of his chief concerns was what he considered the methodological weakness of much of the legal and policy literature, with insufficient sensitivity shown to the extent to which scholars and others "are emphasizing tested hypotheses about the consequences of alternative actions and the extent to which they are utilizing their individual judgements about untested propositions" (ibid., 4). McKean was very much of the mind that economic analysis could provide clarification on these fronts and, in doing so, greatly solidify the basis of legal analysis (ibid., 5) – in essence, put it on a more scientific footing (ibid., 40). His goal, however, was not to identify "optimal policies," but to lay out the consequences – derived from economic reasoning – of different legal arrangements and to assess the extent to which these consequences had been recognized in the literature. This perspective was motivated in part by the view that the economist had little or nothing to say about fairness, but even more by McKean's sense that concepts such as fairness and equity, and of the appropriate trade-offs between them, are necessarily ambiguous (ibid., 5).

McKean devoted the first part of his analysis of products liability to an overview of its case history. This history will not concern us here other than to note three points. First, McKean was convinced that competitive forces had been and continued to be a significant check on the sale of defective products (McKean 1970, 7–8). Second, he saw the laws governing products liability through the previous century as reflecting efficient adaptation of law to the circumstances of the case and the moment in time (ibid., 7–8), a position hinted at by Coase (1960) and soon after to be more fully developed by Richard Posner and others. Third, he did not view the recent trend toward producer liability for product-related injuries as representing a particularly large break with the past or as a cause for great concern (McKean 1970, 22).

1.1 The toolkit: economics for the legal scholar

McKean next turned his attention to the future and the analysis of the economic implications of alternative systems of products liability being proposed in the legal literature. However, being aware of his audience, McKean began his discussion by providing an elementary economics tutorial – laying out, in essence, the basic elements of a toolkit for legal-economic analysis. McKean identified four concepts that he considered to be crucial elements of such a toolkit, two for positive (or descriptive) analysis and two for normative (or prescriptive) analysis:

> To understand the economic implications of alternative policies, one has to have clearly in mind the notions of cost and comparative advantage; and to see how much and how little economics can guide one regarding policy, it is important to understand what is meant by "economic efficiency" and "Pareto optimality."
>
> (McKean 1970, 25)

He then proceeded to discuss these concepts and their application, utilizing a simplified world consisting of three producers and two commodities. While much of this discussion is unremarkable and thus will not concern us here,[11] the efficiency discussion is another matter, for it was directed squarely to the Coase theorem and as such is illustrative of the important place that Coase's result played in the development of an economic analysis of law, as well as (as we shall see) in McKean's own subsequent analysis in this article.

1.1.1 The benefits of voluntary exchange

McKean's working definition of "economic efficiency" was that of a set of points at which it is impossible to expand the output of one good without reducing the output of the other (McKean 1970, 27). At such a point, all goods are being produced according to the dictates of comparative advantage. McKean pointed out that the presence of inefficiency implies that there is scope for a deal that is advantageous to both parties and shifts the production of each good to the agent(s) having a comparative advantage in that activity (ibid., 28). McKean suggested that such an exchange "*will occur*" if the gains exceed the transaction costs, and thus that an efficient outcome will obtain. Moreover, he said, if the transaction costs associated with the exchange process are greater than the gains, we can assume that the outcome, though in one sense inefficient, "may be the most efficient *attainable* solution" (ibid., 28–29, emphases added). A system of voluntary exchange, then, will lead to an outcome that satisfies the dictates of Pareto optimality. While the government *may* be able to provide a lower-cost resolution of the problem, McKean noted, we cannot be confident of this because "one cannot be sure what the costs of the action are" and the government's actions are "almost certain to involve compulsion for some persons" (who are losers by the deal) (ibid., 29).

For McKean, the key link between the concept of efficiency and products liability came via the negotiation result that Ronald Coase had demonstrated in his analysis of alternative liability assignments in "The Problem of Social Cost" (1960) – a result that George Stigler (1966, 113) had subsequently labeled the "Coase theorem." McKean stated the result this way:

> If there were zero transaction costs, and if people eschewed the use of coercion or interference with voluntary exchanges, it would not matter, as far as resource allocation is concerned, how the legal rights or liabilities were initially assigned, that is, whether the purchaser or the manufacturer was liable for injuries and damages.
>
> (McKean 1970, 30)

Though the Coase theorem tells us that efficiency is guaranteed in a world of zero transaction costs, McKean was not willing simply to lay the Coase theorem onto the real world and pronounce the outcomes efficient. "[It] should be stressed," he said, "that [the theorem's] assumptions are *fairly heroic*" (ibid., 31, emphasis added). However, he contended, this did not make Coase's result non-useful, for it offered the ability to "*simplify and think about products liability*" (ibid., 31, emphasis added). Coase's negotiation result, for McKean, was "a point of departure," but his position was that "it pays … to understand the oversimplified situation before trying to make judgments about the more complex ones" (ibid., 31) – a standard defense of the use of abstraction in economic analysis, made to an audience that might naturally recoil at such abstraction, given the practical bent of so much of legal analysis.

Nor, said McKean, are these insights regarding the workings of the voluntary exchange process confined to the realm of "products," as traditionally defined. He pointed out that risk bearing and injury prevention, central to products liability, are "outputs" that can be (and often are) exchanged in the marketplace. The issue, from an efficiency perspective, is one of placing risk and the injury prevention activities on the party who has a comparative advantage in dealing with them, and the Coase theorem, he said, tells us that, if transaction costs are zero, voluntary exchange in the marketplace will do precisely that, just as it does for, say, wheat (McKean 1970, 36).

Though McKean characterized Coase's negotiation result as a point of departure, he did not consider it empirically empty. Citing the case of injuries from automobile accidents, McKean argued that transaction costs might be low enough to allow the appropriate combination of automobile safety features and driver precaution to emerge through the marketplace. The precise characteristics of this (efficient) outcome would depend upon the public's willingness to pay for safety features (or the amount that they would accept in payment to undertake precaution) and their level of risk aversion. The parties have an incentive to negotiate the placement of damage prevention with that party who has the comparative advantage in producing it, though he allowed that the magnitude of transaction costs "may vary greatly" in these different situations. Because transaction costs are not always zero and in fact may be "large" in certain instances, the appropriate (efficient) assignment of liability "is not completely clear" (ibid., 34).

1.1.2 Questioning efficiency

Having shown that, in a frictionless world, efficient negotiated solutions will emerge as a matter of course and that the costs associated with such negotiations in the real world may leave us in an inefficient position – one that may or may not be amenable to an efficiency-enhancing intervention by government – McKean turned to the larger normative question of whether "economic efficiency" or "Pareto optimality" are, as he put it, "good," and thus appropriate objectives for law. In doing so, he identified four issues which suggested to him that there may not be widespread agreement on making efficiency concerns "the supreme objective" of law (McKean 1970, 36).

First, he said, the meaning of "economic efficiency" is not given or even obvious. It could be measured in a static sense, or it could take account of long-run factors of various sorts, including changes in technology and information availability, and pursuing either long-run or short-run efficiency considerations could very well come at the expense of the other. Second, there may not be broad-based agreement that the set of Pareto optimal points are the *best* points or outcomes – the ones that society should pursue. Third, efficiency as characterized in a Coasean negotiation/Pareto optimality context privileges the process of voluntary exchange as against other forms of decision-making, such as majority rule, and a particular measure of output rather than some other "output" goal, such as social stability, that society may prefer. Finally, individuals may prefer inefficient outcomes that are more, or potentially more, beneficial to themselves because such outcomes afford a less costly road to achieving additional benefits, such as wealth for themselves (McKean 1970, 36–38).

McKean sensed that it was distributional concerns – fairness or equity – that lay behind the move by courts toward strict liability. However, those who advocated the "deep pockets" approach, he said, appeared to minimize the impact of other consequences, such as those on the efficiency front, relative to placing liability on the party with the greatest ability to pay, often reasoning that the manufacturer can spread these costs over the general public as part of the cost of doing business (McKean 1970, 39).[12] Though McKean was very much in favor of an increased focus on efficiency as against its perceived neglect by legal scholars, he was not arguing – at least explicitly – that efficiency is the only appropriate goal for law or even the most important one, nor for a sort of disciplinary imperialism that would have economic considerations completely *supplant* the traditional normative concerns of law. However, he also emphasized that though the economist's efficiency analysis "at best yields indecisive clues" regarding the appropriate course for policy decisions, this did not imply that economics should be "discarded casually" as an aid to reasoning about the impacts of policies. Indeed, the natural sciences can do no better, contended McKean (1970, 40, quoting Blum and Kalven 1965, 65). However, like economics, he continued, they can and do provide information for policymakers that can assist in the determination of the most appropriate course of action, even though "they cannot tell us *what ought to be done*" (McKean 1970, 40, emphasis added).

1.1.3 The most basic of all laws: demand

The tool that McKean urged upon law as the centerpiece of legal impact analysis was perhaps the most basic of all of the economist's tools: the law of demand. The problem, he asserted, is that "[n]on-economists often doubt this proposition" (McKean 1970, 40), effectively assuming that people often do not respond to variations in prices. When it comes to dangerous products, McKean said, the law of demand doubters argue that the risk to life and limb is a sufficient deterrent – that monetary penalties would not increase the amount of precaution taken.[13] He granted that the deterrent effect would indeed be zero if the risk at issue was a "100 percent probability of death or serious injury," but that, he insisted, is not generally the level of risk in question (ibid., 40). What is really at stake in most products liability situations, he pointed out, are modifications in behavior that have a small impact on the probability of injury or death. Nor, he said, are these types of trade-offs in any way unique: individuals make them constantly in their everyday lives – whether crossing busy streets at dusk, driving in rainy weather, or accepting a job that entails a slightly greater risk of injury or death in return for a higher wage. "Only heroes or suicides," he said, "take actions that involve extremely high probabilities of death or serious injury, but all of us trade modest increases in those probabilities for modest savings or rewards" (ibid., 55).

Given this, legal rules that add a financial penalty to this risk factor could, in fact, have a significant effect on the expected costs that individuals associate with risky behaviors and so make a noticeable impact on the level of precaution taken against accidents and injury (McKean 1970, 41). While there may be some question regarding the sensitivity (or elasticity) of the behavioral response to the change in price, the inverse relationship between price and quantity demanded is "one of the few *tested propositions* we have" (ibid., 42, emphasis added) and, for McKean, was perhaps the brightest symbol of what economics could add to law to set straight the legal reasoning process. It was also the "basic tool" that he would use in tracing the implications, or consequences, of alternative products liability schemes (ibid., 41).

The critic of the economic approach, McKean reasoned, might object that if fairness or equity is all that is relevant, there is no point in being able to trace these behavioral consequences. However, McKean did not see matters this way. The consequences revealed by the law of demand and other economic concepts matter, he said, because even though "few persons" consider economic efficiency to be "their overriding objective or criterion," people do care about efficiency in some basic (as opposed to technical economic) sense – we "have enough to gain" from it "to keep it in mind as one of our important objectives" (McKean 1970, 42). The fact that people *actually do* pay attention to efficiency, then, means that it is important to have an analytical system – such as that offered by economics – which reveals the likely consequences of alternative courses of action (ibid., 42).

1.2 Applying the tools: the economics of products liability alternatives

Having laid the theoretical groundwork for making an economic analysis of products liability law, McKean turned his attention to the question of what economics could add to the evaluation of various possible assignments of liability for product-related harms.

1.2.1 Caveat emptor

Given the recent emphasis on strict liability in the literature and case law, it is interesting that McKean chose to begin his own analysis with a discussion of its near antithesis, "complete *caveat emptor*" – making customers liable for losses that occur when using a product. Such a system, he argued, would generate an efficient allocation of resources in a world of zero transaction costs,[14] in that each product, including safety, would be produced by those having a comparative advantage in that activity: "Purchasers of products would hire producers to reduce product hazards and 'bribe themselves' to exercise care as long as these actions paid" (McKean 1970, 43). An identical logic was applied to third parties – e.g., innocent bystanders – who might be harmed as a result of the use of the product. For McKean, all of this was clearly revealed by the Coase theorem's logic.

McKean was well aware that this conclusion relied upon the "fairly heroic" assumption of zero transaction costs, but he was of the mind that customer liability will make "certain transaction costs ... relatively low." McKean's reasoning was that "established market relationships" for products facilitate the ability of consumers to register their preferences for safety features, safety information, and the like in the marketplace via their willingness to pay for these things. Drawing on his lesson regarding the law of demand and the associated propensity to substitute in response to price changes, McKean argued that if consumer preferences on the safety front are frustrated in the marketplace, one of two things will happen. Some, when faced with the choice between greater risk at a lower price and less risk at a higher price, will choose the former, while others will substitute alternative products or versions of the same product produced by other sellers. The marketplace, for its part, will generate a response to this expression of preferences. Producers who lose through this process, meanwhile, will respond by increasing safety quality or adjusting prices (McKean 1970, 44–45). While none of this eliminates injuries from or disappointments with products, the end result is that the registration of preferences through the market and the availability of alternatives in a competitive system gives consumers what they want, in the sense of the package of "outputs" – including risk and injury prevention – that they prefer to purchase.

McKean recognized, however, that this outcome can only be achieved in the presence of a sufficiently competitive market and, in light of this, was very explicit in allowing that none of this promises a panacea. *Caveat emptor*, he said, offers "some consequences that many persons would regard as desirable, but it will not by any means yield a tidy economy free of misery and possible inefficiency" (McKean 1970, 47). The question, then, was how these possible imperfections stacked up against various systems of producer liability.

1.2.2 Producer liability

As McKean noted at several points in his analysis, producer liability would be as efficient as *caveat emptor* if transaction costs were zero, per the Coase theorem, but the reality of transaction costs suggested the likelihood of significantly greater inefficiencies under a producer liability system[15] – cost-related inefficiencies overlooked in the legal literature but made crystal clear, he believed, by economic analysis.

Seller liability with fault and privity of contract – the system that had long governed products liability – created a more complex web of issues, McKean argued, than that found under *caveat emptor*. The incentive for consumers to exercise precaution or strike cost-reducing bargains with producers would be higher than under strict liability, though less than under *caveat emptor*, and the demonstration of fault is itself a costly process (McKean 1970, 48). A system of seller liability with fault would also have the effect of inducing most buyers and sellers to carry liability insurance, but insurance companies will then increasingly resort to litigation to protect their interests, which further increases transaction costs relative to *caveat emptor* (ibid., 49). Continued movement in the direction of producer liability would only serve to "intensify" the issues, McKean reasoned, with absolute producer liability – where fault is irrelevant (a rule that McKean labeled "rather extreme"; ibid., 50)[16] – offering something approaching the worst of all possible worlds.[17] Economic analysis, he said, suggests that:

> one might expect more court cases and court costs, greater difficulties in bribing purchasers to be careful, more efforts to publish warnings and instructions (perhaps disclaimers, if they still gave producers some shelter), a greater tendency to produce safer products, a smaller range of product choice for poor consumers, and a partially offsetting tendency to neglect safety because of the more extensive use of liability insurance.
>
> (ibid., 50)

Goods prices would rise due to the higher accident prevention costs incurred by manufacturers. Litigation costs would also be higher than under *caveat emptor* because of the need to properly establish the extent of injuries (which would otherwise be overstated by victims) (ibid., 51). The crux of the problem, McKean argued, is that under producer liability there will be a dramatic increase in those instances in which accident prevention is not undertaken by the party having the comparative advantage in undertaking such activities (ibid., 51).[18]

None of this, for McKean, *demonstrated* that *caveat emptor* is the proper course of action. The problem, in his opinion, was that the discussion in the legal literature had largely ignored these very real efficiency-related consequences – consequences that were plainly revealed by the application of economic logic. One may legitimately prefer a compensation without fault or social insurance plan, he said, but economic analysis reveals that a free lunch is not to be expected: "When considering a change that would bring gains, one cannot neglect the costs or disadvantages against which the gains must be weighed" (McKean 1970, 56).

1.3 Legal reasoning, from the perspective of economics

Though McKean's assessment of alternative products liability options had not overtly passed judgment on their relative efficacies, it is difficult to come away from his discussion feeling other than that he believed that *caveat emptor* had much to recommend it, in efficiency terms. As this perspective put him very much at odds with recent legal trends, it comes as no surprise to find McKean suggesting that, with only a few exceptions, the implications of alternative legal rules were not "being discussed critically" or "with much precision and care" (McKean 1970, 56) – the implication being, of course, that economic analysis offered a road to greater precision.

Perhaps the "most striking" feature of the legal literature, for McKean, was that "equity or fairness" were emphasized, "almost to the exclusion of any other considerations" (McKean 1970, 56). Indeed, in many cases, he said, the writers seemed concerned with little more than which party deserves our sympathy (ibid., 58). While admitting that his impressions might be idiosyncratic, his evaluation was that those writing in the legal literature seemed to be "thinking uncritically in the same sense that the six blind men were being uncritical in the old poem about feeling the elephant" (ibid., 56).[19] McKean again took pains to emphasize that he did not consider fairness unimportant; his point was simply that "one should at least ask himself whether or not there are other relevant considerations in designing products liability policy" – touch other parts of the elephant, so to speak – including, at a minimum, thinking about the link between the distribution of wealth and the amount of wealth to be distributed (the latter of which, of course, goes to efficiency) (ibid., 57). If these commentators truly believed that there are no efficiency-related implications associated with the positions they were advocating, they should be explicit about saying so in order that others might assess the validity of their claims (ibid., 57).

However, even if fairness is all that matters, McKean continued, an important role for economic analysis remains in terms of assessing the quality of the analysis and the validity of the positions taken. Economic analysis, he asserted, reveals that fairness-related judgments are neither simple nor obvious (McKean 1970, 56). For example, he said, while fairness considerations may dictate compensation to consumers or third parties, the effect of such a rule may be a reduced selection of products for consumers, including the inability to buy a lower-quality product and assume additional risk of injury in return for a lower price – something that would hit the poor particularly hard. Fairness for victims, then, may come at the expense of fairness to the poor. Though he considered fairness issues both "important" and worthy of debate, McKean was convinced that it would be "more helpful" if the discussions were undertaken with "increased caution and accuracy" – toward which end the application of economic reasoning had much to offer (ibid., 59).

Though McKean was troubled by the lack of attention to considerations that economists would identify as important, he was equally bothered by the quality of the economic analysis employed when such issues *were* treated in the products liability literature. Too often, he said, "the generalizations are too broad or the inferences that are drawn are not warranted." At times, claims are made which are factually incorrect. In other instances, authors "make strong assertions" regarding economic effects "where there is little evidence or even where there are tested hypotheses that (until refuted) cut in the opposite direction" (McKean 1970, 60). Finally, said McKean, the literature is replete with statements that speak to the costs associated with one course of action but neglect to examine the costs associated with the alternative course of action (ibid., 61). Thus, we find claims that consumers cannot influence producer behavior, that consumers could not possibly prefer to trade off quality for price, that no company would go out of business because of increased tort liability, and that the failure to exercise precaution increases accident-related costs (ignoring the fact that the precaution itself may have a greater cost-increasing effect) (ibid., 59–61). Bad economics, in short, was a recipe for bad law.

All of this being said, McKean did not consider the legal literature a complete wasteland so far as economic considerations were concerned. As he noted in the concluding paragraph of his article,

> the products liability literature contains quite a few analyses which do recognize considerations other than equity (such as economic effects), the inherent difficulties of agreeing on what is equitable, the significance of permitting or of interfering with individual choice, the relevance of the law of demand and deterrence, and the possible impacts of products liability laws on total costs and output. A few writers have even drawn on the Coase theorem, recognizing that with zero transaction costs and voluntary exchange the assignment of liability would make no difference to resource allocation, but also recognizing that in the real world the implications of liability assignment are often complex and uncertain.
>
> (McKean 1970, 62–63)[20]

Yet, he clearly did not consider this the heart of the products liability literature, nor did he believe that the trends in the legal discussion were moving the field in the direction implied by solid economic analysis. In sum, there was much to be gained, in McKean's mind, from bringing the economic methods and perspective to bear on the law.

2 The responses

With McKean having made his case for what economic analysis could add to the subject of products liability, the floor passed to the four scholars enlisted to provide commentaries on his efforts. Two economists, James Buchanan and Robert Dorfman, and two law professors, Guido Calabresi and Grant Gilmore, had been enlisted – each a scholar of significant stature in his field. Though at this stage we have no evidence to shed light on why or how these four particular respondents were selected, both an examination of their respective backgrounds and a reading of the comments on McKean's analysis suggest that one of the economists and one of the lawyers were chosen because they would be expected to be at least somewhat sympathetic to McKean's economic approach and the application of it, and that the other respondents were selected because they could be counted on to provide more critical commentary. That, at least, is how things turned out, as the reactions illustrated the challenges of getting economists and lawyers to buy into the application of economic reasoning – or at least that of the type McKean was selling – to the law.

2.1 Buchanan and the defense of caveat emptor

Buchanan, like McKean, had been educated at the University of Chicago, and he had also published a number of articles dealing with externalities over the course of the 1960s. As such, he was well versed in the externality-related logic – including the Coase theorem – that had a bearing on tort law generally and products liability in particular. Perhaps even more than McKean, Buchanan was convinced of the benefits of voluntary exchange within a market context and of an economic analysis that made voluntary exchange processes a foundational element of the conceptual framework. Indeed, he made no bones about the prejudices with which he approached the problem:

> As an economist who studies market processes, disciplinary prejudice alone suggests to me that departures from caveat emptor should be carefully scrutinized and accepted only after specific argument accompanied by convincing evidence. As an individualist, who places a high value on freedom of exchange, any limitations on the exchange process, either directly or indirectly, arouse my initial skepticism.
>
> (Buchanan 1970, 72)

That he applied to his comment the title, "In Defense of Caveat Emptor," then, should not be an occasion for surprise. His "defense," though, was more in the way of a critique of producer liability than a direct defense of *caveat emptor*.

To illustrate how economic reasoning makes a strong case against producer liability, Buchanan instanced a market for coal, where the coal is available in two qualities – the only difference being that a lump of the low-quality coal has a higher probability of exploding and possibly causing severe damage to persons and property than a lump of the high-quality coal. A move to impose liability on producers here, Buchanan noted, would increase the price of both low- and high-quality coal, but the former much more, in order to account for the high associated liability costs. Consumers thus would be prevented from purchasing lower-quality coal at a much lower price that reflects a willingness to assume the relevant risks (Buchanan 1970, 66). Where McKean had hedged his bets somewhat on the normative front, Buchanan did no such thing. The effects of a move to make producers liable for damages, he said, "are clear and unambiguous." This policy "can be condemned" on both efficiency and equity grounds, the latter because the burden of the law will be "concentrated on the poor" (ibid., 67). Buchanan, then, had used economic analysis to turn the legal profession's equity argument on its head – a result, he believed, of the legal profession's tendency to assume that increases in product quality could be achieved without cost (ibid., 72).[21]

Buchanan, like McKean, was a die-hard proponent of the Coase theorem's utility – in fact, even more so, being of the mind that equally efficient results should obtain even in the presence of transaction costs.[22] If so, then producer liability should be equivalent to *caveat emptor*, at least on the face of it. This presented Buchanan with something of a conundrum and led him to reject McKean's position that the theorem was a useful point of entry into products liability issues. The problem, he said, is that "[t]o be fully applicable, the Coase theorem requires the assumption that there be no prohibitions on any mutually advantageous exchanges that may be made as between potential buyers and potential sellers." Strict manufacturer liability, though, has the effect of imposing such a prohibition because those who would prefer to purchase low-quality products are "effectively ... prevented from 'buying risk'," meaning that the law itself effectively forecloses the exchanges that allow the Coase theorem to work its magic (ibid., 69).

Buchanan was not, however, willing totally to write off the possible beneficial effects of producer liability. A "plausible case can be made," for example, if the accidents resulting from the product impact third parties (Buchanan 1970, 70). However, he continued, this does not imply that a blanket move toward producer liability is justified. Rather, "[a] discriminating approach is required," one that examines the issues "on a product-by-product basis, *taking into account the relevant economic criteria*" (ibid., 71, emphasis added). The problem, he said, is that this puts the economic approach into a bit of conflict with the legal one. "The law, as the law, tends to be general in applicability. If it is not, it ceases to be law in a certain sense." However, the economics involved here, he said, "varies from product to product." The issue, then, is one of how to achieve an appropriate balance in light of this tension. His conclusion? "Commence with some prejudice for caveat emptor and be sophisticated in the application of departures from this principle" (ibid., 73).

2.2 Calabresi's alternative economic analysis of law

Like Buchanan, Yale Law Professor Guido Calabresi, who had been developing an economics-infused approach to accident law for more than a decade, could be expected to be favorably disposed to McKean's position, though not completely so. For while Calabresi was trained in economics and relied heavily on it in his own work, his economics did not lead him to McKean's conclusions.[23] Calabresi, writing with his student, Kenneth Bass III, had "very little argument with" McKean's basic framework of analysis, which, as he noted, "closely parallels" aspects of the analysis set out in his own work – including the grounding of the theory in the Coase theorem,[24] an insight that Calabresi regarded as "crucial" and deserving of "constant attention" (Calabresi and Bass 1970, 77).

However, one of the reasons why the idea is so important, Calabresi said, is the reality of transaction costs which makes it "imperative" that liability be placed upon the party that would bear the costs if the negotiation process contemplated by the theorem were actually feasible – that is, the least-cost avoider (Calabresi and Bass 1970, 77). McKean, he argued, had missed the mark in his conclusions as to who has this "comparative advantage in safety" (ibid., 78). From where Calabresi stood, McKean's analysis of alternative liability systems suffered from the same "overgeneralization and ... insufficient analysis" that McKean had ascribed to those lawyers whose analysis he had found wanting (ibid., 78–79):

> On the one hand, he takes a view from the mountain top and talks as if all product liability situations were the same in terms of the comparative

> advantage of users and sellers in producing safety. On the other hand, he becomes immersed in the trees and talks as if the party on whom the law initially puts the burden of an accident is the party who will ultimately bear the burden and hence will be the party who will make the decisions for or against a safer but potentially more expensive alternative.
>
> (Calabresi and Bass 1970, 79)

"The usual case," Calabresi countered, "is quite different" (Calabresi and Bass 1970, 79), and he went on to examine three situations – transfusion hepatitis, cosmetics reactions, and drug allergies – that pointed to very different liability rules based on the least-cost avoider criterion. Moreover, he said, "to the extent general conclusions about products liability are useful, economic theory indicates that consumers do not in practice have a comparative advantage in safety" (ibid., 88).

Though Calabresi disagreed strongly with the lessons that McKean drew from his economics, he was equally adamant that economics had much to offer to legal reasoning – a position that should come as no surprise given that Calabresi had spent more than a decade attempting to ground an approach to accident law in economic logic and to persuade other legal scholars to follow his lead. Lawyers and legislators alike, he said,

> are, in fact, making such choices quite unintelligently all the time in accident law. We are making choices without even asking the right questions, as is indicated by the disturbing tendency in products liability cases to look simply for the best loss spreader as if that were the only goal of accident law, and by the inclination to resort to collective fiat to accomplish primary accident cost reduction.
>
> (Calabresi and Bass 1970, 90–91)

Economics, for Calabresi, offered a way forward, and he was of the mind that "[a]nalyses like Professor McKean's, tentative and incomplete though they may be, are crucial first steps toward asking the right questions. Taken as that they are very valuable indeed" (Calabresi and Bass 1970, 90).

2.3 Dorfman's hesitancy about an economic analysis of law

While Buchanan and Calabresi were in agreement with McKean's basic thesis regarding the utility of economics for legal reasoning – even if not with the particulars of his analysis – McKean's optimism was challenged by the other two commentators, economist Robert Dorfman of Harvard, and University of Chicago Law Professor Grant Gilmore. If their objections have a familiar ring to them, it is because they seem to harken forward to the objections made over the next two decades both by some economists and by those in other disciplines to the extension of the boundaries of economics.

Dorfman, an economic theorist with a strong interest in environmental economics, certainly could not be counted as one who saw an essential unity in the social sciences, with economics providing the unifying framework. Economics and law, he said, "are such very different disciplines that lawyers and economists are bound to have difficulty in understanding each other's approach to this or any other topic," largely because they "approach problems in very different ways and with different purposes in view" (Dorfman 1970, 92). The lawyer's point of view, he suggested, is "pre-eminently practical":

> Whether the lawyer sits on the bench or stands before it, his business is to make social decisions. In making those decisions he has many things to take into account, and, in particular, he has to apply the standards of ethics and justice and mutual obligation that are inscribed in the law.
>
> (Dorfman 1970, 92)

The economist, in contrast, "is not concerned with reaching decisions at all." Instead, said Dorfman,

> His business is part that of a scientist and part that of a social critic. His task is to describe the way the world operates and if possible to describe it so well and so profoundly that he can infer how the world would operate if conditions were somewhat altered, that is, so he can predict the consequences of following different policies.
>
> (Dorfman 1970, 92)

One, then, is very much in the fray, while the other stands above, or at least apart from, it. Given their very different positions, Dorfman found it "no wonder that economists tend to feel that lawyers have inadequate understanding of the facts of life and employ low standards of analytical rigor, while lawyers regard economists as wildly impractical types." The reality of the matter, he said, is that "[w]e both are right" (Dorfman 1970, 92–93).

When it came to the particulars of McKean's analysis, Dorfman felt compelled to preface his response by noting that "it will become amply evident in the next few minutes that Professor McKean and I adhere to different schools of economics." These differences, he stressed, "are fundamental and are at bottom methodological" (Dorfman 1970, 94). The distinction, for Dorfman, was well exemplified by McKean's reliance on the Coase theorem. McKean, he said,

> believes that you can get to the economic essence of a problem by simplifying it in a particular way. I believe in simplification, too, because I recognize that the human mind can deal with only very simple problems, but I am very suspicious of the particular kind of simplification that Professor McKean regards as being most illuminating.
>
> (Dorfman 1970, 94)

The Coase theorem world assumed by McKean "is a very useful conceptual device for some economic problems," he said, but for others it "can be grossly distorting" (Dorfman 1970, 94). Products liability analysis, for Dorfman, fell into the latter category. Though he disagreed with McKean's assumptions, Dorfman was willing to see where the analysis that was based upon them would lead and to accept the resulting conclusions given the assumptions being made. The problem, according to Dorfman, was that McKean had not followed his assumptions "to their logical conclusions" (ibid., 94).

The assumptions "placed before us," Dorfman reminded his audience, are that the costs of information, of transacting, and of enforcement are zero. Given this, he said, "I feel obliged to draw a conclusion from them – namely, that in such a world, the problem of products liability would hardly exist" (Dorfman 1970, 95). Dorfman then proceeded to make the case for his conclusion. He pointed out that two types of defects may give rise to products liability: defects in design and defects in manufacture. The latter, he asserted, "would not exist at all in a world in which information was free," since the existence and nature of the defect would be known to everyone from producers to end users. If such goods were sold at all, they would be sold as defective products, and at a discount, and there thus would be no associated liability issues. A related logic would apply to design defects in such a context (ibid., 95–98). If you knowingly purchase a product that is defective, and at a price that fully accounts for this, you get to live with the consequences. A world of costless information that McKean had assumed, then, would put us in a world in which, as Dorfman put it, "the significance and occurrence of products liability would be so wildly different from what they are in the real world that the assumptions are hardly a first approximation to the actual problem that confronts us at this meeting" (ibid., 98). Because issues of products liability arise out of a context of costly and thus inadequate information, said Dorfman, "*We cannot abstract from these costs without changing the problem in an essential manner*" (ibid., 98, emphasis added).

Dorfman also took McKean to task for assuming that markets are perfectly competitive and then failing to follow through on *that* assumption as well. While acknowledging the economists' (including his own) love of such a world and their "addiction" to thinking about markets in these terms, with the associated "beautiful theorems," one must fully understand, and allow for, he said, what all of this implies. The chief issue of present concern, for Dorfman, was that in this perfectly competitive environment there is no "brand" identity and thus no way to catch out producers of defective products or for consumers to identify higher-quality versions of the product for substitution purposes. Conversely, if the details are known, the market in question must not be competitive, and none of the competitive adjustment features on which McKean's *caveat emptor* system relied, including Pareto optimal adjustments, can be invoked (Dorfman 1970, 98–99). "Neither Professor McKean nor I can have it both ways," Dorfman argued (ibid., 99). The relevant questions, he said, bear on the economic consequences of products liability in a world of imperfect markets, and McKean's analysis had provided no insights on that score (ibid., 101).

As severe as these critiques of McKean's analysis were, Dorfman's final one – a "warning," as he called it – was perhaps the most damning of all. McKean had made much of his analysis hang on the law of demand and Dorfman noted that this was an element of the "economist's creed" with which he agreed "wholeheartedly" (Dorfman 1970, 101). The problem, as Dorfman saw it, was that "this undoubted doctrine is *largely irrelevant* to predicting the effects of changing the law of products liability" (ibid., 101, emphasis added). What was it that would cause a believing economist to doubt the application of something as fundamental as the law of demand in this context? At issue, for Dorfman, was the "psychology of responses to changes in the penalties that will be incurred in very remote contingencies":

> I do not question that people will eat more apples if the price falls. I do question that they will eat more apples or wash them less carefully if the Northwest Apple Growers Association should undertake to indemnify all people who suffer arsenic poisoning from eating apples ... It does not seem plausible to me that anyone without a suicidal bent, who is tempted by a succulent unwashed apple, would say to himself, "The chances that it will make me sick are very small, and besides, if it does so, the law of products liability is there to console my heirs and assigns."
>
> (Dorfman 1970, 102)

Nor did Dorfman contest McKean's contention that people will trade off money for risk. However, he said, "there appears to be a threshold beyond which people's reactions change, because even sophisticated minds find it impossible to envisage or appreciate contingencies with very low probabilities of occurrence" (Dorfman 1970, 102), as evidenced by the fact that people seem to treat a one in 1,000 risk identically to one in 100,000.

Dorfman, then, was questioning the very premise of applying the economic model of behavior to law, both in terms of its ability to unify the methods of these disparate fields of inquiry and as respects its utility for effective problem-solving outside its traditional domain. Though he did not go so far as to insist on the correctness of his objection to relying on the law of demand here, he did maintain that McKean was "on unsafe ground" in so confidently asserting its applicability, arguing that this is "an empirical question, not a logical consequence of the law" (Dorfman 1970, 102).

2.4 Gilmore: the lawyer's dissent

While Dorfman remarked that he felt a bit out of place in commenting on products liability law, University of Chicago Law Professor Grant Gilmore seems to have felt himself even more so. Gilmore declared that he "was known to know nothing – either about economic theory or about recent legal trends in what has come to be called Products Liability," and thus took as his task "to comment, in childlike fashion, on these mysteries – more or less, it may be, like the little boy who, rightly or wrongly, cried out in mid-procession that the Emperor had no clothes on" (Gilmore 1970, 103).[25] However, as he did not have a horse in the race, so to speak, on either the economics or legal fronts, he provides us with a useful set of insights into how the typical legal scholar might have reacted to McKean's analysis, as well as to the broader question of bringing economic analysis to bear on the law.

Gilmore took as McKean's thesis the idea that:

> clarity of thought will be promoted if lawyers and judges, instead of concentrating on such fuzzily defined or undefined or perhaps undefinable concepts as 'fairness,' 'equity' and 'justice,' take into account the presumable economic effects of the various possible schemes of loss allocation on costs, prices, output and the use of resources.
>
> (Gilmore 1970, 104)

The problem, he said, is that despite "having, by his remarkable analysis, immeasurably clarified our thought," McKean did not appear "to have any convictions or even any particularly strong feelings about what we should do next" – about how cases should be decided or what statutes should be adopted (ibid., 104). This, he continued, posed a particularly difficult problem for the lawyer and the judge, in that they are "professionally required" to provide answers to such questions, even though lawyers understand that they cannot be certain that the answers they are providing are the correct ones. Economics, then, did not seem to Gilmore to offer much utility as a problem-solving device for law.

Like each of the other commentators, Gilmore latched onto McKean's repeated invocations of the Coase theorem, an idea to which he was no stranger given that Coase was his colleague on the Chicago law faculty. While Gilmore accepted the theorem as "true" (Gilmore 1970, 105), he took issue with McKean's contention that it is an appropriate starting point for the analysis, suggesting that it is so highly over-simplified that it simply could not be expected to provide relevant insights. Not only are there always transaction costs, said Gilmore, but also it will generally be impossible to predict how large they may be in any given situation or to ascertain how large they were even after the fact. Moreover, he added, the "truly voluntary" – that is, non-coerced – exchanges that are at the heart of the theorem have "become almost an extinct species in the all too real world" (ibid., 105). The issue that this raises, he said, is that we cannot move in straightforward fashion from the simplified Coase theorem world to the complexity of reality, as the real world is not "recognizably of the same family as the simple model." The reductionism that McKean saw as a strength of the economic approach was, for Gilmore, a fatal weakness, as conclusions drawn for the real world based upon such a model, he said, will either be false or will be true only by accident (ibid., 105–106).

All of this, for Gilmore, pointed to the dangers for law inherent in setting one's course based on the Coase theorem. Such a theory "conditions our thinking about the real world," as, indeed, "it is meant to do" (Gilmore 1970,

106). The conclusion to which it leads is that, since the direction of liability does not matter, products liability should cease to evolve further in the direction of producer liability, as it at least does no good to proceed further in that direction. This, for Gilmore, was the "ultimate counsel" of McKean's analysis. However, given the distance between the world of the Coase theorem and the world in which we live, Gilmore was not willing to accept the conclusion that the evolution of products liability law should cease, or that it would be a good thing to turn back the clock (ibid., 107). In fact, Gilmore cast doubt on the entire enterprise of using economic analysis to inform legal reasoning, and his comments on this score are worth quoting in full:

> The Coase theorem, truly understood, may prove conclusively that all the changes which have taken place in our society since 1850 – or 1750 or 1650 – have been, from the point of view of economic theory, undesirable. The realization that this was so would sadden me but I would not see that much could be done about it or that anything should be done about it. I do not for a moment dispute or deny the great value of abstract theoretical analysis or the pure delight of engaging in it. I doubt that it gives us, or is meant to give us, guidance in handling real problems in the real world.
>
> (Gilmore 1970, 116)

Nor, it seems, did Gilmore see any prospect that economic reasoning would ultimately prove to be influential in reforming law:

> It is entirely within the realm of possibility that the forces which have determined the shift in risk allocation which has been going on for the better part of a hundred years will cease to operate and that we will reverse our course – or set out in a new direction never before dreamed of. But that, if it were to happen, would have nothing to do with the fact that we had at last learned to think clearly about what was going on. The wit of man is from the start overmatched by the weight of circumstance and, except in the simplest situations, thinking clearly and acting sensibly really have nothing to do with each other.
>
> (Gilmore 1970, 116)

If Gilmore was the representative black letter man of law, the prospects for an economic analysis of law seemed grim indeed.

3 Conclusion: boundary crossing, or scientific imperialism?

It hardly needs stating that in the decades following the products liability conference, the economic analysis of law evolved into the most important new movement in legal scholarship in the last half-century – though there is plenty of room for argument over the extent to which the insights of economics have influenced the law itself.[26] That said, if influence on subsequent scholarship is the measure of importance, the products liability symposium organized by the Joint

Committee of the AEA and the AALS is a relatively minor blip in the history of law and economics. The papers from the symposium are seldom cited, and historians seem largely unaware of its existence or that of the Joint Committee.[27] McKean's perspective on the potential fruits from extending the economic paradigm eventually won the day among substantial groups of both lawyers and economists, but it was Calabresi and Posner (1973), rather than McKean, who persuaded lawyers to begin to take economics seriously as a tool of analysis, and scholars such as Becker and William Landes who convinced economists to do the same.[28]

Yet, the symposium is an important moment in the history of economic analysis of law for several reasons. First, it reflects an overt attempt to *create* an economic analysis of law. Economic analysis of law did not just happen; a small community of scholars made a conscious effort to bring it into being as a replacement for (in the minds of some) or complement to (in the minds of others) traditional methods of legal analysis. The fact that the Joint Committee's effort did not succeed does not make the concerted effort any less real, particularly when one realizes that certain of the key players – notably Manne – were engaged in similar efforts on other fronts.[29]

Second, even among those who agreed that economic thinking could profitably inform legal analysis – in the present instance, McKean, Buchanan, and Calabresi – there was no uniformity of view on what the analysis entailed or implied. Differently put, economics did not offer a "science" of law in the sense of suggesting singular determinate optimal solutions to legal questions. While its proponents preached clarity of thought, there was no real clarity on what that thought should be, as evidenced from the very different conclusions regarding liability for product-related harms reached, and the justifications offered, by these three economics-favoring protagonists.

Third, and perhaps most obviously, the symposium provides the first real insight into the challenges that economic analysis of law would confront in establishing a toehold within both economics and law. Gilmore's commentary is illustrative of the barriers that economists faced in marketing their product to lawyers, even within the walls of the University of Chicago Law School, while Dorfman's critique made it clear that convincing leading economic theorists that their models could be exported outside their traditional context was not going to be an easy road.[30]

It is fair to say that no small amount of the explanation for the critical responses of Gilmore and Dorfman lay in McKean's decision to ground his analysis in the Coase theorem.[31] Viewed from the perspective of the present, McKean's move seems a logical one. After all, the theorem is one of the core ideas of the economic analysis of law and the basis for much of the reasoning that flows from it. The symposium witnessed no dissent regarding the theorem's validity as a proposition in economic logic. The issue, instead, was relevance and whether the reductionism that the economist brought to the table had any currency in the legal realm. The clash between the messy practicalities of law and the "all is for the best in the best of all possible worlds" nature of the Coase theorem could not have been more stark. This is not to say that McKean made a mistake making the theorem the starting point for his analysis. Indeed, one could argue that Posner did the same in *Economic Analysis of Law* (1973), and to great effect. The problem was that McKean approached the subject with an economist's mentality, allowing abstraction to be the driver of his analysis. If economic analysis was going to make serious inroads into legal analysis generally, the game would have to be played on the legal scholar's turf, even if using the economist's ball. In the end, that turn was the key to its eventual success.

This brings us directly to a final lesson to be drawn from the products liability symposium: the difficulties attending attempts to fasten the "imperialism" moniker to boundary crossings involving economics. One of the great myths surrounding the history of economic analysis of law – one perpetuated by economists as well as by critics of the economic approach – is that it developed out of an exporting or colonization effort by economists. This early effort was not, despite McKean's centrality in the story, one of economists attempting to take over law. While McKean himself may be seen, without violence, as engaged in imperialism, Manne, as a lawyer, could not have been, and McKean was a tool enlisted to help reshape legal thinking in a manner that Manne and others he had recruited to the cause found congenial. The reality is that the scholarly and institution-building pioneers of the subject – Calabresi, Manne, and Posner, lawyers all – were *importing* economic analysis into the legal arena. Economists were relative latecomers to the party, Becker's and Landes' contributions notwithstanding.[32]

The term "imperialism" denotes the effort of *A* to exert power or influence over *B*.[33] The early history of economic analysis of law, though, represents anything but this. A small group of lawyers decided that it would be useful for their discipline to draw more heavily on the methods of economics. Yes, some economists slowly began to join the chorus. Yes, with time, economic analysis came to inform legal reasoning in a significant way. And yes, a significant number of economists have come to believe that economics offers the best grounding for legal reasoning. Yet, the economics profession has, to this point, exerted no power or influence over law.[34] Instead, the influence of economics on legal reasoning is the result of an increasingly large group of legal scholars finding utility in the methods of economic analysis and adding it to their box of tools in their research and in the classroom, and of law school deans making the decision to place economists or lawyers with substantial economics training on their faculties and add "law and economics" courses to the curriculum.

The process at work here, then, was not "imperialism" but boundary crossing – of which imperialism is a subset. Boundary crossing, including with economics, has been at the heart of American jurisprudence since the early 20th century,[35] though the economics of this earlier period was of a decidedly different flavor and, ironically (but not coincidentally), gave rise to some of the legal principles called into question by the modern economic analysis of law and defended vociferously by critics of it. It is all well and good to question whether the tools of the economist are appropriate for the analysis of legal relationships, but to do so within the context of imperialistic rhetoric serves both to mask and fundamentally misunderstand the historical processes through which the economic approach to law came to be, and to garner a measure of acceptance within the legal academy.

Acknowledgments

The author wishes to thank Roger Backhouse, Uskali Mäki, Alain Marciano, Manuela Fernández Pinto, Adrian Walsh, and participants in the workshops on "Interdisciplinary Social Science" (ENS-Cachan) and "Scientific Imperialism" (TINT) for instructive comments related to the present analysis. The financial support of the Earhart Foundation, the National Endowment for the Humanities, and the Institute for New Economic Thinking is gratefully acknowledged.

Notes

1 Distinguished Professor of Economics, University of Colorado Denver, USA. Email: steven.medema@ucdenver.edu.
2 However, see Mitchell (1999) and Medema (2000) for an analysis of the development of public choice analysis and the work of Pedro Teixeira (e.g., Teixeira 2009, 2014) on the development of human capital analysis. The Introduction by Mäki, Walsh, and Fernández Pinto offers a sketch of the various extant perspectives on scientific imperialism.
3 Medema (2013) provides an overview of the formation and early history of the Joint Committee.
4 See e.g., Calabresi (1961, 1965a, 1965b, 1970).
5 Marciano (2012) and Medema (2014a, 2014b) provide discussions of the treatment of Coase's negotiation result – better known today as the "Coase theorem" – by Calabresi and in the legal literature during the 1960s.
6 Two significant pieces of evidence for its solidity within the economics profession are its recognition within the *Journal of Economic Literature* field codes, and the presence of "law and economics" courses in the undergraduate economics curriculum across the United States.
7 It goes almost without saying that "law and economics" has a long heritage in legal analysis, its heyday coming during the legal realist period of the 1920s, 1930s, and 1940s, but that law and economics was of a very different type from the modern economic analysis of law. See e.g., Hovenkamp (1990), Duxbury (1995), and Medema (1998) for analyses of the earlier and later law and economics movements.
8 By "traditionally 'non-economic' areas of law" we exclude antitrust, regulation, etc., where economists and legal scholars had worked side by side and even hand in hand for many decades. See e.g., Duxbury (1995) and Hovenkamp (1990).
9 The same, as it happens, is true of the history of the application of economic analysis to political science.
10 This goal is consistent with what Mäki (2013, 334) has characterized as "imperialism of style."
11 It is worth noting, however, that McKean referred the reader to Armen Alchian and William Allen's *University Economics* (1967) for an elaboration of the points raised in his tutorial, rather than, say, a more popular textbook such as Samuelson's *Economics* (1967). Alchian and Allen had been McKean's colleagues at UCLA, and their text is generally considered to have a more market-oriented flavor than Samuelson's.

12 McKean cites *Escola v. Coca-Cola Bottling Co.* as an illustration of this attitude. See *Escola v. Coca-Cola Bottling Co.*, 24 Cal. 2d 453, 462, 150 P.2d 436, 441 (1944) (Traynor, J., concurring).

13 Although McKean did not note it in his essay, this very sentiment was at the heart of Blum and Kalven's (1964) critique of Guido Calabresi's (1961) analysis of deterrence, a line of argument that was very much informed by his extensive training in economics.

14 McKean defined zero transaction costs as a situation in which there are zero costs of negotiating and enforcing agreements, and "zero costs of information about products' characteristics." Such a world, he said, is not one "of complete certainty in this analysis, for such a world would have no accidents, unforeseen events, or doubts about liability" (McKean 1970, 43, n.108).

15 See e.g., McKean 1970, 50–51.

16 It is curious that McKean saw fit to label absolute producer liability "extreme" but did not do the same for absolute consumer liability (*caveat emptor*).

17 Still worse, though, McKean argued, would be a no-fault system with *governmental* compensation, which some were advocating, as this would remove the burden of accident costs from producers as well as consumers, leaving both parties with reduced incentive to engage in precautionary activities (McKean 1970, 52).

18 McKean went on to note that prices would rise in industries producing *safe* products as well, as consumers of products whose prices had risen substituted other products. There would also be secondary distributional effects, as returns to inputs in industries producing unsafe products would fall, while owners of inputs in safe products industries would rise (McKean 1970, 51). Interestingly, these effects would also accompany a system of *caveat emptor*, but McKean did not note these as a black mark against that system.

19 The reference here is to John Godfrey Saxe, "Six Blind Men and the Elephant," in *The Poems of John Godfrey Saxe* (Boston, MA: Ticknor and Fields, 1868).

20 McKean cited as illustrations of such work Franklin (1966), Keeton (1967), Plant (1955), Blum and Kalven (1964) and, notably, Calabresi (1965a, 1970).

21 It is not surprising to see Buchanan make an appeal to fairness as well as efficiency. His 1974 essay reviewing Posner's *Economic Analysis of Law* finds Buchanan far more comfortable with the use of price theory to assess the impacts of legal rules than with efficiency-driven jurisprudence (Buchanan 1974).

22 The logic, according to Buchanan, is that transaction costs which preclude further exchange make the existing position efficient, in that it is not cost-effective to move away from it.

23 Calabresi noted that his paper had been prepared as a reaction to an original draft by McKean, and that McKean had since made some modifications of emphasis that might cause his own remarks to seem a bit strong (Calabresi and Bass 1970, 74).

24 See e.g., Calabresi (1965a, 1970).

25 Gilmore's most significant contributions to law include his book, *The Death of Contract* (1974), and his role as one of the architects of the Uniform Commercial Code. This raises the question of why Manne did not secure the services of someone who had written on products liability but was not attached to the economic approach, as Calabresi was, and thus could be suitably critical of McKean's analysis. Unfortunately, the archival records contain no information on this score.

26 That law itself has been more immune from the influence of economics than legal scholarship could be said to validate Gilmore's prediction.

27 This author is currently engaged in a study of the history of the Joint Committee's origins, activities, and eventual demise in the mid-1970s.

28 Landes was Becker's student at Columbia in the 1960s and appears to have been the first economist to establish a research agenda in the economic analysis of law, one

that emphasized empirical analysis as a necessary complement to theory. Landes began a long and fruitful research collaboration with Posner in the mid-1970s.

29 The most influential of these other efforts may be the institutes, organized by Manne, that taught economic analysis and its application to law to law professors and judges. See e.g., www.law.gmu.edu/about/history (accessed January 28, 2016).

30 Even Ronald Coase, widely credited as one of the founders of modern law and economics, was skeptical of the application of economics outside the traditional realm of the subject – a position that put him at odds with his law school colleague, Richard Posner. See e.g., Coase (1978, 1993) and Posner (1993).

31 One finds a similar set of attitudes in the conference discussion that followed the five paper presentations – the analysis of which here is precluded by space limitations. However, see Manne (1970).

32 Indeed, when Landes presented some of his early research on the subject at Chicago in 1967, he was told by a senior Chicago economist that he should instead focus on subjects that were of more than "marginal interest" to economists (Landes 1997, 34).

33 See e.g., the *Oxford English Dictionary* entry on "imperialism," www.oed.com/view/Entry/92285?redirectedFrom=imperialism#eid (accessed January 27, 2016).

34 In fact, an argument can be made that a goodly number of the economists working on legal topics have little interest in law or the legal reasoning process, but instead have mined this territory because of the publication possibilities it offered (Medema 2015).

35 See e.g., Duxbury (1995) and Hovenkamp (1990).

References

Alchian, Armen A. and William R. Allen. 1967. *University Economics*. Second edition. Belmont, CA: Wadsworth Publishing Company.

Becker, Gary S. 1968. "Crime and Punishment: An Economic Approach". *Journal of Political Economy* 76(2): 169–217.

Becker, Gary S. 1976. *The Economic Approach to Human Behavior*. Chicago, IL: University of Chicago Press.

Blum, Walter J. and Harry Kalven, Jr. 1964. "Public Law Perspectives on a Private Law Problem: Auto Compensation Plans". *University of Chicago Law Review* 31(4): 641–723.

Blum, Walter J. and Harry Kalven, Jr. 1965. *Public Law Perspectives on a Private Law Problem: Auto Compensation Plans*. Boston, MA: Little, Brown and Company.

Buchanan, James M. 1970. "In Defense of Caveat Emptor". *University of Chicago Law Review* 38(1): 64–73.

Buchanan, James M. 1974. "Good Economics. Bad Law". *Virginia Law Review* 60(3): 483–492.

Calabresi, Guido. 1961. "Some Thoughts on Risk Distribution and the Law of Torts". *Yale Law Journal* 70(March): 499–553.

Calabresi, Guido. 1965a. "The Decision for Accidents: An Approach to Nonfault Allocation of Costs". *Harvard Law Review* 78(4): 713–745.

Calabresi, Guido. 1965b. "Fault, Accidents and the Wonderful World of Blum and Kalven". *Yale Law Journal* 75(2): 216–238.

Calabresi, Guido. 1970. *The Cost of Accidents: A Legal and Economic Analysis*. New Haven, CT: Yale University Press.

Calabresi, Guido and Kenneth C. Bass III. 1970. "Right Approach, Wrong Implications: A Critique of McKean on Products Liability". *University of Chicago Law Review* 38(1): 74–91.

Coase, Ronald H. 1960. "The Problem of Social Cost". *Journal of Law and Economics* 3(October): 1–44.

Coase, Ronald H. 1978. "Economics and Contiguous Disciplines". *Journal of Legal Studies* 7(2): 201–211.

Coase, Ronald H. 1993. "Law and Economics at Chicago". *Journal of Law and Economics* 36(1): 239–254.

Dorfman, Robert. 1970. "The Economics of Products Liability: A Reaction to McKean". *University of Chicago Law Review* 38(1): 92–102.

Dupré, John. 1994. "Against Scientific Imperialism". *PSA: Proceedings of the Biennial Meeting of the Philosophy of Science Association* 2: 374–381.

Dupré, John. 2001. *Human Nature and the Limits of Science*. Oxford: Oxford University Press.

Duxbury, Neil. 1995. *Patterns of American Jurisprudence*. Oxford: Oxford University Press.

Fine, Ben. 1999. "A Question of Economics: Is it Colonizing the Social Sciences?" *Economy and Society* 28(3): 403–425.

Fine, Ben. 2002. "Economic Imperialism: A View from the Periphery". *Review of Radical Political Economics* 34(2): 187–201.

Fine, Ben and Dimitris Milonakis. 2009. *From Economics Imperialism to Freakonomics: The Shifting Boundaries Between Economics and Other Social Sciences*. Abingdon: Routledge.

Franklin, Marc A. 1966. "When Worlds Collide: Liability Theories and Disclaimers in Defective-Product Cases". *Stanford Law Review* 18(6): 974–1020.

Gilmore, Grant. 1970. "Products Liability: A Commentary". *University of Chicago Law Review* 38(1): 103–116.

Gilmore, Grant. 1974. *The Death of Contract*. Columbus: The Ohio State University Press.

Hovenkamp, Herbert. 1990. "The First Great Law & Economics Movement". *Stanford Law Review* 42(April): 993–1058.

Joint Committee of the American Economic Association and the Association of American Law Schools. 1967. "Report of the Joint Committee of the American Economic Association and the Association of American Law Schools". *American Economic Review* 57(2): 715–717.

Keeton, Robert E. 1967. "Is There a Place for Negligence in Modern Tort Law?" *Virginia Law Review* 53(4): 886–898.

Landes, William M. 1997. "The Art of Law and Economics: An Autobiographical Essay". *The American Economist* 41(1): 31–42.

Lazear, Edward. 2000. "Economic Imperialism". *Quarterly Journal of Economics* 115(1): 99–146.

Mäki, Uskali. 2013. "Scientific Imperialism: Difficulties in Definition, Identification, and Assessment". *International Studies in the Philosophy of Science* 27(3): 325–339.

Manne, Henry G. 1970. "Edited Transcript of AALS-AEA Conference on Products Liability". *University of Chicago Law Review* 38(1): 117–141.

Manne, Henry G. and Gerald M. Meier. 1970. "Foreword". *University of Chicago Law Review* 38(1): 1–2.

Marciano, Alain. 2012. "Guido Calabresi's Economic Analysis of Law, Coase and the Coase Theorem". *International Review of Law and Economics* 32(1): 110–118.

McKean, Roland N. 1970. "Products Liability: Trends and Implications". *University of Chicago Law Review* 38(1): 3–63.

Medema, Steven G. 1998. "Wandering the Road from Pluralism to Posner: The Transformation of Law and Economics in the Twentieth Century". *History of Political Economy* 30(Supplement): 202–224.

Medema, Steven G. 2000. "'Related Disciplines': The Professionalization of Public Choice Analysis". *The History of Applied Economics: History of Political Economy Annual Supplement* 32: 289–323.

Medema, Steven G. 2013. "Bringing Law and Economics Together". Working Paper, University of Colorado Denver.

Medema, Steven G. 2014a. "Debating Law's Irrelevance: Legal Scholarship and the Coase Theorem in the 1960s". *Texas A&M Law Review* 2(2): 159–213.

Medema, Steven G. 2014b. "Juris Prudence: Calabresi's Uneasy Relationship with the Coase Theorem". *Law and Contemporary Problems* 77(2): 65–95.

Medema, Steven G. 2015. "From Dismal to Dominance? Law and Economics and the Values of Imperial Science". In Aristides N. Hatzis and Nicholas Mercuro (eds). *Law and Economics: Philosophical Issues and Fundamental Questions*. London: Routledge.

Mitchell, William C. 1999. "Political Science and Public Choice: 1950–1970". *Public Choice* 98(3–4): 237–249.

Plant, Marcus L. 1955. "Strict Liability of Manufacturers for Injuries Caused by Defects in Products – An Opposing View". *Tennessee Law Review* 24: 938–945.

Posner, Richard A. 1973. *Economic Analysis of Law*. Boston, MA: Little, Brown.

Posner, Richard A. 1993. "Ronald Coase and Methodology". *The Journal of Economic Perspectives* 7(4): 195–210.

Samuelson, Paul A. 1967. *Economics*. Seventh edition. New York: McGraw-Hill.

Stigler, George J. 1966. *The Theory of Price*. Third edition. New York: Macmillan.

Stigler, George J. 1984. "Economics: The Imperial Science?" *Scandinavian Journal of Economics* 86(3): 301–313.

Teixeira, Pedro N. 2009. *Jacob Mincer: The Human Capital Labour Economist*. Oxford: Oxford University Press and IZA.

Teixeira, Pedro N. 2014. "Gary Becker's Early Work on Human Capital – Collaborations and Distinctiveness". *IZA Journal of Labor Economics* 3(1): 1–20.

6 Rational choice as neo-decisionism

Decision-making in political science and economics after 1945

Nicolas Guilhot and Alain Marciano

Introduction

Usually associated with the work of a small set of postwar economists (essentially Kenneth Arrow, Duncan Black, James Buchanan, Anthony Downs, William Riker, Thomas Schelling, or Gordon Tullock), rational choice theories in political science are considered to be one of the most successful examples of economics imperialism over a neighboring discipline (see e.g. Green and Shapiro 1994; Lazear 2000; Monroe 2001). Interestingly, the notion that economics was taking over the study of politics was not only entertained by the advocates of new economic approaches to political problems; their opponents and critics took it for granted too, only in order to oppose what they saw as a regrettable trend. Critical political scientists or sociologists have since recycled the notion that the application of rational choice theory in the social sciences is essentially the "enlarged and modernized application of the 'economic approach' to social problems" (Calhoun and Wacquant 1989, 41).

In the following, we want to qualify this view, and to propose a slightly different interpretation of the rise of rational choice approaches after World War II. It is a fairly common view that the rise of "rational choice" in political science is the primary manifestation of the imperialism of economics. Little is known about the specific pathways of this disciplinary imperialism. It is possible, following for instance Mäki (2013), to develop a more fine-tuned picture, in particular by taking into account both the internal and the external factors that made it possible for seemingly "economic" methodologies to spread within political science. Indeed, political scientists have often looked up to economics as a model to be emulated in the study of politics.[1] David Easton suggested that economics was possibly the only discipline that could analyze the notion of equilibrium satisfactorily, something that political science was not yet in position to do (Easton 1953). In the subfield of international relations, the major innovation of the 1970s was the rise of a systemic theory of international politics that explicitly modeled itself after economics (Waltz 1979). One could thus describe a process of methodological imperialism driven in part by factors internal to political science. As Mäki suggested:

> [e]conomics imperialism is a matter of a persistent pursuit to increase the degree of unification provided by rational choice theory by way of applying it to new types of explanandum phenomena that are located in territories that are occupied by disciplines other than economics
>
> (Mäki 2002, 238)

It seems indeed that this is what happened to political science after 1945.

While not incorrect, this description is too schematic, partly because it results from a rational and retrospective reconstruction associated, rightly or not, with the hegemony of the Chicago school. Rational choice, we argue, did not define economics in the postwar years, nor even in the 1950s: it represented a relatively recent and still marginal development in the discipline and it did not yet have the capacity to redefine economics, let alone political science. Stating the problem only in terms of imperialism or disciplinary turf battles prevents us from developing a better historical grasp of this development.

Our basic argument is quite straightforward, and parallels the argument made by S.M. Amadae in her contribution to this volume. Rational choice approaches cannot be considered as a straightforward case of imperialism of economics[2] over political science, at least in the immediate postwar years (although things would change later and are different today). These approaches signaled a methodological rupture within economics itself and there was a joint transformation of these disciplines as they struggled to deal with the emerging topic of decision-making. Indeed, "[t]he set of methods that came together to form rational choice theory," S.M. Amadae rightly suggested in a prior work, "did not flow out of mainstream economics" (Amadae 2003, 148) – a point largely confirmed by the historians of economic thought or the historians of science. As decision-making became a central focus of political research in the postwar years, it triggered the development of new methodologies that either ignored disciplinary boundaries or affected economics and political science in equal measure. The notion of "rationality" that was developed in these disciplines was indeed tightly associated with decision theory, and the expression "rational choice" bears witness to this close association. Economics registered this impact just as much as political science. We first want to establish that decision-making became important both in economics and in political science, and served as the central concept around which economic analyses of political phenomena were developed. While the analysis in terms of "imperialism" is certainly useful, we want to complement this internalist account with an externalist perspective that points at the *joint transformation* of these disciplines in response to important cultural and political transformations.

Once this has been established, the task of the disciplinary historian is to understand what caused this turn: why did decision-making become the central focus of political science and economics from the mid-1940s onwards? What kind of problem did rational choice provide an answer to? Which previous theories of political decision-making did it seek to replace? Why did establishing rationality criteria for decisions become so important after 1945, thus giving economics a place of pride in the analysis of political questions?

We believe that economists and political scientists were driven to rethink their approaches to decision-making because of a generalized anxiety about the future of liberal politics. Rational choice became a new language that was used to legitimate both infringements to liberal constitutionalism and its re-foundation on a more authoritative basis. In domestic and international matters, traditional modes of decision-making were called into question. Our second line of argument is that rational choice was an effort at "taming," as it were, exceptional political decisions by establishing rationality criteria that would make them acceptable in the absence of other forms of legitimation, in particular democratic ones. The initial context of the preoccupation with decision-making was a diffuse fear that liberal institutions and liberalism were not capable of meeting the challenges of the postwar world. The crisis of the 1930s and World War II had destroyed anything that may have been left of prior beliefs in historical progress toward increasingly rational forms of social life. The democratization of politics, observers such as Walter Lippmann or Hans Morgenthau claimed, had not brought about a greater rationality of social life but had, on the contrary, unleashed the irrationality of the masses, often in the form of utopian and messianic nationalism.

The onset of the Cold War gave a new lease of life to the notion that liberal democracies were ill-equipped to defend themselves against enemies willing to use liberal freedoms and rights in order to seize power and abolish them. It also gave traction to the idea that decision-making capacities unconstrained by constitutional provisions were sometimes in order if such outcomes were to be prevented (Friedrich 1957; Rossiter 1948; see also Greenberg 2014). Finally, the atomic bomb presented new dilemmas and called for political and strategic decisions taken in the complete absence of historical precedents and preexisting normative frameworks (Ghamari-Tabrizi 2005). It required a complex command-and-control structure that made it paramount to specify unambiguously decisional processes (Schlosser 2013). It also profoundly changed the nature of governance by concentrating decision-making power at the top and giving rise to what Elaine Scarry has called "thermonuclear monarchy," fundamentally altering the self-perception of the American republic (Scarry 2014). To put it differently, following World War II, political decisions were increasingly sovereign "decisions upon the exception" that pronounced upon fundamental issues while remaining detached from existing normative frameworks and historical precedents. "Decisionism," which referred in the 1920s and 1930s to the anti-liberal legal doctrine popularized by Carl Schmitt, was now the dominant mood in several quarters of political science, not least international relations and strategic studies.[3]

Yet, the justification for these unaccountable forms of authority no longer took the form of a conservative attack on liberal traditions or a rejection of scientific and political rationalism: it now paraded under the guise of a scientific, rationalistic approach to the problem of political decision-making.[4] Exceptional decisions were no longer expressing some mystical notion of sovereignty, but logical calculations that could be justified on the basis of their economic (formal) properties. As it was understood after 1945, "rationality" was summoned in order to "tame the terrors of decisions too consequential to be left to human reason alone, traditionally understood as mindful deliberation" (Erickson et al. 2013, 2). Rational choice, we argue, is better understood as a form of "neo-decisionism" that thrived on the crisis of the traditional modes of legitimation of political decisions in liberal democracies. "Rationality" thus became a substitute for the curtailment of democracy in political decision-making as well as a justification for it. It also developed on the ruins of previous hopes placed in the role of "Reason" in human affairs and in history, inherited from the Enlightenment. In the process, it defined a new normativity for political decision-making, no longer defined by legal norms but by "scientific" ones.

1 The rise of decision-making in political science and economics

It may be useful to start by putting a bit of empirical flesh on the bones of our argument. As can be seen from Figure 6.1, "decision-making" appears to be a distinctly postwar phenomenon as a focus of research in political science and economics. While in absolute terms the number of articles that included "decision-making" in their title remains quite low, it should not cast doubt on the significance of its manifold increase in the decades immediately following World War II. From being non-existent before 1939, it became an up-and-coming topic both in economics and in political science.

Although absolute numbers remained quite low, this trend was not insignificant since it was clearly identified by political scientists, for whom it provided nothing short of the methodological identity of the discipline. In 1962, political scientist William Riker claimed that "the subject studied by political scientists is decision-making" (Riker 1962, 11). He was echoed a few years later by Herbert Simon, who felt justified in saying that decision-making was "not … some highly special aspect of the political process, but … its central core" (Simon 1966, 15).

The reasons for this collective infatuation with decision-making are many – and we will explore some of them in detail below – but those that political scientists were most explicit about were essentially of a methodological nature.

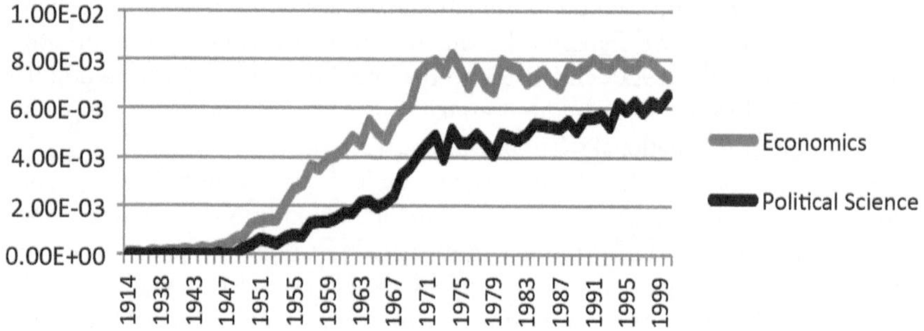

Figure 6.1 "Decision Making" in Political Science and Economics Journals, 1914-200
Source: JStor data for research.

Studying decisions was considered the royal way toward establishing political science on a more scientific and less historical basis. Already in 1950, Yale political scientist Harold Lasswell noted that:

> [i]n the immediate future it is probable that political scientists can improve their results by paying more attention to the construction of theoretical models" and, in this regard, "[a]ttention needs to be given to newly developed models of the decision making process".
>
> (in Simon et al. 1950, 423–425)

For him, these models were those developed in von Neumann and Morgenstern's theory of games: this "first mathematically well-developed theory of rational judgment … confronts the political scientist with the problem of determining whether the terms used in the mathematical notation can be given operational meaning in the situations relevant to political science" (ibid., 424). Although he took his work in a direction that would eventually become critical of rational choice, Herbert Simon was also among the first to stress the importance of "decision-making" in the analyses of administrative organization. His 1947 book *Administrative Behavior* was widely reviewed in political science journals and some reviewers similarly pointed out that his approach could provide a new theoretical ground to political science (e.g. Cook 1950).

The emergence of decision-making as a feature of disciplinary identity was thus associated with struggles over the nature of political science at a time of uncertainty over its position within the changing landscape of the social sciences. The study of decision-making in terms of the formal properties of the decision itself made it possible to study politics without any reference to a historical or normative context, and as such it appealed to those within the profession who wanted to turn political science into a "science" and detach it from history and law. These "professional mourners" (Germino 1963, 437) eager to put an end to the study of historical institutions and legal systems were not so many, but they voiced their uneasiness or dissatisfaction so strongly and it "captured so well the mood of a sizeable body that it spread through […] the profession […] with appalling swiftness" (ibid., 437).[5] Some of their colleagues still insisted that political theory should make more room for the study of the *foundations* of political regimes, and therefore take in ethics, morals, and history. This was, for instance, the view defended by Alfred Cobban, who urged his colleagues to "take up again the tradition of Western political thought and, in doing so, to resume that "continuous transformation of morals into politics" (Cobban 1953, 337). By the mid-1950s, however, this was a rearguard battle, unable to contain the surge of behavioral social science.[6] The growth of "behavioralism" in political science, with its emphasis on political process and dynamics rather than formal documents or institutions, was thus correlated with the increasing importance of decision-making as a key focus of the discipline (Solovey 2013). The same process occurred in international relations, where the traditional emphasis on diplomatic history gave way to attempts at formalizing decisional processes – such as in the volume published by Snyder, Bruck, and Sapin, *Decision-making as an Approach to the Study of International Politics* (1954) and more generally in the rise of game theory and systems theory in the field.

The postwar *Methodenstreit* in political science thus tended to generate new alliances cutting across disciplinary boundaries, and in particular alliances between political scientists and economists. A number of the former have looked up to economics as a disciplinary model for tackling essentially similar questions, i.e. questions of choice and decision. This, in turn, has given credence to the notion that economics was an "imperialist" discipline, colonizing the neighboring field of political science. Indeed, a number of political scientists have often used formulations that seemed to confirm this view. Yet, these pronouncements were mostly strategic, and they did not reflect accurately the evolution of political science, nor, indeed, of economics. For in this latter discipline, the same trend can be observed, albeit more pronounced, with the same point of departure.

In economics too, decision-making was not a central issue before World War II and became a core focus of the discipline only after 1945. While the graph does not contain the answer to the question of whether or not rational choice was a concern imported from economics into political science (and indicates a slight priority of economics in registering the trend of research on decision-making), a wealth of circumstantial and anecdotal evidence makes its interpretation unequivocal. Without exceptions, the early practitioners of game theory who have written about the early days of rational choice suggest that economists were largely unfazed by these new methodological gimmicks. A 1946 review of von Neumann and Morgenstern's *Theory of Games and Economic Behaviour* in the *Review of Economic Studies* gave the tone: the book in no way represented the new departure – a "revolution" as Kaysen put it in the title of his essay – for economics that its authors claimed it was (Kaysen 1946–47). In his reminiscences about game theory at Princeton, for instance, Martin Shubik suggested that economists were initially skeptical, while political scientists were much more receptive (Shubik 1992).[7]

The few economists initially interested in the theory of games and rational choice were often interested in political phenomena, intended essentially as a matter of decision-making. For them, any division or distinction in terms of scope between economics and political science – and even more broadly, between the social sciences in general – had to be abandoned. Disciplinary boundaries were unnecessary. They defended the idea of an explanatory unification between the social sciences, which should be integrated or unified around the use of a "scientific method" (see e.g. Rice et al. 1950) – which meant essentially that their propositions had to pass the test of mathematization in order to achieve scientific status. The most explicit formulations of this methodological program can be found in the writings of Duncan Black, who published in 1950 an article titled "The Unity of Political and Economic Science," in which he suggested that a "satisfactory" theory of politics could be entirely based upon a "Theory of Committee Decisions" (Black 1950, 506–507). Black was interested in developing a science of politics with his work on committees.[8] The science of politics he was proposing and which he saw as the equivalent of the theories developed by Walras in his *Élements d'Économie Pure* or by Pareto in his *Manuel* (Black 1950, 506), consisted in an economic (mathematical and statistical) analysis of how groups of individuals (committees) made decisions and choices. In other words, "a Theory of Committee Decisions" was "a sufficient means to construct a Theory of Politics" and, conversely, "a Theory of Politics [...] c[ould] be developed out of the theory of committees" (ibid., 506–507), which itself could be entirely formulated in the terms of "Economic Science," as "a set of formal or mathematical proposition" (Black 1950, 506–507).[9] Yet, Black did not simply advocate the use of a specific method imported from economics. He called for the abolition of any differentiation between economics and political science, on the ground that political systems too were means to make decisions. Economics and politics really "form[ed] two branches of the same subject" (Black 1950, 512), as they sought "to answer practically the same question" (Black 1950, 512): "each relates to *choosing* of some kind" (Black 1950, 514, emphasis added). The convergence between the disciplines was thus advocated on the basis of the identity between the social subsystems they were supposed to deal with – the economy and politics.[10]

That decision theory became the pivot for important disciplinary and methodological realignments can also be seen in the work of Kenneth Arrow. Like Black, Arrow assumed a fundamental "analogy between economic choice and political choice" (Arrow 1951, 5) as a starting point for his argument about democracies, the unrestricted domain of preferences and dictatorship. His *Social Choice and Individual Values* famously begins with the claim that "[i]n a capitalist democracy there are essentially two methods by which social *choices* can be made: voting [...] and the market mechanism" (Arrow 1951, 1, emphasis added). Both these choices, however, could be made by using a particular "machinery" (Arrow 1951, 8): the social welfare function, which was supposed to aggregate individual utility functions representing preferences considered to be *given*.

Another case of this conflation between economics and politics can be found in the work of Anthony Downs (1956, 1957a, 1957b). The first version of Downs' work, published in 1956 by the Department of Economics of Stanford University for the Office of Naval Research, was titled *An Economic Theory of Government Decision-Making in a Democracy*. An economist, Downs had nonetheless written a text that spoke directly to political scientists. In his review, Charles E. Lindblom claimed that Downs' "definition, assumptions, and propositions are on the whole familiar to political scientists" (Downs 1956, 242) because the "central critical hypothesis" of his model – namely that "a political party is an organization for making nominations and contesting elections in order to win control of the government machinery" (ibid., 242) – was nothing else than "a re-formulation of a proposition often encountered in political science" (ibid., 242). Despite the fact that Downs seemed to suggest that he was proposing a theory of democracy, he actually reduced democracy to decision-making and restricted the role of the political system to performing economic tasks only, to the extent that political decisions meant for him essentially decisions about the allocation of resources, i.e. budgetary decisions (see e.g. Downs 1960). As one of the book reviewers noted, Downs proceeded to "the assimilation of politics to economics" (Diamond 1959, 208).

Along the same lines, one can mention Gary Becker's short article about democracy published in 1958 in the *Journal of Law and Economics*. This piece warrants some attention, despite the fact that politics has never been a central line of research for Becker, because the arguments he proposes are very close to Downs', Arrow's, or Black's. To him, the government exclusively engages in economic activities, geared toward the allocation of scarce resources. Significantly, he defined democracy as an "institutional arrangement for arriving at political *decisions*" (Becker 1958, 106, emphasis added).

Last but not least, *The Calculus of Consent* (Buchanan and Tullock 1962), which we analyze in greater detail below, was an attempt by economists to tackle issues that until then were the purview of constitutional lawyers. A seminal text of the school of Public Choice and of what would later become known as "constitutional economics," the book offered an economic analysis of both decision-making rules and of decisions upon decision-making rules, i.e. constitutional decisions. When in 1963 Buchanan and Tullock first gathered a few colleagues – economists, political scientists, and even a political philosopher – to discuss how political choices are made and how institutions affect those choices, they chose to call the group the "committee on non-market decision making." It would eventually evolve into the Public Choice society and develop significant clout, but it was initially a development taking place on the outer margins of economics.

To summarize, the tendency toward explanatory unification that took place in the 1950s between economics and political science did not simply result from the trespassing on the disciplinary border by economists or the subjection of political science to another discipline. Political scientists were also attracted by what they considered to be "economic" methods. Both statistical and circumstantial elements thus suggest that interest in approaches oriented toward decisions emanated as much from political science as they did from economics. Rational choice, on the other hand, did not represent a mainstream development in the discipline of economics. Its rise in the social sciences was the result of an overlap of interests that led both political scientists and economists to search for formal or mathematical methods for the study of political decisions. Why this focus on political decision-making drove methodological innovations in both disciplines is the question to which we now turn.

2 The context and the focus on decision-making

The first economic analyses of political phenomena, whether they came from political scientists or from economists, did not simply consider politics as a decision-making activity: they sought to provide a new benchmark for legitimate politics under the form of "rationality." As the hypothesis of an external and methodological imperialism of economics seems less convincing, we have to look for a more nuanced and fine-grained contextualization of the rise of rational choice theories.

First, the development of rational choice theory did not reflect an unabashed optimism in the capacities of human reason reflected in economic science but, on the contrary, a deep anxiety about the ability to uphold standards of rationality in political matters (Heyck 2015). It is significant, for instance, that David Easton, who was one of the main advocates of a "scientific" approach to politics and repeatedly gestured toward economics as an epistemic model, described his project as a reaction against what he saw as rising anti-rationalist trends in political thought. *The Political System* was explicitly directed against the "mood in Western civilization directed against the use of scientific reason" (Easton 1953, 6). "The present epoch," Easton wrote, "may well be the beginning of a long period of the decline of men's faith in reason" (ibid., 7). The Enlightenment faith "in the ability of the kind of reasoning implied by scientific method to solve social problems, empirical and even moral" (ibid., 7) was under attack. Easton was particularly worried by a "movement back to theology" that emphasized spiritual ideals detached from empirical reason, and by the kind of "traditionalism" for which "reason must yield to the greater wisdom of prejudice, tradition, and accumulated experience knowable largely through history" (ibid., 21). While Easton had in mind Michael Oakeshott's attack on political rationalism, there was no dearth of examples closer to home, starting with the emerging realist movement in international relations theory. It may be that a number of political scientists were certainly driven by economics envy in terms of disciplinary prestige and credibility – what Mäki (2013) would call an imperialism of "standing" – but it is clear that there were a number of intellectual reasons unrelated to economics that explain the appeal of rational theories of decision at the time.

One was certainly an attempt at salvaging a notion of "rationality" from the generalized discredit in which previous notions of "Reason" and hopes placed in progress found themselves. Rational choice approaches were based on a thin, formal or procedural rationality that was very different from the substantive "reason" allegedly embedded in history or human practice and which informed the Enlightenment tradition. The diffusion of "rationality" in the postwar social sciences, as Easton clearly indicates, was associated with a more modest and less secure conception of reason, confronted with a return of spiritualism and historicism in political thought and various forms of anti-Enlightenment reactions. Rational choice was thus part of a wider attempt in the social sciences at not giving up completely on rationalism as the entire Enlightenment tradition was under attack.[11]

That rational choice methods were intimately connected to this historical background is further suggested by the success of Herbert Simon, who capitalized on this *Zeitgeist* more than any other theorist. Compared to other rational choice approaches such as game theory (or more generally what Simon called "subjective expected utility"), his approach posited an even thinner notion of rationality, based on vicarious learning and adaptation rather than extensive calculative capacities. Rationality, for him, was essentially defined by its limitations, and the "rational chooser" was a creature of limited rationality, often fumbling in the dark and advancing through trial and error, emulation or restriction of her options. Behind Simon's "bounded rationality," there was a historical context in which positing any strong notion of human rationality was no longer possible because the crisis of the 1930s had exposed its essential weakness and frailty: as he put it much later, "reason was not, could not have been, our principal shield against Nazism" (Simon 1983, 10–11). In this respect, it was developments in psychology, much more than in economics, that informed a number of perspectives on rational decision. This is exactly what one of the first reviewers of Simon's *Administrative Behavior* perceived and stressed: "[i]f any 'theory' is involved, it is that decision-making is the heart of administration, and that the vocabulary of administrative theory must be derived from the logic and psychology of human choice" (Angel 1947, 185).

A second, related factor was the need to re-legitimate democracy and liberal institutions, as S.M. Amadae has pointed out in her pioneering work. In the wake of the 30-year crisis, democratic processes and liberal institutions could no longer be counted upon to deliver rational outcomes or to make sound decisions in a timely fashion. For a variety of disillusioned liberals, this suspicion also extended to democratic publics, which could not be trusted to exercise government in a rational and effective manner. "Where mass opinion dominates the government," Walter Lippmann wrote in 1955, "there is a morbid derangement of the true functions of power. The derangement brings about the enfeeblement, verging on the paralysis, of the capacity to govern" (Lippmann 1955, 15). The rational character of the democratic political process and institutions could no longer be taken for granted. It had to be made secure through explicit specifications of what constituted "rationality."

Diverse as they were, rational choice methods were thus set in an ambivalent relation to liberal institutions. On the one hand, they offered new legitimation criteria for decision-making that no longer relied on processes such as vote or deliberation; on the other hand, they were premised on a number of notions – individual preferences and freedom of choice, in particular – that were germane to the liberal democratic imagination. William Riker perfectly understood that the great value of game theory was the promise of reconciling the predictability of decisions with the idea of free choice: while traditionally the social sciences' emphasis on generalization and determinacy seemed incompatible with human freedom, "game theory offer[ed] a way out of this dilemma by combining the possibilities of generalization and free choice" (Riker 1992, 210). Rational choice thus took stock from the critiques of political rationalism that had flourished in the previous decades while seeking to reformulate a more modest understanding of rationality in political decision-making and develop new legitimation criteria for political decisions that were no longer based on democratic process but on the formal properties of the decision itself.

Third, in the wake of the 1929 crisis and during the New Deal, there were powerful incentives that led economists to devote more and more attention to the question of political decision-making. It is during this period that economists developed notions of radical uncertainty that unsettled the capacity to make choices: Frank Knight's "unmeasurable uncertainty" is a product of its time, and bears witness to deep anxieties about a future that could no longer be made amenable to calculation and prediction (Katznelson 2013, 33). The crisis was also the historical event that established the recognition that markets could fail, and with them the capitalist economy (see Backhouse 2015).[12] The idea that markets could be left to their own devices and be expected to regulate themselves had lost all credibility, along with the notion that the allocation of resources was spontaneously efficient. The question of what should be allocated through the market and what should be allocated through authoritative decisions became a pressing issue, and one that took the form of a fundamental political decision. Starting in the 1930s, a number of these decisions were made by the US Supreme Court, which in some matters sought "to establish the perimeters of governmental authority in the field of economic regulation" (Maidment 1991, 2). This judicial activism was resented in particular by economists attached to the operation of the free market, who then turned their attention to constitutional issues in an unprecedented way and tried to place limits on politics by turning economic rationality into the normative foundation of the social order.

It is thus no coincidence that, starting in the late 1930s, political decision-making became a crucial topic not only for political scientists, who had to revise a number of assumptions of classical democratic theory, but also for economists, forced to come to terms with the political nature of the economic order. As the liberal democratic and the capitalist economic order could no longer be expected to be self-sustaining but required political determination, what could ensure that political decisions remained compliant with these orders? One of the defining features of the nascent neoliberal movement was indeed that it recognized the "primacy of politics" (Friedrich 1955). In an age of ideological battles and economic crisis, the legal framework within which economic transactions took place became an object of analysis in its own right for a new generation of economic thinkers. At Chicago, the first programs in "law and economics" were launched right after World War II (in 1946), with the objective of analyzing the regulation of the economy by the state and the US Supreme Court, but also of ensuring that the "rationality" of judicial decisions essentially meant that they remained in line with the fundamental principles of a market economy. This focus on "law and order," which distinguished neoliberalism in general (Foucault 2004), delineated a new space of reflection where the lines between economics and political theory became blurred. The question of what was to be allocated through market mechanisms and what was to be allocated authoritatively through political power was, in fact, a question about the nature of the polity or, to put it differently, a constitutional decision. The various answers it could receive delineated a wide range of possible regimes going from libertarian capitalism to socialism, with various forms of social democracy in between. The fact that decision-making became so prominent in postwar political science and economics, in our view, is directly related to the fact that the fundamental issues of the time could all be seen to some extent as exceptional decisions that were no longer bound by the strictures of liberal constitutionalism.

3 Rational choice as Cold War constitutionalism

"In the loose employment of the term 'Cold War,'" Carl Friedrich warned in 1950, "there lurk very real dangers for the future of constitutionalism" (Friedrich 1950, 582). A German émigré scholar who had written extensively on constitutional politics and witnessed from afar the collapse of the Weimar Republic, Friedrich was convinced that the survival of democracies required the their defense not be bound by the strictures of liberal politics. Left to its own devices, liberalism was incapable of ensuring its own survival and was bound to be undone by its enemies, external or internal (who happened to be the same during the Cold War). This triggered a search for the acceptable conditions under which exceptional powers required by the international situation and its domestic ramifications could be exercised without endangering a liberal social and political order. The resurgence of thinking about exceptional powers was a characteristic feature of postwar political science. It took place, however, in a liberal political context viscerally attached to constitutional guarantees. Some authors like Friedrich, his former student Frederick Watkins, or the political scientist Clinton Rossiter took a hard look at the pre-liberal political tradition and developed a reflection about emergency powers in democracies that found some inspiration in that tradition. To some extent, they sought to rehabilitate a decisionist approach to politics in emergency situations. Others, however, flinched at the prospect of such illiberal perspectives. A whole current of economic and political thinking sought to re-legitimate liberal constitutionalism by establishing it on a stronger basis, in order to repel any "organic" visions of sovereignty positing the existence of the "state" as a supra-individual entity (Amadae 2003, 16). Rational choice approaches in economics did just that.

Beyond their differences, however, what these two intellectual projects had in common was an effort to bring sovereign decisions in line with rationality criteria. If political decisions had to be taken in the absence of historical precedent, normative warrant, or popular mandate, political scientists and economists sought to articulate guidelines that would provide a new justification for these decisions. Rational choice, in that sense, represented a new normative framework of politics and, at a more fundamental level, a new form of constitutionalism adapted to the contingencies of the Cold War. We call "Cold War constitutionalism" the attempt at requalifying exceptional, sovereign decisions in terms of their rationality, and at replacing the traditional legal or democratic legitimation of decisive authority with a "scientific" one. The main problem that Cold War constitutionalism sought to address was to prevent the emergence of organic visions of the state and yet accommodate forms of exceptional authority bypassing the deliberative procedures of democracy. Its main theorists thus sought to make room for exceptional decisions, yet exclusively in the context of a strengthened liberalism. The main problem for Cold War constitutionalism was not so much to come up with new political solutions, but essentially to justify existing political practices through a new normative framework. This framework was essentially redefining liberalism by subordinating it to security imperatives, of which the prisoner's dilemma would become the ultimate symbol (Amadae 2016).

While this trend was rather obvious in political science, it also became significant in economics. The question of the exceptional decision wormed its way into the discipline as a reflection about decision-making under conditions of radical uncertainty. In the wake of the Depression, some economists started developing new understandings of risk that were no longer amenable to computation. Frank Knight introduced the notion of "unmeasurable uncertainty" to refer to situations in which the possible range of outcomes did not simply take the form of a statistically distributed risk that was calculable but of a structural unknown (Knight 1921, 232). This new, realist understanding of social life would thrive in the postwar years and influence a number of economists. It placed the emphasis on the uneven distribution of information, on the cost of acquiring it (Downs 1957a, 1957b) and of making decisions (Buchanan and Tullock 1962), and on the fact that individuals were ignorant. Uncertainty, lack of information, ignorance, and decision-making costs became staples of postwar economic thought, in a development that paralleled the diminishing faith in individual and collective rationality that was taking hold in political science and political theory. Just as political scientists were struggling to define procedural criteria of rationality detached from any content and "largely independent from personality or context" (Erickson et al. 2013, 3), economists sought to stipulate how rational decision-making could obtain when individuals were ignorant, information costly, and decisions painstaking. Far from being antithetical to rational decision-making, these notions were now taken as the building blocks of a new procedural rationality. By assigning a cost to information and decision-making, economists were able to bring back decisions taken in unique or exceptional situations within the fold of rationality. Once information is given a cost, ignorance no longer indicates a lack of rationality: on the contrary, ignorance becomes a *rational* response to the scarcity and the cost of information (e.g. Stigler 1962) – just as in politics, a decision taken in the absence of knowledge and normative warrant can be rational because of the costs entailed in delaying or forsaking it.

That both political scientists and economists were concerned with the need to craft a new language for justifying critical and risky political decisions in terms of their rationality appears clearly in one of the most important manifestos of Cold War constitutionalism, James Buchanan and Gordon Tullock's *The Calculus of Consent*. The book offered an economic analysis of constitution-making, i.e. of decisions about decision-making rules. These fundamental political decisions were also decisions over the nature of the polity: Buchanan and Tullock considered the constitution a decision over the dividing line between social and individual choice or, to put it differently, between the public sector and the market. Was it possible to stipulate formal, rational rules for determining which decisions should be made through some collective mechanism, and which should remain purely individual decisions? *The Calculus of Consent* appeared *prima facie* as a reconstruction of the state from the ground up on the basis of a radical individualism and a purely economic rationality. Constitutional economics emerged as an attempt at establishing the rationality of an individualistic, libertarian political order against any possibility of prioritizing some notion of "common good" above individual preferences, and this was achieved by locating economic rationality at the very core of constituent power. Building upon the previous work of Kenneth Arrow and his impossibility theorem, Buchanan and Tullock conceived of rationality as a property that could only be attributed to individuals, never to collectives.[13] Yet, it is important to point out that they did not simply substitute economics for classical political theory and the doctrine of the state, or politics with market mechanisms. Despite their intention to do away with notions of superior entitlement, they did not get rid of the notions of constituent power or sovereign authority, but on the contrary they proceeded to give them the *imprimatur* of economic rationality. Just as Carl Friedrich's *Constitutional Reason of State* was an effort at justifying the exercise of prerogative powers in terms that were germane to the tradition of liberal constitutionalism, *The Calculus of Consent* sought to legitimize new forms of authority in the language of economic rationality. A full analysis of the book cannot be proposed within the confines of this chapter, but here we will focus specifically on the way in which Buchanan and Tullock handle decisions that seem to depart from the requirement of reflecting all individual preferences in an optimal way. In particular, they describe such decisions as the expression of a rationality that undergirds the entire constitutional order, not as an exception to it.

Buchanan and Tullock's main objective was indeed to offer a justification of a liberal constitutional order that was entirely premised upon individual freedom, yet one that was also capable of accommodating the exercise of exceptional authority in some situations. Their constitution allowed for decision-making processes that were suboptimal, in the sense that they did not conform with Pareto optimality and therefore were neither "rational" nor democratic.

To do so, Buchanan and Tullock developed a theory of restricted decision-making under liberal conditions that introduced the notion of cost and cost-calculation in the definition of decision-making. Decision-making is costly, which is why individuals may choose to remain ignorant, not to partake in decision-making, or to decide in the absence of certainty. It is by referring to Buchanan's mentor, Frank Knight, that they introduced the problem of decision under times of constraint and uncertainty: "Professor Frank H. Knight has often posed the question: When should an individual rationally stop considering the pros and cons of an issue and reach a decision?" (Buchanan and Tullock 1962, 97). The situation of decision under time constraint or in situations of radical uncertainty as Knight (or Knight as a mouthpiece for Buchanan and Tullock) formulates it is a *locus classicus* of any decisionist argument: in such situations, although decisions are costly, irresolution and the absence of decision take an even higher toll.

By ascribing a cost to decision-making, Buchanan and Tullock displaced the discussion of political decision-making onto a new terrain. It meant that individuals would settle for less-than-optimal decision-making procedures: while unanimity ensures that every individual is part of the decisive group and therefore cancels any external cost potentially inflicted upon individual preferences, it is also the decision-making procedure that entails the highest decision costs, which is why simple majority rule has become the primary democratic procedure in a number of democratic contexts.[14]

Since calculation itself has a cost, it becomes rational to interrupt it at some point if the cost of acquiring information and reaching a decision increases beyond the costs of a bad or suboptimal decision. The expedient decision now appears to be entirely in line with the rational principles that apply also to "normal" decisions and undergird established and inclusive decision-making procedures. Acting decisively in uncertain situations is just as rational as acting according to existing laws in normal situations.[15] The sovereign decision is not a pure decision that takes exception to normal procedures, but it is squarely located within the calculative continuum that underlies the entire constitutional order. It balances external versus decision costs and as such it is not an exception to normal economic rationality; it is only the resolution of specific values in the cost function. The same calculative rationality applies to established decision-making procedures just as much as to exceptional decisions that step into the unknown. *The Calculus of Consent* ended up suggesting that it was possible to maintain authoritative decision-making capacities and to justify them in the terms of a liberal, individualistic normative vocabulary as expressions of a common economic rationality. To put it differently: Buchanan and Tullock were busy developing in economics the same kind of authoritarian or robust liberalism that a number of political scientists and political theorists were theorizing in their discipline.

Constitutional economics was thus part of a wider trend in postwar political theory that sought to "rationalize" the disquieting figure of the sovereign decision that expressed itself in its constituent form or in the exceptional decision taken in the presence of an existential threat. This is best understood by comparing *The Calculus of Consent* with a contemporary text by Carl Friedrich that must be quoted at length:

> In light of this ever-present contingency, one might argue that a rational decision is one which most nearly can be defended in reasoning upon all the facts and values involved in a given situation. But if this is done, the decision-maker is cast in the mould of a Hamlet-like worrier who postpones action in waiting for the ongoing search of additional "data." The need of deciding in time, the deadline, the *kairos*, are aspects of all decision-making which, although themselves part of the rationality process, are in turn contingent and do not permit of reasoned elaboration beyond a certain point. The tremendous role of authority in all political decision-making comes into play at this point. When understood as the capacity for reasoned elaboration, it can be understood as an integer by which a great deal of potential rationality is introduced into the equation of the decision

> in condensed form, as it were. The much criticized doctrine of the secrets of empire, the *arcana imperii*, is thereby revealed to be no more (no less) than *the recognition of the role of authority in political decision-making – not as an alternative to rationality, but as an integral part of such rationality*.
>
> (Friedrich 1964, 80, emphasis added)

Like the economists of the time, Friedrich was reformulating a traditional notion of sovereign decision as an instance of "rational" choice and as a matter of "costs." Friedrich was clear that he sought to rehabilitate reason of state and sovereignty: in Friedrich, as in Schmitt, whose influence on Friedrich had been strong in the 1930s, there is an overarching "authority" that is qualitatively different from other social forces. Friedrich was thus rebuilding sovereignty on a new basis, which was no longer tradition or natural rights, but rationality.

The articulation of expedient decision and rationality appears more clearly in the polemic against welfare economics that runs through *The Calculus of Consent*. One of Buchanan and Tullock's main arguments is that it can be rational for individuals to settle for politically suboptimal (i.e. restricted) decision-making procedures because decision-making costs increase exponentially as the size of the decisive group increases: "even the most rationally constructed constitution will allow some decisions to be made that are 'nonoptimal' in the Pareto sense" (Buchanan and Tullock 1962, 94). At this extremity of the spectrum of possible decision-making procedures, dictatorship stands as the most cost-efficient method of political choice, and if its goals are shared by the collective, i.e. if the external costs it could impose on individuals are low or non-existent, then it is also the most rational. Buchanan and Tullock are clear that if there is a well-established notion of "common good," then it is rational to dispense with the costly procedures and institutions of the democratic process and delegate "all effective decision-making power to a single decision-maker" (Buchanan and Tullock 1962, 99–100). This is what, in their view, points at the danger of such notions as "public interest" or "common good." As Jacob Talmon had put it, "where full unanimity is postulated, there is no escape from the imposition of a single will" (Talmon 1952, 207). Yet, even they recognize that some decisions are characterized by unanimity, *in primis* the constitutional decision itself. Logically, and although they do not dwell on this issue, *The Calculus of Consent* implies that dictatorial measures taken for safeguarding the constitution are perfectly rational, and compatible with the libertarian constitutional framework they propose, since they are the most economic way to act in a situation of unanimity of purpose. Even though at first sight Buchanan and Tullock were concerned by organic visions of authority and sought to prevent their emergence, they did not bar the possibility of exceptional forms of authority, reduced to a restricted group or to a dictator. The result was a constitutional theory that was entirely built upon individualistic foundations, yet which included the decisionist figures of the dictator or of the general will, now justified in terms of economic rationality. These figures were now legitimate not because they were inherited from some pre-liberal tradition, but because they were specific cases of a general cost function and, as such, belonged to a logically consistent liberal order. The sovereign decision was no longer defined and limited by legal provisions, but by economic calculations, which seemed capable of constraining sovereignty in a way that the legal order could not. It was rational because it was embedded within a wider economics of choice. One of the most radical attempts at re-founding liberal constitutionalism, Public Choice theory reformulated a decisionist perspective under the guise of rational choice.

4 Conclusion

The notion of economic imperialism over political science applied to "rational choice" is more a taken-for-granted idea than an established fact. There is no doubt that rational choice was a "trading zone" (Galison 1997) between the two disciplines and provided a common methodological toolbox for (some) economists and political scientists. However, it certainly was not a bundle of economic methods imported into or imposed on political science. That it eventually came to be perceived as such, when political science was pressured into adopting rational choice models and formalistic methodologies, does not change the fact that rational choice initially emerged on the periphery of both disciplines, in an effort to address the question of political decision-making in new ways. Rather than entertaining or refining the notion of "economics imperialism," we have located the impetus for the growth of rational choice not only in disciplinary turf battles over methodological grounds – although the diffusion of rational choice also took this form – but more fundamentally in a specific historical context and a set of problems of direct concern to both economists and political theorists. For many of them, the postwar crisis of liberalism required an effort to think afresh about the conditions under which a liberal economic and political order could survive. In particular, this entailed a need to justify exceptional and authoritative political decisions dictated by the exigencies of the Cold War, yet in terms congruent with the basic tenets of liberalism. These efforts aimed at legitimizing in new and seemingly "scientific" terms forms of authority that no longer relied on the democratic deliberation. "Rationality" provided this new legitimation basis, precisely at a time when the rationality of the publics of democracies had become a sham that the social sciences had dispelled and that nobody entertained. The formal logic of decisional architectures – whether organizations or choice schematics – replaced the deliberative and ethical powers of the individual. Of course, once rational choice had taken hold it became more readily identified with economics, in part because of the greater internal diversity of political science, giving way to notions of "economics imperialism." To the extent that it did not bear an explicit relation to the historical conjuncture it initially reacted to – not least because of its intrinsic formalism – it became discussed exclusively in terms of disciplinary identities and methodological merits. However, that is another story …

Acknowledgments

We have benefited from the comments of Steve Medema, and we are very grateful to him. We also thank the editors of the volume for their invitation to contribute.

The research leading to these results has received funding from the European Research Council under the European Community's Seventh Framework Programme (FP7/2007–2013) Grant Agreement no. [284231].

Notes

1 Political scientists were in their large majority "hostile to the emerging field of rational choice in which the approach and tools of economics are applied to politics" (Mitchell 1999, 237), or, at best, "indifferent" to economics (Mitchell 1999; Medema 2000). A good illustration of how political scientists then perceived economics is provided by the organization of a conference at Northwestern University on "Approaches to the Study of Politics," in which no economists were asked to talk about politics (see also Riker 1959, 209; Riker 1961, 911; Medema 2000, 6). In other words, interactions, joint works, and collaborations were even less frequent than the attempts made by scholars to cross the boundaries of their respective discipline.

2 Another instance of ambiguous imperialism is law and economics (see Medema, this volume).

3 See for instance the observations of Judith Shklar in Shklar 1964b.

4 As Judith Shklar acutely pointed out at the time: Shklar 1964a.

5 Thus, David Easton stresses "the decline of modern political theory" (Easton 1951, 1953), and Alfred Cobban, with a broader characterization, speaks of the "decline of political theory" (Cobban 1953). Focusing on the specific situation in his home country, Catlin notes that "England is in a poorer way as a fount of political ideas than she has been for centuries" (Catlin 1951, 641). Commenting on Easton's (1953) *The Political System*, Hacker says that the book "is indicative of the methodological trauma through which political science is now passing" (Hacker 1954, 775). A few years later, Peter Laslett (1956) and Robert Dahl (1958) would state that the discipline is "dead." Rodney Grey concludes his review of a collective volume entitled *The Policy Sciences: Recent Developments in Scope and Methods*, edited by Daniel Lerner and Harold Laswell, that: "The absence of anything important by a political scientist in such a collection is a comment on the present state of that discipline" (Grey 1953, 548).

6 For a detailed analysis, see Gunnell 2005, 2013.

7 Other important contributors to what would become a game theoretical analysis of decisions came not from economics, but from mathematics, like Howard Raiffa.

8 Black was influenced by "the class lectures of Professor A.K. White on the possibility of forming a pure science of Politics" (Black 1958 [1987], xi). Black completed in 1947 a first version of his book on *The Theory of Committees and Elections* (1958) in 1947. At that time, from what he wrote in a 1948 article, the forthcoming book was "entitled The Pure Science of Politics" (Black 1948a, 23).

9 Black belongs to the Rochester positivist approach to politics (Mitchell 1988, 102–106), the more technical and mathematical of the schools in Public Choice.

10 One finds the same type of perspective in an article by Joseph Spengler (1950, 374): "[t]he forms of behaviour of which each treats, while not identical, flows largely from out distinguishable related kinds of scarcity. Both sciences therefore are much concerned with problems of allocation and collective choice."

11 The seminal text on this is Shklar 1957.
12 On the difference between market failures and the failure of markets, see Medema and Marciano 2015.
13 See the discussion in Amadae 2003.
14 There is a difference between these authors around the notion of rationality. Arrow, by contrast with the others, was interested in the social choice. Rationality is a criterion used to evaluate the outcome of the decision. He therefore meant that a dictatorial social choice is rational, whatever this choice is. Downs assumed individual rationality. Whatever its outcome, a political decision is always rational. For Becker, the role of rationality is less clear. Ideal democracies – that work like free and competitive markets – do not seem to need rationality; the most efficient political parties survive in a way that is not different from what Alchian (1950) or Friedman (1953) had argued (see also Becker 1962). In actual democracies, rationality seems to play a larger role precisely because those evolutionary pressures have disappeared.
15 Buchanan and Tullock indeed present rules as automated decision-making devices that lower the costs of decisions. From that perspective, there is not substantive difference between exceptional decisions and normal ones.

References

Alchian, Armen. 1950. "Uncertainty, Evolution, and Economic Theory". *Journal of Political Economy* 58: 211–221.

Amadae, S. M. 2003. *Rationalizing Capitalist Democracy: The Cold War Origins of Rational Choice Liberalism*. Chicago, IL: University of Chicago Press.

Amadae, S. M. 2016. *Prisoners of Reason: Game Theory and Neoliberal Political Economy*. Cambridge: Cambridge University Press.

Angel, Arthur D. 1947. "Review of Administrative Behavior: A Study of Decision-Making Processes in Administrative Organization by Herbert A. Simon". *Annals of the American Academy of Political and Social Science* 254: 184–185.

Arrow, Kenneth J. 1951. *Social Choice and Individual Values*. New York: Wiley and Sons.

Arrow, Kenneth J. 1964. "Control in Large Organizations". *Management Science* 10: 397–408.

Backhouse, Roger. 2015. "Economic Power and the Financial Machine: Competing Conceptions of Market Failure in the Great Depression". *History of Political Economy* 47(annual suppl 1): 99–126.

Becker, Gary S. 1958. "Competition and Democracy". *Journal of Law and Economics* 1: 105–109.

Becker, Gary S. 1962. "Irrational Behavior and Economic Theory". *Journal of Political Economy* 70: 1–13.

Black, Duncan. 1948a. "On the Rationale of Group Decision-making". *Journal of Political Economy* 56: 23–34.

Black, Duncan. 1948b. "The Elasticity of Committee Decisions with an Altering Size of Majority". *Econometrica* 16: 262–270.

Black, Duncan. 1948c. "The Decisions of a Committee Using a Special Majority". *Econometrica* 16: 245–261.

Black, Duncan. 1949a. "The Theory of Elections in Single-Member Constituencies". *Canadian Journal of Economics and Political Science/Revue Canadienne d'Économique et de Science Politique* 15: 158–175.

Black, Duncan. 1949b. "Some Theoretical Schemes of Proportional Representation". *Canadian Journal of Economics and Political Science/Revue Canadienne d'Économique et de Science Politique* 15: 334–343.

Black, Duncan. 1950. "The Unity of Political and Economic Science". *Economic Journal* (September): 506–514.

Black, Duncan. 1958 [1998]. *The Theory of Committees and Elections*. Dordrecht: Kluwer.

Buchanan, James M. and Gordon Tullock. 1962. *The Calculus of Consent*. Ann Arbor: Michigan University Press.

Calhoun, Craig and Loïc Wacquant. 1989. "Intérêt, rationalité et culture: à propos d'un récent débat sur la théorie de l'action". *Actes de la Recherche en Sciences Sociales* 78: 41–60.

Catlin, George. 1951. "Contemporary British Political Thought". *American Political Science Review* 46(3): 641–659.

Cobban, Alfred. 1953. "The Decline of Political Theory". *Political Science Quarterly* 68: 321–337.

Cook, Thomas I. 1950. "Review of Administrative Behavior, a Study of Decision-Making Processes in Administrative Organization by Robert A. Simon". *Review of Politics* 12(3): 405–407.

Dahl, Robert. 1958. "A Critique of the Ruling Elite Model". *American Political Science Review* 52(2): 463–469.

Diamond, Martin. 1959. "Review of Anthony Downs' 'An Economic Theory of Democracy'." *Journal of Political Economy* 67: 208–211.

Downs, Anthony. 1956. *An Economic Theory of Government Decision-Making in a Democracy*. Technical Report No. 32, prepared for the Office of Naval Research, Stanford, CA, Stanford University, Department of Economics. (Mimeographed; Armed Services Technical Information Agency, Knott Building, Dayton, Ohio.)

Downs, Anthony. 1957a. "An Economic Theory of Political Action in a Democracy". *The Journal of Political Economy* 65: 135–150.

Downs, Anthony. 1957b. *An Economic Theory of Democracy*. New York: Harper.

Downs, Anthony. 1960. "Why the Government Budget is Too Small in a Democracy". *World Politics* 12: 541–563.

Easton, David. 1951. "The Decline of Modern Political Theory". *The Journal of Politics* 13: 36–58.

Easton, David. 1953. *The Political System: An Inquiry into the State of Political Science*. New York: Alfred A. Knopf.

Erickson, Paul, Judy L. Klein, Lorraine Daston, Rebecca Lemov, Thomas Sturn and Michael D. Gordin. 2013. *How Reason Almost Lost its Mind: The Strange Career of Cold War Rationality*. Chicago, IL: The University of Chicago Press.

Foucault, Michel. 2004. *Naissance de la biopolitique. Cours au Collège de France 1978–1979*. Paris: Gallimard-Seuil.

Friedman, M. 1953. *The Methodology of Positive Economics. Essays in Positive Economics*. Chicago, IL: University of Chicago Press.

Friedrich, Carl J. 1950. *Constitutional Government and Democracy: Theory and Practice in Europe and America*. Boston, MA: Ginn and Company.

Friedrich, Carl J. 1955. "The Political Thought of Neo-Liberalism". *The American Political Science Review* 49: 509–525.

Friedrich, Carl J. 1957. *Constitutional Reason of State: The Survival of the Constitutional Order*. Providence, RI: Brown University Press.

Friedrich, Carl J. 1964. "On Rereading Machiavelli and Althusius: Reason, Rationality, and Religion". In Carl J. Friedrich (ed.). *Rational Decision*. New York: Prentice Hall, pp. 177–196.

Galison, Peter. 1997. *Image and Logic: A Material Culture of Microphysics*. Chicago, IL: The University of Chicago Press.

Germino, Dante. 1963. "The Revival of Political Theory". *Journal of Politics* 25(3): 437–460.

Ghamari-Tabrizi, Sharon. 2005. *The Worlds of Herman Kahn: The Intuitive Science of Thermonuclear War*. Cambridge, MA: Harvard University Press.

Green, Donald P. and Ian Shapiro. 1994. *Pathologies of Rational Choice Theory: A Critique of Applications in Political Science*. New Haven, CT: Yale University Press.

Greenberg, Udi. 2014. *The Weimar Century: German Émigrés and the Ideological Foundations of the Cold War*. Princeton, NJ: Princeton University Press.

Grey, Rodney. 1953. "Review of *The Policy Sciences: Recent Developments in Scope and Method* by Daniel Lerner; Harold D. Lasswell; Harold H. Fisher; Ernest R. Hilgard; Saul K. Padover; C. Easton Rothwell; Ithiel de Sola Pool". *Canadian Journal of Economics and Political Science/Revue canadienne d'Economique et de Science politique* 19(4): 546–548.

Gunnell, John G. 2005. "Political Science on the Cusp: Recovering a Discipline's Past". *American Political Science Review* 99: 597–609.

Gunnell, John G. 2013. "The Reconstitution of Political Theory: David Easton, Behavioralism, and the Long Road to System". *Journal of the History of the Behavioral Sciences* 49: 190–210.

Heyck, Hunter. 2015. *Age of System: Understanding the Development of Modern Social Science*. Baltimore, MD: Johns Hopkins University Press.

Katznelson, Ira. 2013. *Fear Itself: The New Deal and the Origins of our Time*. New York: Liveright.

Kaysen, Carl. 1946–47. "A Revolution in Economic Theory?" *Review of Economic Studies* 14: 1–15.

Knight, Frank H. 1921. *Risk, Uncertainty and Profit*. Boston, MA and New York: Houghton Mifflin Company.

Lazear, Edward P. 2000. "Economic Imperialism". *Quarterly Journal of Economics* 115: 99–146.

Lindblom, Charles E. 1957. "In Praise of Political Science". *World Politics* 9: 240–253.

Lippmann, Walter. 1955. *Essays in the Public Philosophy*. Boston, MA: Little, Brown and Company.

Maidment, Richard A. 1991. *The Judicial Response to the New Deal: the US Supreme Court and Economic Regulation, 1934–1936*. Manchester: Manchester University Press.

Mäki, Uskali. 2002. "Symposium on Explanations and Social Ontology 2: Explanatory Ecumenism and Economics Imperialism". *Economics and Philosophy* 18: 235–257.

Mäki, Uskali. 2009. "Economics Imperialism: Concept and Constraints". *Philosophy of the Social Sciences* 39: 351–380.

Mäki, Uskali. 2013. "Scientific Imperialism: Difficulties in Definition, Identification, and Assessment". *International Studies in the Philosophy of Science* 27: 325–339.

McKean, Roland N. 1965. "The Unseen Hand in Government". *American Economic Review* 55: 496–506.

Medema, Steve. 2000. "'Related Disciplines': The Professionalization of Public Choice Analysis". *The History of Applied Economics, History of Political Economy Annual Supplement* 32: 289–323.

Medema, Steve and Alain Marciano. 2015. "Market Failure in Context: Introduction". *History of Political Economy* 47(6): 1–19.

Mitchell, William C. 1961. "Politics as the Allocation of Values: A Critique". *Ethics* 71: 79–89.

Mitchell, William C. 1988. "Virginia, Rochester, and Bloomington: Twenty-Five Years of Public Choice and Political Science". *Public Choice* 56: 101–119.

Mitchell, William C. 1999. "Political Science and Public Choice: 1950–70". *Public Choice* 98: 237–249.

Monroe, Kristen Renwick. 2001. "Paradigm Shift: From Rational Choice to Perspective". *International Political Science Review/Revue internationale de science politique* 22: 151–172.

Monroe, Kristen Renwick. 2004. "The Chicago School: Forgotten but Not Gone". *Perspective on Politics* 2: 95–98.

Rice, Stuart A., Francis G. Wilson and Thomas I. Cook. 1950. "Generalists versus Specialists in Social Science: Discussion". *American Political Science Review* 44(2): 380–393.

Riker, William H. 1959. "Review of Approaches to the Study of Politics, edited by Roland Young, Midwest". *Journal of Political Science* 3: 207–210.

Riker, William H. 1961. "Voting and Summation of Preferences: An Interpretive Bibliographic Review of Selected Developments during the Last Decade". *American Political Science Review* 55: 900–911.

Riker, William H. 1962a. *The Theory of Political Coalitions*. New Haven, CT: Yale University Press.

Riker, William H. 1962b. "Review of *The Calculus of Consent* by James M. Buchanan and Gordon Tullock". *Midwest Journal of Political Science* 6: 408–411.

Riker, William H. 1992. "The Entry for Game Theory into Political Science". In E. Roy Weintraub (ed.). *Toward a History of Game Theory*. Durham, NC: Duke University Press, pp. 207–223.

Robbins, Lionel. 1932. *An Essay on the Nature and Significance of Economic Science*. London: Macmillan.

Rossiter, Clinton. 1948. *Constitutional Dictatorship: Crisis Government in the Modern Democracies*. Princeton, NJ: Princeton University Press.

Scarry, Elaine. 2014. *Thermonuclear Monarchy: Choosing Between Democracy and Doom*. New York: W.W. Norton & Co.

Schelling, Thomas. 1958. "The Strategy of Conflict. Prospectus for a Reorientation of Game Theory". *Journal of Conflict Resolution* 2: 203–264.

Schelling, Thomas. 1960. *The Strategy of Conflict*. Cambridge, MA: Harvard University Press.

Schlosser, Eric. 2013. *Command and Control: Nuclear Weapons, the Damascus Incident, and the Illusion of Safety*. New York: Penguin.

Shklar, Judith N. 1957. *After Utopia: The Decline of Political Faith*. Princeton, NJ: Princeton University Press.

Shklar, Judith N. 1964a. "Decisionism". In Carl J. Friedrich (ed.). *Rational Decision*. New Brunswick, NJ: Transaction, pp. 3–17.

Shklar, Judith N. 1964b. *Legalism: Law, Morals, and Political Trials*. Cambridge, MA: Harvard University Press.

Shubik, Martin. 1992. "Game Theory at Princeton, 1949–1955: A Personal Reminiscence". In E. Roy Weintraub (ed.). *Toward a History of Game Theory*. Durham, NC: Duke University Press, pp. 161–163.

Simon, Herbert A. 1966. "Political Research: The Decision-Making Framework". In David Easton (ed.). *Varieties of Political Theory*. Englewood Cliffs, NJ: Prentice Hall, pp. 15–24.

Simon, Herbert A. 1983. *Reason in Human Affairs*. Stanford, CA: Stanford University Press.

Simon, Herbert A., Max Radin, George A. Lundberg and Harold D. Lasswell. 1950. "The Semantics of Political Science: Discussion". *American Political Science Review* 44(2): 407–425.

Snyder, Richard, H.W. Bruck and Burton Sapin. 1954. *Decision-making as an Approach to the Study of International Politics*. Princeton, NJ: Foreign Policy Analysis Project, Princeton University.

Solovey, Mark. 2013. *Shaky Foundations: The Politics-Patronage-Social Science Nexus in Cold War America*. New Brunswick, NJ: Rutgers University Press.

Spengler, Joseph J. 1950. "Generalists versus Specialists in Social Science: An Economist's Views". *The American Political Science Review* 44: 358–379.

Stigler, George J. 1962. "The Economics of Information". *Journal of Political Economy* 69: 213–225.

Talmon, Jacob L. 1952. *The Origins of Totalitarian Democracy*. London: Secker & Warburg.

Waltz, Kenneth. 1979. *Theory of International Politics*. Long Grove, IL: Waveland Press.

7 Economics imperialism reconsidered

S.M. Amadae

Introduction

This chapter uses Uskali Mäki's (2009) concepts of "good" and "bad" imperialism to investigate the "economics imperialism" thesis. If science expands by offering (a) consilience, and (b) epistemological and ontological unity – that is, it explains more phenomena with greater parsimony – then this is good scientific expansion. Economics imperialism is only bad if the methodology of economics expands outside its domain without increasing understanding in the above manners.

I take from Steve Clarke and Adrian Walsh the view that the burden of proof is on the shoulders of an analyst claiming to identify bad imperialism because she needs to account for "why imperialistic ideas are liable to succeed in the disciplines they attempt to colonise, despite their lack of explanatory virtue" (Clarke and Walsh 2009, 202). This chapter argues three points. First, alleged "economics imperialism" is better described as "rational choice imperialism" because there is a disciplinary break between the earlier schools of classical and neoclassical economics versus game theory. Second, following Mäki, the advances and imperializing power of rational choice is *derivational*, rather than epistemological, conciliatory, or ontological. Third, the stakes of accurately identifying and even resisting rational choice imperialism follow from it potentially reflecting a type of politically motivated imperialism that may not contribute to scientific understanding. Clarke and Walsh recognize that some forms of disciplinary imperialism dismiss alternative manners of inquiry, and potentially also inform common-sense views of valid or appropriate social ontologies.

1 Good and bad "scientific imperialism"

The case of scientific imperialism focal to this chapter is that of economics imperialism which refers to starkly polarized territory in which many theorists are critical of the increasing encroachment of economic assumptions on domains of inquiry formerly operating with alternative approaches (Sen 1977; Heath 2008), and other researchers are celebratory (Becker 1978; Posner 1984; Buchanan 1975). However, before discussing economics imperialism specifically, it is useful to follow Mäki in developing tools to investigate this so-called "imperialism" in neutral terms. Mäki recommends having an open mind as to whether disciplinary imperialism is "good" or "bad," and therefore articulates a means to discover whether disciplinary expansion is consistent with scientific unification and greater explanatory power. Suppose, for example, that a scientific theory such as Isaac Newton's theory of gravitation is able to explain the movement of objects both on Earth and throughout the heavens, then although this modern physics colonizes astronomy with the same analytic structure applied to the motion of Earth-bound corporeal bodies, no credible philosopher of science will denounce this expansion as contrary to the principles of effective scientific inquiry. As Clarke and Walsh (2009) point out, in fact it seems likely that if a body of theory is applied to a new domain of study and those theorists find it useful, then *prima facie* it would seem that this exercise shares common ground with the widespread and self-sustaining acceptance of Newtonian physics. Thus Mäki notes that, "unification provides a norm that is firmly embedded in the institution called science" (Mäki 2009, 262). This type of expansion fits comfortably with the belief in scientific progress.

Mäki spells out features of a successful extension of a scientific method from one domain to others. Unification of inquiry should be effective with respect to ontology, pragmatics, and epistemology. For the first, Maki states that, "ontological unification is a matter of redescribing large classes of apparently independent explanandum phenomena as forms or manifestations of a common system of entities, causes, and mechanisms" (Mäki 2009, 364). The core idea is that a scientific theory may expand credibly if it applies to a broader ontology akin to Newtonian gravitation pertaining to motion both within and beyond Earth's atmosphere. As I will explore below, in the case of the expansion of economic science, it would exhibit the property of greater ontological relevance if its mode of explanation successfully explains not only traditional market transactions, but, for example, also all rational decision-making. Economists propose to achieve this aim (Hausman 2012).

With respect to pragmatism, Mäki refers to two types of consilience: subsumption and cardinality. If one theory more successfully explains both phenomena to which it initially was applied and in addition those previously explained by another theory, then the broader theory has more explanatory power. The Copernican model of celestial mechanics explained both the Earth's movement vis-à-vis the sun and that of the other planets' orbits around the sun, and over time was seen to offer a more synthetic perspective than that of the Ptolomaic system with its well-known reliance on epicycles (Kuhn 1962). Similarly, if one theory is able to explain more bundles of facts than another, then that more encompassing theory seems obviously more effective. Mathematical models of waves can explain these patterns of energy transfer in multiple substances to include radio waves, water waves, and sound waves. Again, looking ahead to the theme of economics imperialism, if game theory can explain both markets and crime whereas neoclassical economics could only explain markets, then game theory seems to offer greater explanatory power.

With respect to epistemology, Mäki points out that scientists prefer parsimony in theory choice, typically deferring to theories making the fewest assumptions. Thus, if empirical tests are used to determine the accuracy of hypotheses generated by theories, and the data support two different theories, the one with greater parsimony will likely be preferred. Again, in reference to the expansion of economics beyond its borders, if rational choice and Freudian psychology both explain individuals' consumptive choices, but rational choice has fewer assumptions, then likely theorists will find it to be the more epistemically effective.

In his discussion of features of scientific theories that make them more compelling and able to expand their domains of relevance, Mäki also introduces an alternative means by which theories can increase their range of application. He calls this a type of unification that proceeds through the power of derivation as opposed to greater explanation of ontological phenomena. The idea is that a theory may provide the means to derive many conclusions analytically, and hence generate an impressive quantity of results that cover a range of phenomena in domains formerly addressed by alternative theories and methods. Yet such a theory is limited to derivational unification if its formal results are not vindicated by empirical verification. Below I will argue that the concept of imperialism through derivational unification is particularly relevant to the type of expansion characterized by rational choice.

To summarize, then, I concur with Mäki that the expansion of a scientific theory from one domain to another appears to be consistent with the aims of science. Hence it is valuable to identify features of scientific unification that are consistent with the enterprise of science, and the celebration of increased ontological, practical, and epistemological applicability is thus warranted. However, Mäki does alert us to a type of unification through the multiplication of analytic derivations without empirical vindication that could characterize a manner of expansion that may not enhance the explanatory power of scientific inquiry.

2 Revisiting the "economics imperialism" thesis

George Stigler's "Economics – The Imperial Science?" (1984) discusses how the discipline of economics "has been aggressive in addressing central problems in a considerable number of neighboring social disciplines without any invitations" (Stigler 1984, 311). He describes how theorists have made forays in modeling diverse social phenomena like criminal activity and family planning (e.g., Becker 1978) using concepts and methods from economics. Thus, the exercise of economics imperialism exhibited by Anthony Downs (1957), James M. Buchanan and Gordon Tullock (1962), Thomas Schelling (1960), Michael Taylor (1976) and Russell Hardin (1982) is generally accepted to represent applying economic arguments to democracy, the social contract, law, conflict and conflict resolution, cooperation, and collective action. Before interrogating this understanding of the type of imperialism exhibited by game theory in the next section, the goal here is to provide a charitable description of what is normally referred to by the phrase "economics imperialism" (Lazear 2000; Miller 1997, 1181). This task is helpful due to the common currency of the phrase and the imagery that this common lore narrative evokes.

The essence of the economics methodology being expropriated by other fields, which include political science, international relations, jurisprudence, public policy, arbitration, conflict resolution, sociology, and psychology, is that rational agents maximize utility in all decision-making independent of its context. Thus individual choice can be modeled according to whether actors behave consistently given that outcomes are usually realized in accordance with probabilities rather than certainties. Additionally, interactions are modeled assuming that individuals maximize individual gain competitively. The assumptions underlying the disciplinary unification are parsimonious and only demand that actors have complete and consistent preferences over outcomes (Hargreaves Heap and Varoufakis 2004, 8); and that, from these assumptions, utility functions can be constructed that pertain to diverse choice situations from markets to politics (Mueller 2003; Pettit 2002), and love and truth-telling (e.g., Becker 1978; Lewis 1969) to evolutionary science (Trivers 1971; Smith 1982; Dawkins 1979). Nothing seems to be outside the confines of this application of bare bones individualistic maximization to every conceivable domain of rational judgment (Hausman 2012). Even the sympathy characterized by Adam Smith's (1982 [1759]) impartial spectator has been subjected to the assumption that all individuals maximize some type of gain in every decision and interaction (Pettit 2002). If actors demonstrate altruism, then this is merely another strategy by which they benefit, either by aiding the promotion of their genetic identity through kin, or seeking immortality and longevity through reputation (Dawkins 2006).

In considering these examples of economics imperialism it is evident that one particular brand of economic theorizing underlies this trend: rational choice theory. This is a point acknowledged by Mäki (2002, 239). Here the fundamental concepts are narrow self-interest and comporting with the axioms of expected utility theory (Mäki 2002, 237). Economics imperialism is not that of Adam Smith's supply/demand analysis; it is not the neoclassicals' formalization of diminishing marginal utility; nor is it Keynesianism or macroeconomics. The economic method underlying the late 20th-century domination of economics over other disciplines is specifically the game theory revolution which encompasses expected utility theory as an intrinsic part (von Neumann and Morgenstern 1953 [1944]). Although Mäki also refers to "cost benefit analysis" being essential to economics imperialism (Mäki 2013, 332; see also Clarke and Walsh 2009, 196), it is unclear whether individual expected utility maximization is a class of cost–benefit analysis, vice versa, or if they are derived from the same fundamental theory.

So far I have discussed Mäki's classification of "good" and "bad" imperialism with the former representing effective scientific unification that reflects progressive understanding of phenomena, and the latter involving "new types of explanandum phenomena [that] are located in territories that are occupied by [different] disciplines … and where [the colonizing discipline] presents itself hegemonically as being in possession of superior theories and methods, thereby excluding rival theories and approaches from consideration" (Mäki 2009, 274). Economics imperialism is synonymous with rational choice imperialism, and it remains to be determined whether the multidisciplinary adoption of game theory (encompassing expected utility theory, which serves as the basis of decision theory applicable in parametric contexts without any interactive partners) represents a good or bad form of science expansion.[1] If game theory helps to unify our understanding of human behavior with parsimonious assumptions that encompass a greater range of social phenomena, which are subject to empirical verification, or offers greater efficacy of explanation than alternative theories, then it qualifies as constructive expansion.

The next section scrutinizes the rational choice revolution first as a species of economics imperialism, and draws attention to how it may be more appropriate to view this theoretical transformation as a distinctive rupture contrasting with prior economic theorizing. In fact, game theory first established itself in fields external to economics before finally colonizing economics itself. The following section interrogates whether rational choice offers epistemic unification through either encompassing a greater ontological range of experience, or through offering greater empirical verification. This section raises the possibility that the unification and expansion offered by rational choice is better understood as falling into the derivational category that Mäki argues is not a legitimate species of scientific progress.

The stakes of accurately understanding the nature of economics imperialism, as stated by Clarke and Walsh, is that if one method of scientific inquiry is adopted in another field, either voluntarily with the hopes of professional success, or coercively due to the allocation of opportunities for funding and prestige, then it is possible both that alternative theories may be lost, and alternative ontologies or fields of value may recede from sight (Clarke and Walsh 2009). This is because alternative theories may afford forms of explanation that focus on individual identity in terms of types, without being able to speak to experiential data that pertain to specific individuals (Mäki 2002, 253). Furthermore, I will suggest that possibly subjecting individuals to environments in which they become data points for the type of behavioral decision-making analysis typical of rational choice (either in experiments, or in the process of designing institutions, public policies, and legal frameworks in accordance with the assumptions and findings of rational choice) may significantly alter their self-expression to comport with the colonizing means of scientific explanation.

3 Is the "rational choice revolution" really "economics imperialism"?

It is clear that in many, if not most, theorists' minds, economics imperialism and the rational choice revolution are synonymous, dating to the initial publication of John von Neumann and Oskar Morgenstern's *Theory of Games and*

Economic Behavior, and gradually becoming first a solidified canon of literature including Duncan Luce and Howard Raiffa's *Games and Decisions*, Anthony Downs' *An Economic Theory of Democracy* (1957), James Buchanan and Gordon Tullock's *Calculus of Consent* (1962), Russell Hardin's *Collective Action* (1982), and Gary Becker's *Economic Approach to Human Behavior* (1978). This understanding is independent from whether theorists support or criticize the widespread acceptance of rational choice. Viewing rational choice as a product of economics, rather than as post-World War II economics transformed by game theory, may have been encouraged by Morgenstern's vision of revolutionizing the study of economics in terms of strategic interactions (Leonard 1992). Moreover, the early application of game theory to the dual constraint problems encountered by military planners related gaming, linear programming, and military worth (cost–benefit analysis) as an exercise of achieving efficiency in strategy in the late 1940s and 1950s (Erickson 2015; Thomas 2015).

Thus, in this section I question the standard understanding with the aim of better appreciating the nature of colonization entailed by rational choice. I will argue that rather than represent the phenomenon of "economics imperialism," the rational choice revolution is better thought of as a distinctive break in the understanding of human agency, from an instrumentalism of using means most efficiently to achieve ends, to strategic competition among actors to satisfy their preferences exemplified by complete and consistent rankings over all conceivable end states. Moreover, I further explore the hypothesis that the misrepresentation of the rational choice revolution as *economics imperialism* falsely conveys to the disciplinary ascendance of game theory a pedigree of legitimacy that economics and instrumental rationality achieved starting with Adam Smith and augmented by the neoclassical economists' formalization of general equilibrium theory and economic efficiency. I will argue that in fact the rational choice revolution is better understood as a new methodology that defines "rational action" as having consistent preferences over all possible end states, ranking them on a single and finite scale, and making consistent choices among potential end states treated as lotteries of outcomes. This view of rational choice is distinct and at odds with the neoclassical model of rational action that copies the mathematical formalism of rational mechanics requiring rational actors to obey the law of diminishing marginal utility (Mirowski 1989). Even if the rational choice school of thought has imperializing tendencies, these transformed the discipline of economics itself which previously had only achieved the recognition of existing as a free-standing science by deliberately demarcating its territory from psychology and politics (Amadae 2003, 220–248).

Mäki describes how rational choice seems to reflect a continuity with preceding forms of economic analysis: "It is well known that the [Lionel] Robbinsian 'definition' of economics in terms of ends and scarce means … is powerfully scope-expanding" (Mäki 2009, 358). Here Mäki alludes to the observations of Jack Hirshleifer, who notes that, "[w]hat gives economics its imperialist invasive power is that our analytic categories – scarcity, cost, preferences, opportunities, etc. – are truly universal in applicability [1985, 53]" (ibid., 359). The idea is that the neoclassical economists, who used calculus-based constrained optimization in their models, formalized the formerly discursive concepts of scarcity, cost, preferences, etc., and that game theorists took these same formal concepts and then applied them to "reputation, sex, status, eternal salvation, the meaning of life, and a good night's sleep" (Hirshleifer 1985, 53; Mäki 2009, 358). Mäki's list of the colonized fields is lengthy: "politicians' and bureaucrats' behavior, voting and law, crime and punishment, racial discrimination and slavery, marriage and divorce, pornography and prostitution, religion and suicide, drug addiction and abortion, sport and gambling, rock 'n' roll and science" (Mäki 2009, 358).

Even if it were true that rational choice theory is used to model these diverse social interactions and phenomena, it is not true that this was by extending the methods of neoclassical economists to new fields of inquiry and areas of human behavior outside market transactions characterized by ratios of exchange of scarce and fungible goods. Two points about the nature and scope of economic science, to which Robbins (1935) dedicated a monograph by that name, are evident. First, in his endeavor to carefully delimit the boundaries of economic science, Robbins made clear that the appropriate domain of economic analysis is the efficient use of scarce means which are traded against each other in specific ratios depending on individuals' goals and market prices. The neoclassicals' domain of economics is that of economic value, usually expressible in monetary denominations. Here, money is simply one more commodity that is traded for other goods in precise ratios that must uphold the law of diminishing marginal utility – if consumers and producers are rational. Second, with this in mind, possibly countering Karl Marx's thesis that capitalists end up pursuing monetary gain as a valuable end in itself, Robbins (1935) clarified that money can never be an end, but only a mere means. Thus for Robbins, economic science is strictly limited to the domain of the efficient obtainment of means, which he treated as finite tangible resources that are useful in precise ratios. The trade-offs referred to in neoclassical economics are those captured in a two-good problem: increasing consumption of, for example, bread, must come at the price of giving up some wine and vice versa (Robbins 1935). Economic science is strictly the domain of the efficient allocation of means, which are exchangeable, fungible, scarce resources. Agents have preferences over commodity bundles that are suitable for representation in a graphic model of an Edgeworth box.

In order for rational choice theory to model diverse considerations ranging from sex and sleep to suicide and addiction, it was necessary that it wholly abandon the neoclassical framework of constrained maximization which borrowed the mathematics of the path of least action from physics (Mirowski 1989; Amadae 2003). The early game theorists were fully aware of, and celebrated, their clean break free from the perspective of constrained maximization (von Neumann and Morgenstern 1953 [1944]; Luce and Raiffa 1957). Instead of individuals ranking commodity bundles or *finite means*, each exhibiting unique causal properties consistent with mass and energy, and chemical properties and compound substances characteristic of periodic elements, the game theoretic agents rank *outcomes*. Sometimes outcomes can be dollar denominations, or lotteries of varying probabilities of dollar denominations, or possibly being at war or at peace, or winning or losing an election. Thus, for the rational choice revolution to be able to expand the domain of economics beyond the constrained maximization of utility under a budget constraint, its theoretical foundation first had to reformulate the nature of choice to redefine rationality from the efficient use of means to the consistent ranking of ends, or outcomes themselves. Von Neumann and Morgenstern, as well as Luce and Raiffa, discuss this new formalization of expected utility in careful detail. They boast that their bold initiative cuts free from the neoclassicists' limited thinking of constrained optimization under a budget constraint. Not only do they reformulate the notion of utility, but they also model strategic interactions in which individuals maximize their pursuit of anticipated utility against other like-minded players (Leonard 2010).

Thus, game theory developed largely outside economics from the 1940s until roughly 1970. It is well known that von Neumann, a mathematical physicist, and Morgenstern, a less well-known Austrian economist, developed *Theory of Games*, a mathematically dense 600-plus-page volume that was almost entirely ignored by economists (Leonard 1992) until the late 1970s and early 1980s. In the late 1940s and 1950s, early work in game theory was pursued for military strategy (Erickson 2015), mainly at the Santa Monica-based RAND Corporation (Amadae 2003, 2016). Perhaps the first economist who can truly be described as an "economics imperialist," Thomas Schelling, initially used game theory to model nuclear deterrence and conflict before going on to model the segregation of neighborhoods and collective action (Schelling 1960, 1973, 1978; Amadae 2016). Von Neumann and Morgenstern's initial theory, which mainly pertained to two-person zero-sum game theory, seemed much more useful for conflict than applicable to economic exchange in which all actors were supposed to gain rather than one individual's gain being someone else's loss (Leonard 2010). Game theory was a staple in nuclear deterrence (Aumann et al. 1967) in the 1960s. It was also developed in evolutionary game theory as early as the late 1960s (Trivers 1971). It represented a paradigm of powerful derivational import and its sparse assumptions could be used to model any interaction in which agents were understood to have consistent preferences over outcomes (Axelrod 1970).

It is difficult to sustain a thesis of "economics imperialism" if many of the initial trailblazers had disciplinary homes outside economics (Amadae 2003). Kenneth J. Arrow is an obvious exception, but even he accepted that the maximization used in social choice is distinct from the neoclassical concept of maximization under a budget constraint (Arrow 1963; Amadae 2003, 83–132). Economists as a community of theorists did not embrace game theory until the late 1970s and 1980s when the Nash mutual-best-reply equilibrium became the accepted solution concept, in contrast to von Neumann and Morgenstern's more limited solution concept based on extending zero-sum theory to non-zero-sum games (Mirowski 2002; Binmore's introduction to Nash 1996).

The manner of formulating utility, developed by von Neumann and Morgenstern, called "expected utility theory," displaced the formal models of diminishing marginal utility and over time became the standard treatment of utility in economics. However, economists only caught up with work achieved in other disciplines such as political science (Riker and Ordeshook 1973; Taylor 1976), nuclear deterrence (Schelling 1960; Schelling and Halperin 1961), international relations (Jervis 1978), public policy analysis (Stokey and Zeckhauser 1978) and evolutionary game theory (Trivers 1971) after these other disciplines had already standardized canonical game theoretic literature (Myerson 1991). Even behavioral economics, which uses rational choice theory as the standard normative theory of rationality against which to test individuals' systematic deviations, originated from the work of Daniel Kahneman, a psychologist, and Amos Tversky, a mathematics-trained cognitive scientist and psychologist (1979). These facts about which theorists, in what disciplines, played key roles in establishing rational choice theory early on permit us to stand back and reappraise the economics imperialism thesis, and lead us to the conclusion that the rational choice revolution is not a species of economists colonizing other fields so much as the extension of a new formal method to multiple disciplines that followed from its original statement in *Theory of Games and Economic Behaviour* (von Neumann and Morgenstern 1947 [1944]) and application by von Neumann himself to submarine warfare (Leonard 2010).

Before going on to assess the scientific value contributed by game theory to understand a vast array of human behavior, I would like to draw attention to the clear means by which game theory purports to apply to all decision-making. In the words of von Neumann and Morgenstern themselves, they clearly state that their concept of a solution, or the identification of an outcome rational players converge on, "is plausibly a set of rules for each participant which tell him how to behave in every situation which may conceivably arise" (von Neumann and Morgenstern 1953 [1944], 31). Thus, *every conceivable* decision a rational individual may make falls within the purview of expected utility theory and game theory. Moreover, when it comes to agents' expected utility functions, or their complete ranking of all possible outcomes, *everything* of relevance to actors' appraisal of value is argued to be in their preference orderings (Myerson 1991, 7–8). Preference rankings over outcomes "must be a total [meaning exhaustively comprehensive] ranking, incorporating every factor agents take to influence their choices" (Hausman 2012, 34). These default orthodox expected utility rankings in game theory only consider actors' evaluation of outcomes (Hausman 2012, 53), and yet at the same time assume these rankings are exhaustively comprehensive (Hausman 2012, 35). Game theoretic expected utility functions both incorporate these assumptions and then operationalize a mathematics that provides the means to treat individuals' choices according to these terms. Rational choice can imperialistically explain all choice because from its original formulation, expected utility theory was designed to represent all cognitive information necessary to explain rational behavior. Thus, in assembling the points put forward in this section, rational choice is a new methodology for understanding individuals' rational choices that shifts attention from market choice under a budget constraint to consistent rankings over all possible end states, and taking appropriate strategic choices given competition against others who similarly seek to secure rational expectations.

In the next section I suggest that this wider perspective, which is open-minded about the type and source of imperialism, not only is helpful for maintaining a neutral vantage point for evaluating the nature of the rational choice revolution, but also rejects the thesis that the successful and already vindicated methodology of economic science was exported from economics into other disciplines on the promise that it could offer greater explanatory power to achieve scientific unification and consilience. Once rational choice theory is viewed, as its founders recommend, as a disciplinary rift with neoclassical economics rather than a further refinement and expansion of this pre-existing theory, then it becomes easier to evaluate its successes assessed in terms of ontological and epistemological unification, as well as pragmatic progress in capturing more data with fewer assumptions. In moving ahead with this analysis I am grateful for Mäki's helpful distinction between "derivational" and "ontological" expansion (Mäki 2009, 263–64).

4 Subjecting the "rational choice revolution" to Mäki's criteria for good and bad imperialism

According to Mäki, and I think unarguably, the expansion of one method of explanation to a broader range of empirical phenomena fits within the understanding of scientific inquiry. Yet, even given this tacit recognition that the envelopment of one discipline by another may well be a positive development leading to greater explanatory success, Mäki distinguishes criteria that characterize constructive versus destructive imperialism. Obviously, if a theory (a) can satisfactorily explain a larger range of ontological substrata or properties, such as either decisions in markets or elections and choices by humans or single-celled organisms, then this is consistent with science. Similarly, if a theory (b) can either explain more facts than another theory, or cover the entire domain of a competing theory in addition to its prior domain, much like identifying DNA provides a unifying mechanism that sustains both plant and animal life, then this too would seem to be a happy outcome of scientific unification. Furthermore, if a theory were able to (c) predict either more or novel empirical data, such as Albert Einstein's general relativity theory predicts gravitational waves, then this empirical means of verification would endorse this type of scientific expansion to encompass more fields of investigation. The question is, does rational choice achieve (a), (b), or (c)?

In his discussion of economics imperialism, Mäki paves the way to answering this question by also identifying "derivational unification." This refers to "derivations of conclusions from a set of premises" which can also be a feature of scientific expansion but, if unmet with verification by empirical tests, may provide a means to identify a type of scientific imperialism that does not necessarily advance understanding. Writing in 1985, game theorist Robert Aumann acknowledges that rational choice theory offers derivational unification which Mäki describes as "a matter of deriving large classes of explanandum sentences from a parsimonious set of theoretical sentences or inferential patterns." Crucially, "theories are regarded as logical formulae, possibly devoid of truth-value, serving the task of generating implications and saving the phenomena" (Mäki 2009, 363). Aumann directly accepts that game theory is powerful because it offers the means of derivational unification: "the validity of utility maximization does not depend on its being an accurate description of the behavior of individuals." He goes on to note that "rather, it derives its being the underlying postulate that *pulls together most of economic theory*." However, Aumann's crucial acknowledgment follows because he discloses that rational choice's basis in the hypothesis of expected utility maximization cannot be experimentally verified:

> While attractive as hypotheses, there is little theory built on them, *they pull together almost nothing*; they have few interesting consequences. In judging utility maximization, we must ask not 'Is it plausible?' but '*What does it tie together, where does it lead?*'"
>
> (Aumann 1985, 46, quoted by Mäki 2009, 364)

This sense in which rational choice offers an elementary set of choice axioms that define rational action, and can be used by derivation in accordance with game theory to model any situation reducible to these assumptions, points toward the almost endless power of the theory to present the means to derive countless formal models. Game theoretic models are typically tested for their analytic consistency. When they are applied in empirical studies, rather than verifying the accuracy of the models, these experiments test whether actors behave rationally given the assumptions built into the model (e.g., Rapoport and Chammah 1965; Rand et al. 2012). Nobel Prize-winning game theorist Roger B. Myerson admits that there is no way to empirically test the validity of game theory; instead, as he puts it, "one can only ask whether a person who understands the model would feel that he would be making a mistake if he did not make decisions according to the model" (Myerson 1991, 22). In their skeptical assessment of game theory's contributions, Donald P. Green and Ian Shapiro (1996) accuse game theorists of generating formal models but with little value for actually explaining or understanding social contexts as basic as voting or collective action.[2]

In part the derivational force of rational choice stems from the fact that its veracity is entirely analytic. Ken Binmore acknowledges that the theory of rational choice is in essence tautological; it asserts that actors maximize expected utility but this essentially boils down to observing that in every choice, actors by definition choose what they most prefer (Binmore 1994, 169). The model imputes rationality to agents, and defines precisely the characteristics of preference and strategically rational choice that must obtain in order for individuals to qualify as rational and for the formal models to pertain to concrete contexts.

Expected utility theory is the preferred theory of instrumental action from the 1980s to the present, and as such is normative in defining the rules that agents must comport with (Luce and Raiffa 1957). It is also prescriptive in providing an analytic means to design institutions, legal frameworks, and public policies on the assumptions that citizens and consumers are either rational actors, or if they fail to comport with the dictates of the theory, are prone to systematic failures of rational choice that should be remedied by the interventions of policymakers and choice architects (Buchanan 1975; North 1990; Thaler and Sunstein 2009). So long as rational choice cannot be falsified, and insofar as actors who fail to exhibit consistent choice on its terms are labeled irrational, it is difficult to argue against Aumann's point that an important and perhaps chief quality of game theory is its derivational ability to generate models that could be applied to diverse social situations.

Here I briefly introduce the case of US nuclear deterrence which demonstrates both of the tendencies that I have striven to document. On the one hand, the type of modeling that was central to nuclear deterrence in the 1960s is considered a form of "economics imperialism" and yet is better understood to be an expression of the post-World War II rationality project (Thomas 2015; Erickson 2015). On the other hand, at least one well-positioned contemporary defense intellectual, Michael Desch, vehemently argues that the formal modeling approach characterizing deterrence theory, which came to characterize the mainstream approach adopted throughout the US academic profession of political science in the 1980s, is purely analytic without any practical purchase on real problems confronting military planners (Desch 2018).

Desch takes it for granted that the new wave of security studies generated by the academic strategists to implement nuclear deterrence exemplifies economics imperialism insofar as this research can best be characterized as "increasingly abstract and methodologically-fixated … late Golden Age economic theories of nuclear strategy" (Desch 2018, 7). Desch argues that nuclear deterrence theory of the 1950s and 1960s typifies "the methods and approaches of Economics" (ibid., 8). Because his main goal is to contend that this formal approach can lead to endless analytically derived formal models that are of no use for steering strategic policy, his goal is not to exhume the historical origins of this "Economic approach." Yet even in articulating his concern over the emptiness of economic modeling applied to policies of nuclear deterrence, Desch acknowledges the point I make throughout this chapter: the body of thought characterizing rational deterrence is game theory. Desch quotes the World War II- and Cold War-era British operations researcher P.M.S. Blackett, who complained about the limitations of the rational strategists' approach, that it did not "clothe the skeleton conflicts of the theory of games with the complex flesh and blood attributes of real nations; hence the bizarre nature of some of their practical conclusions" (Blackett 1961, 16).

Yet despite his disinclination to excavate the origins of the theory that he refers to as the Economic approach, Desch does much to document the mushrooming formal analytic work on the one hand, paired against its increasing lack of policy applications on the other between 1980 and 2012. Formal quantitative models increased to become approximately 50% of the research presented in top international relations journals during these three decades. Yet, according to Desch, by 2010 the relevance of this research to practical problem-solving in policy venues dropped from full applicability in 1980 to as low as no policy relevance today. Thus, in focusing on the case of nuclear deterrence as a paradigmatic example of "economic imperialism" that would better be designated "rational choice imperialism," or possibly "analytic formal imperialism," Desch provides a sustained and convincing argument that this expansion of scientific method is derivational but not successful in achieving greater ontological scope, increased powers of explanation, or enhanced predictive power.

Yet despite this tautological quality of much of the theoretical work underneath the rational choice umbrella, and the additional fact that behavioral economists use canonical rational choice theory as the standard of pure rationality that human subjects deviate from (Kahneman and Tversky 1979), via its undeniable and much discussed imperialism, rational choice has come to have a lot of impact on the structure of institutions, laws, and policies (e.g., Stern 2006). Thus, if the imperializing power of rational choice stems from its ability to serve as the basis for deriving endless formal models under the assumption that individuals maximize the expected utility reflected in the models, then the cautionary notices put forward by Mäki, and by Clarke and Walsh are significant.

Mäki warns that a scientific theory may be spread, as Ronald Coase warns us about rational choice, with theorists' hopes that they may be able to extract increasing insights into the operation of phenomena without actually attaining any actual ontological or epistemological unification from this derivational capacity (Mäki 2009, 367; Coase 1978, 209). Mäki also worries that the type of explanation that rational choice affords, even if it does turn out to be empirically sound, may tell us something about types of individuals in some categories of circumstances without helping us to understand particular individuals with specific identities. Even if a model were produced to predict the number of suicides under a specific change in public policy under the assumptions of rational agency, this theory would still be useless in predicting the actual suicide of any one individual, or shedding light on that agent's decision-making prior to the indelible act.

Clarke and Walsh further worry that alternative ways of understanding social phenomena and of accounting for human values may be eroded, or even lost entirely. When challenged that they seem to imply a view of science that either upholds progress as a telos along a singular developmental path, or must accept a relativism of explanation, the authors defend the view that some theories can progressively aid in offering superior explanations without also assuming that a particular developmental path is a forgone conclusion. Clarke and Walsh defend their view that despite the fact that imperialism seems to be self-vindicating, because arguably scientists would not adopt theories unless they were motivated by good reasons, still scientific expansion may offer unwelcome restrictions on understanding by eliminating alternatives rather than maintaining an effective intellectual marketplace for ideas.

Here they finally hit on one of the primary reasons that "economics imperialism" could be both abolishing potentially fruitful veins of discovery and stifling creative wherewithal to engage in open exploration. In developing these ideas, they point to the fact that imperialistic tendencies can follow from the exercise of power rather than the self-vindicating nature of the scientific method with expansionist tendencies. Thus, politically motivated rather than intellectually vetted expansion could follow from the superior prestige of a field that thus holds attraction to researchers in another discipline who seek to borrow its pedigree in order to cater to audiences and funding opportunities. For example, the prestige enjoyed by game theorists working on nuclear deterrence helped to make game theory attractive to researchers in other fields (Amadae 2016).

In their final conclusion, Clarke and Walsh alight on another consideration to ponder in evaluating the pros and cons of scientific imperialism. Let us suppose that derivational expansion and success is not a sign of bad imperialism, although by itself it does not entail either ontological or epistemological unification. Thus, the rational choice school could be a step in the constructive expansion of a scientific approach. Perhaps, as Herbert Gintis (2009) suggests, that at some later stage in the future, the rational choice revolution may achieve the unification of the social and behavioral sciences because cognitive science may reveal the truth-value of the elementary claims put forward by game theory, that human brains are constructed to rank all outcomes on a single scale of reference. However, even before this point were carefully established, in the meantime expected utility theory could transform the social world in its image if its basic tenets were absorbed throughout popular culture to account for the nature of rational action and consumer choice (Clarke and Walsh 2013, 349–350; Dixit 2010).

This suggestion is not far-fetched because research shows that individual behavior can be changed to conform to the predictions of rational choice under any of three circumstances. Students at elite institutions exposed to game theory can become normalized to accept its understanding of rational self-interest (Frank et al. 1993; Miller 1999). Not only are economics students more prone to cheating, but also game theory actively condones cheating as rational (McCabe and Trevino 1993, 1995; Nonacs 2013). Individuals who function in institutions that only reward behavior with an incentive structure matching the motivational structure suggested by rational choice will, rather than treat action choices as intrinsically meaningful or worthwhile, tend to conform to incentives relinquishing other motivational sets (Mazar and Ariely 2006). Encountering actors whose behavior or self-understanding conforms to expected utility theory and strategic rationality can also encourage action indistinguishable from that predicted by non-cooperative game theory, and therefore can serve to fragment bonds of trust among members of a society (Amadae 2016, 69–98). Given the ongoing assertions by game theorists that they can capture everything of relevance throughout the social sciences and even, increasingly, the humanities, within the scope of rational choice (e.g., Brams 2012), Clarke and Walsh's concern deserves to be heard.

5 Convergence of appraisal of the social significance of the rational choice revolution

In this concluding section I will reflect briefly on the significance of the rational choice revolution that impacted diverse fields throughout the social sciences and professional programs. I urge rejecting the "economics imperialism" thesis suggesting the increasing colonization of diverse disciplines using the template of neoclassical economics. Not only do I argue that the former is a much more accurate way to understand the development of rational choice as a distinctive method that replaced neoclassical economics, but furthermore I suggest that this broader perspective helps at least to raise the question of whether the late 20th-century shift in the social and behavioral sciences to the rational actor model may signal a wider transformation of scientific approach to the study of human society. Here I have in mind the scale of disciplinary transformation that Michel Foucault argued best typifies the development of the human sciences during the Enlightenment that was concurrent with the rise of panoptic disciplinary techniques (Foucault 1979). He further suggested, toward the end of his scholarly career, that a similarly proportioned transformation may be entailed in the late-modern shift to neoliberal political economy (Foucault 2010).

I invoke Foucault to make the case that scientific development is relative to the aims and goals of scientists, and their intersection with the cultural milieu in which their science must gain an institutional footing and demonstrate its relevance. I am also not worried about the familiar charge of relativism that referencing Foucault can induce. This is because I hold that even though, for example, the Newtonian and Einsteinian scientific programs are incommensurable in their elementary assumptions and represent clear and distinct research programs that study the same ontological substrata, still each is clearly valuable and serves as a paradigmatic case of sound scientific theory. Incommensurability does not necessarily imply that good science and bad science cannot be differentiated because all are relative to the cultures and interests of scientists. Let us suppose that scholasticism and the tactics of inquisition, spectacle, and divine right of kings yielded to contradistinctive democratic will formation, surveillance, and disciplining to achieve modern institutional efficiencies. It is similarly possible that post-modern social science may now be predicated on a divergent approach to human agency that dissolves the relevance of separate disciplines because it demands that actors exhaustively rank world states, and strategically compete against each other, in all domains of action. The brilliance of Foucault's work on regimes of knowledge production and the organization of power is that he interrelates the authority of human scientists achieved through scientific practices with their role in governing and normalizing behavior.

In *Prisoners of Reason* (2016) I argue that orthodox rational choice entails four unique assumptions about human behavior which, by being postulated as normative and prescriptive, have played a role in refashioning individual subjectivity to comply with its demands. These unprecedented imperatives are: (a) that only outcomes have value, and not the means or processes by which outcomes are achieved (Hausman 2012, 53); (b) that the total value available for consideration is finite and can be monetized (Giocoli 2006); (c) that individuals must compete against each other strategically without the ability for working together in teams (Bacharach 2006); and (d) that impartial or disinterested action is prohibited under the assumption that every choice must further agents' goals (Sen 1985, in Sen 2002, 206–224). Although this concluding section is not the place to explore these basic assumptions underlying the imperialistic rationality project, it is easy to touch on how radical these assumptions are. They deny the possibility of commitment, or the type of moral action consistent with the classical liberal no-harm principle or Kantian-style side constraints on action requiring that actors avoid interfering with one another (Heath 2008). Game theory requires that the entire range of experiential value available to agents be of one commensurable, finite type that must encompass aesthetic beauty, tranquility, fulfillment, friendship, loyalty, and trust, in addition to fungible rewards such as cash prizes or incentives. It rejects the possibility that actors can act out of solidarity, and it invalidates the type of selfless charity or beneficence characterizing the imperfect duties and positive virtues stipulated by the classical liberal theorists Adam Smith and Immanuel Kant.

From this perspective, then, Clarke and Walsh's warning that rational choice imperialism may seep down into the common person's worldview, authoritatively demarcating available ontologies for social expression and informing agents' self-understanding of traditional social rites and institutions is pertinent (Clarke and Walsh 2009, 350). Rational choice imperialism is inconsistent with the classical political economy of Adam Smith, which accepts negative and positive virtue, the possibility of impartial judgments, and that markets must be bounded by the rule of law (Smith] 1982 [1759; Amadae 2003, 205–219; Amadae 2008), and is incommensurable with the neoclassical marginalists' definition of rational action as obeying the law of diminishing marginal utility and pertaining only to efficient means rather than the consistent and exhaustive ranking of ends (Amadae 2003). Thus, it behooves us to step back and appraise the full scope and dimensions of the transformation in social theory and practice entailed by game theory.

I close by introducing Paul Forman's article, "On the Historical Forms of Knowledge Production and Curation: Modernity Entailed Disciplinarity, Postmodernity Entails Antidisciplinarity" (2012), because through a distinct method he arrives at almost identical conclusions to those put forward here (and in Amadae 2016). He argues that there has been a comprehensive shift in scientific inquiry that has entailed both the methods and conclusions that map directly onto the 1970s epistemic solidification of the rational choice approach. Where philosophers of economics discuss "economics imperialism," and the increasing colonization of rational choice of formerly disparate disciplines, Forman, a historian of science, notices a falling away of disciplines altogether as formerly distinctive branches of knowledge are now unified under one singular method. Interestingly, Forman identifies the same four distinctive attributes of rational choice versus the traditional human sciences (including economics) that preceded it. One is the transformation from the study of means, such as Robbins discusses as the pertinent domain of economics, and legitimate processes to a comprehensive focus on ends and outcomes (Forman 2012, 59). I concur with Forman that Western European modern and liberal societies and human sciences focused on fair and impartial means and procedures to generate outcomes, and that this practice and institutional arrangement has fallen away under the new rational choice appraisal of outcomes (see e.g., John Rawls as a modern theorist, discussed by Forman (2012, 74)). Even knowledge production itself ceases to be the result of disinterested processes wherein the integrity of procedures is a mark of the effective and veridical establishment of truth-claims. Whereas it was a key attribute of knowledge production throughout modernity that scientists were disinterested, the rational choice approach holds that no action can be disinterested, and even views the generation of truth-claims as signaling techniques whereby the decision whether to utter a truth or falsehood is a function of preferences and how the speaker best profits (Forman 2012, 74, 79; Lewis 1969). Forman also notes that whereas solidarity was taken to be constitutive of any healthy and vibrant community, in postmodernity solidarity, or working together with a joint goal, has negligible to no credibility as a meaningful category of action (Forman 2012, 87, 88; Tuck 2008). Forman also points to how the core values underlying modern liberalism were assumed to be universally relevant to all actors, whereas under rational choice, all actors' appraisal of value must be relative to their own personal holdings that accurately identify each individual's opportunity costs for giving up one favorable outcome to achieve a more favorable outcome.

In conclusion, Forman's argument that the modern disciplines that have been coextensive with the modern human sciences have very recently yielded to a new, ends-driven, performance metric-sensitive, individualistic, and self-interested mode of inquiry, correlates with the attributes of the rational choice revolution that I have earmarked as foundational. This convergence helps to render credible the idea that not only is the so-called "economic imperialism" of game theory more indicative of a cross-disciplinary transformation than a vindicated exercise of scientific unification, but also that the copious attention devoted to evaluating the significance of this imperialism is crucial.

Notes

1 Rational choice theory became the designation for the family of theories that grew out of the new approach to decision-making put forward by von Neumann and Morgenstern (1944, 1947), and it includes decision theory (applicable to single agents), cooperative and non-cooperative game theory, social choice, public choice, positive

political theory, and public policy analysis. See Amadae (2003) and Erickson (2015). Luce and Raiffa (1957) wrote the first authoritative text; see also Shubik (1982), which together with von Neumann and Morgenstern, demonstrate that game theory was a new approach to studying social interactions that became central to multiple disciplines throughout the social sciences, including economics. Nicolas Guilhot and Alain Marciano argue in "Rational Choice as Neo-Decisionism: Decision-Making in Political Science and Economics after 1945" (in this volume) that the standard "economics imperialism" thesis is misguided for overlooking the fact that the disciplinary transformation centering on rational decision-making largely occurred outside economics.

2 Jaakko Kuorikoski and Aki Lehtinen (2010) argue that it may be possible to make analytic progress via the means of derivation. However, ultimately this would need to be directly associated with empirical testing to count as a means of making progress in social science.

References

Amadae, S.M. 2003. *Rationalizing Capitalist Democracy: The Cold War Origins of Rational Choice Liberalism*. Chicago, IL: University of Chicago Press.

Amadae, S.M. 2008. "Utility, Universality, and Impartiality in Adam Smith's Jurisprudence". *Adam Smith Review* 4: 238–246.

Amadae, S.M. 2016. *Prisoners of Reason: Game Theory and Neoliberal Political Economy*. New York: Cambridge University Press.

Arrow, Kenneth J. 1963. *Social Choice and Individual Values*, second revised edn. New Haven, CT: Yale University Press.

Aumann, Robert J., John C. Harsanyi, John P. Mayberry, Michael Maschler, Herbert E. Scharf, Reinhard Selton and Richard Stearns. 1967. *Models of Gradual Reduction of Arms*. Arms Control and Disarmament Agency, ACDA/ST-116, Sept. 1, 1967.

Axelrod, Robert M. 1970. *Conflict of Interest: A Theory of Divergent Goals with Applications to Politics*. Markham Political Science Series. Chicago, IL: Markham Pub. Co.

Bacharach, Michael. 2006. *Beyond Individual Choice*. Princeton, NJ: Princeton University Press.

Becker, Gary. 1978. *The Economic Approach to Human Behavior*. Chicago, IL: University of Chicago Press.

Binmore, Ken. 1994. *Game Theory and the Social Contract: Playing Fair*. Vol. 1. Cambridge, MA: MIT Press.

Blackett, P.M.S. 1961. "Critique of some Contemporary Defense Thinking". *Encounter* 16(4): 9–17.

Brams, Steven J. 2012. *Game Theory and the Humanities: Bridging Two Worlds*. Cambridge, MA: MIT Press.

Buchanan, James M. 1975. *The Limits of Liberty: Between Anarchy and Leviathan*. Chicago, IL: University of Chicago Press.

Buchanan, James M. 1976. "A Hobbesian Interpretation of the Rawlsian Difference Principle". *Kyklos* 29(1): 5–25.

Buchanan, James M. and Gordon Tullock. 1962. *The Calculus of Consent*. Ann Arbor, MI: University of Michigan Press.

Clarke, Steve and Adrian Walsh. 2009. "Scientific Imperialism and the Proper Relations between the Sciences". *International Studies in the Philosophy of Science* 23(2): 195–207.

Clarke, Steve and Adrian Walsh. 2013. "Imperialism, Progress, Developmental Teleology, and Interdisciplinary Unification". *International Studies in the Philosophy of Science* 27(3): 341–351.

Coase, Ronald H. 1978. "Economics and Contiguous Disciplines". *Journal of Legal Studies* 7: 201–211.

Dawkins, Richard. 1979. *Selfish Gene*. Oxford: Oxford University Press.

Dawkins, Richard. 2006. *The God Delusion*. Boston, MA: Houghton Mifflin.

Desch, Michael C. 2018. *A Cult of the Irrelevant?* Princeton, NJ: Princeton University Press.

Dixit, Avinash K. and Barry J. Naleuff. 2010. *The Art of Strategy: A Game Theorist's Guide to Success in Business and Life*. New York: W.W. Norton.

Downs, Anthony. 1957. *An Economic Theory of Democracy*. New York: Harper and Row.

Erickson, Paul. 2015. *The World the Game Theorists Made*. Chicago, IL: University of Chicago Press.

Forman, Paul. 2012. "On the Historical Forms of Knowledge Production and Curation: Modernity Entailed Disciplinarity, Postmodernity Entails Antidisciplinarity". *Osiris* 27(1): 56–97.

Foucault, Michel. 1979. *Discipline and Punish*. Trans. by Alan Sheridan. New York: Vintage.

Foucault, Michel. 2010. *Birth of Biopolitics: Lectures at the College de France*. New York: Picador.

Frank, Robert, Thomas Gilovich and Donald Regan. 1993. "Does Studying Economics Inhibit Cooperation?" *Journal of Economic Perspectives* 7: 159–171.

Gintis, Herbert. 2009. *The Bounds of Reason: Game Theory and the Unification of the Behavioral Sciences*. Princeton, NJ: Princeton University Press.

Giocoli, Nicola. 2006. "Do Prudent Agents Play Lotteries: Von Neumann's Contributions to the Theory of Rational Behavior". *Journal for the History of Economic Thought* 28(1): 95–109.

Green, Donald P. and Ian Shapiro. 1996. *Pathologies of Rational Choice Theory: A Critique of Applications in Political Science*. New Haven, CT: Yale University Press.

Hardin, Russell. 1982. *Collective Action*. Baltimore, MD: Johns Hopkins University Press.

Hargreaves Heap, Shaun and Yanis Varoufakis. 2004. *Game Theory: A Critical Text*. Second edition. London and New York: Routledge.

Hausman, Daniel. 2012. *Preference, Value, Choice, and Welfare*. New York: Cambridge University Press.

Heath, Joseph. 2001. *Communicative Action and Rational Choice*. Studies in Contemporary German Social Thought. Cambridge, MA: MIT Press.

Heath, Joseph. 2008. *Following the Rules: Practical Reasoning and Deontic Constraint*. New York: Oxford University Press.

Hirshleifer, Jack. 1985. "The Expanding Domain of Economics". *American Economic Review Special Issue* 75: 53–68.

Jervis, Robert. 1978. "Cooperation Under the Security Dilemma". *World Politics* 30(2): 167–214.

Kahneman, Daniel and Amos Tversky. 1979. "Prospect Theory: An Analysis of a Decision under Risk". *Econometrica* 42(2): 363–392.

Kuhn, Thomas. 1962. *Structure of Scientific Revolutions*. Chicago, IL: University of Chicago Press.

Kuorikoski, Jaasko and Aki Lehtinen. 2010. "Economic Imperialism and Solution Concepts and Political Science". *Philosophy of the Social Science* 40(3): 347–384.

Laezer, Edward. 2000. "Economic Imperialism". *The Quarterly Journal of Economic* 95(1): 99–146.

Leonard, Robert. 1992. "Creating a Context for Game Theory". In E. Roy Weintraub (ed.). *Toward a History of Game Theory*. Durham, NC: Duke University Press, pp. 29–76.

Leonard, Robert. 2010. *Von Neumann, Morgenstern, and the Creation of Game Theory: From Chess to Social Science, 1900–1960*. Historical Perspectives on Modern Economics. New York: Cambridge University Press.

Lewis, David K. 1969. *Convention: A Philosophical Study*. Cambridge, MA: Harvard University Press.

Luce, R. Duncan and Howard Raiffa. 1957. *Games and Decisions; Introduction and Critical Survey*. New York: Wiley.

Mäki, Uskali. 2002. "Explanatory Ecumenism and Economics Imperialism". *Economics and Philosophy* 18: 235–257.

Mäki, Uskali. 2009. "Economics Imperialism: Concept and Constraints". *Philosophy of the Social Sciences* 39(3): 351–380.

Mäki, Uskali. 2013. "Scientific Imperialism: Difficulties in Definition, Identification, and Assessment". *International Studies in the Philosophy of Science* 27(3): 325–339.

Mazar, Nina and Dan Ariely. 2006. "Dishonesty in Everyday Life and its Policy Implications". *Journal of Public Policy & Marketing* 25(1): 117–126.

McCabe, Donald and Linda Klebe Trevino. 1993. "Academic Dishonesty". *Journal of Higher Education* 64(5): 522–538.

McCabe, Donald and Linda Klebe Trevino. 1995. "Cheating Among Business Students". *Journal of Management Education* 19(2): 205–218.

McCabe, Donald, Kennith Butterfield and Linda Klebe Trevino. 2006. "Academic Dishonesty in Graduate Business Programs". *Academy of Management Learning and Education* 5(3): 294–305.

Miller, Dale. 1999. "The Norm of Self-Interest". *American Psychologist* 54: 1–8.

Miller, Gary J. 1997. "The impact of economics on contemporary political sciences". *Journal of Economic Literature* 35(3): 1173–1204.

Mirowski, Philip. 1989. *More Heat Than Light: Economics as Social Physics, Physics as Nature's Economics*. Historical Perspectives on Modern Economics. Cambridge and New York: Cambridge University Press.

Mirowski, Philip. 2002. *Machine Dreams: Economics Becomes a Cyborg Science*. New York: Cambridge University Press.

Mueller, Dennis C. 2003. *Public Choice III*. Cambridge: Cambridge University Press.

Myerson, Roger B. 1991. *Game Theory: Analysis of Conflict*. Cambridge, MA: Harvard University Press.

Nash, John F. 1950. "The Bargaining Problem". *Econometrica* 18(2): 155–162.

Nash, John F. 1996. *Essays on Game Theory*. Cheltenham and Brookfield, VT: E. Elgar.

Nonacs, Peter. 2013. "Cheating to Learn: How a UCLA Professor Gamed a Game Theory Midterm". April 13, https://curious.kcrw.com/2013/04/cheating-to-learn-how-a-ucla-professor-gamed-a-game-theory-midterm#.VtqnHfEw8q8 (accessed March 5, 2016).

North, Douglass Cecil. 1990. *Institutions, Institutional Change, and Economic Performance. The Political Economy of Institutions and Decisions*. New York: Cambridge University Press.

Pettit, Philip. 2002. "Virtus Normativa: Rational Choice Perspectives". In Philip Pettit (ed.). *Rules, Reasons, and Norms*. Oxford: Clarendon Press.

Posner, Richard A. 1984. "Wealth Maximization and Judicial Decision-making". *International Review of Law and Economics* 4(2): 131–135.

Posner, Richard A. 1985. "Wealth Maximization Revisited". *Notre Dame Journal of Law, Ethics, and Public Policy* 2: 85–106.

Rand, David G., Joshua D. Greene and Martin A. Nowak. 2012. "Spontaneous Giving and Calculated Greed". *Nature* 489: 427–430.

Rapoport, Anatol and Albert M. Chammah. 1965. *Prisoner's Dilemma; a Study in Conflict and Cooperation*. Ann Arbor: University of Michigan Press.

Rawls, John. 1971. *A Theory of Justice*. Cambridge, MA: Belknap Press of Harvard University Press.

Riker, William H. and Peter C. Ordeshook. 1973. *An Introduction to Positive Political Theory*. Prentice Hall Contemporary Political Theory Series. Englewood Cliffs, NJ: Prentice Hall.

Robbins, Lionel. 1935. *An Essay on the Nature & Significance of Economic Science*. Second edition. London: Macmillan.

Schelling, Thomas C. 1960. *Strategy and Conflict*. Cambridge, MA: Harvard University Press.

Schelling, Thomas C. 1973. "Hockey Helmets, Concealed Weapons, and Daylight Saving: A Study of Binary Choices with Externalities". *The Journal of Conflict Resolution* 17(3): 381–428.

Schelling, Thomas C. 1978. *Micromotives and Macrobehavior*. New York: Norton & Co.

Schelling, Thomas C. and Morton H. Halperin. 1961. *Strategy and Arms Control*. New York: Twentieth Century Fund.

Sen, Amartya K. 1977. "Rational Fools: A Critique of the Behavioral Foundations of Economic Theory". *Philosophy and Public Affairs* 6(4): 317–344.

Sen, Amartya K. 2002. "Goals, Commitment, and Identity". In *Rationality and Freedom*. Cambridge, MA: Belknap Press.

Shubik, Martin. 1982. *Game Theory in the Social Science: Concepts and Solutions*. Cambridge, MA: MIT Press.

Smith, Adam. 1982 [1759]. *The Theory of Moral Sentiments*. Edited by D.D. Macfie Raphael and A.L. Macfie Raphael. Indianapolis, IN: Liberty Fund.

Smith, Maynard. 1982. *Evolution and the Theory of Games*. New York: Cambridge University Press.

Stern, N.H. 2006. *Stern Review: The Economics of Climate Change*. London: HMS Treasury.

Stigler, George. 1984. "Economics – The Imperial Science?" *Scandinavian Journal of Economics* 86: 301–313.

Stokey, Edith and Richard Zeckhauser. 1978. *A Primer for Policy Analysis*. New York: W.W. Norton.

Taylor, Michael. 1976. *Anarchy and Cooperation*. New York: Wiley.

Thaler, Richard and Cass Sunstein. 2009. *Nudge: Improving Decisions about Health, Wealth, and Happiness*. New York: Penguin.

Thomas, William. 2015. *Rational Action: The Sciences of Policy in Britain and America, 1940–60*. Cambridge, MA: MIT Press.

Trivers, Robert L. 1971. "The Evolution of Reciprocal Altruism". *Quarterly Review of Biology* 46: 35–37.

Tuck, Richard. 2008. *Free Riding*. Cambridge, MA: Harvard University Press.

Von Neumann, John and Oskar Morgenstern. 1947 [1944]. *Theory of Games and Economic Behavior*. Princeton, NJ: Princeton University Press.

8 Crossing boundaries, displacing previous knowledge and claiming superiority

Is the economics of discrimination a conquest of economics imperialism?

Cléo Chassonnery-Zaïgouche[1]

Introduction

Discrimination has been a continuous concern in economics, originally focusing on wage inequalities and segregation. Gary Becker's (1971b) work on discrimination is commonly viewed as an expression of "economics imperialism." In this work Becker makes racial discrimination fit in a trade model, applying price theory to an object previously perceived as "owned" by other social sciences – providing an example of transgression in terms of "property rights" (Becker 1971b, 10). After a contrasted reception of *The Economics of Discrimination* (Swedberg 1990, 27–28; Fleury 2012), analysis of racial discrimination was quickly integrated within labor economics textbooks and academic curricula. Taken as an imperialist conquest, the integration of the economics of discrimination today seems complete: the whole Beckerian project – the "economic approach to human behavior" – considerably changed the study of economics over the past 50 years. This chapter aims at recasting the traditional understanding of economics imperialism by proposing another historical narrative of the economics of discrimination.

According to the received view of the history of discrimination in economics (Prasch 2004), Becker instituted this research program in economics, introducing the concept of "taste for discrimination" in his PhD thesis. After the second edition of *The Economics of Discrimination* (Becker 1971b), Kenneth J. Arrow and Edmund S. Phelps separately developed theories of "statistical discrimination" (Arrow 1972; Phelps 1972), leading to new explanations for discriminatory behaviors, based on beliefs, stereotypes, and signal effects. In a broader perspective, empirical methods – micro-econometric estimations and later laboratory and field experiments – expanded the scope of the economic analysis of discrimination in the early 1970s, from labor to other markets (credit, housing), and non-market areas such as education and health (Chassonnery-Zaïgouche 2014). Three elements from this history – Becker's tastes-based model, Arrow's statistical theory, and a set of empirical works – will illustrate three lessons on the use of economics imperialism as a reflexive concept. In this chapter, two elements define economics imperialism: the crossing of established frontiers between approaches and disciplines, and the intention of substituting one approach for another.

Section one briefly presents the origins of economics imperialism and proposes a two-aspect definition of the concept. The second section contextualizes Becker's model of discrimination as imperialism "within economics" rather than towards other social sciences. Section three characterizes Arrow's statistical discrimination as a "humble" theoretical imperialism, calling for the complementarity of social sciences rather than for the superiority of economics. The final section states that empirical measurements of discrimination in economics are a type of empirical imperialism essentially because these methods replace other social sciences in their contexts of expertise.

Gary S. Becker is a central figure of the traditional account of economics imperialism, a term he proudly uses to describe his general approach. Questioning the imperialist aspect of his work implies assessing whether he was really successful at conquering new territories for economics. Does his theory of discrimination prevail in other social sciences, then and now? Becker's model had a more general impact within economics than outside the discipline, and so this model did not change the way other social sciences studied discrimination. This claim relies on a reappraisal of the recent history of the frontiers of economics. In the mid-1950s, Becker's model was controversial within economics itself. Thus, Becker's first conquest was legitimacy in the discipline. His model of discrimination is not an extension of price theory to non-market domains, but an extension of neo-classical price theory to objects of study previously analyzed by other schools of thought. My historiographical claim in this chapter is that economics imperialism is an ex-post reconstruction of a movement closely related to the change in orthodoxy within economics. By looking at the specific analytical tradition to which Becker belongs, the concept of economics imperialism is extended to encompass the role of smaller units of analysis than a discipline, such as research programs. Beyond the question of the crossing of disciplinary frontiers, the problem thus becomes one of assessing the success and failure of imperialistic perspectives defined at a smaller level.

If economics imperialism is defined as crossing the borders of different territories, is the qualification of imperialism appropriate to evaluate a piece of research's scientific value? In other words, is imperialistic science always "bad science"? The main criticism addressed to economics imperialism concerns reductionism and, as a consequence, the undermining of the influence of other social sciences at the cost of pluralism.[2] Two epistemological issues are at stake in these criticisms: the capacity of economics to build a separate causal explanation – the relative influence of economic factors over a phenomenon or the reduction of all social phenomena to economic factors – and the status of economics as a separate science alongside or above other social sciences. The work of Kenneth Arrow on discrimination, examined in section three, is an example of what I call a "humble" theoretical imperialism. Characterized by a separate economic explanation of discrimination that influences other disciplines, it nevertheless maintains a traditional vision of the division of labor among social sciences. Hence, a certain type of economics imperialism, like the "humble" type, could foster a dialogue rather than a conflict on the frontiers of social sciences.

The last argument concerns economics imperialism's practical significance. Empirical analyses of discrimination are widely used in expertise contexts where other social sciences (would) also have been relevant tools – policy evaluation and calibration, forensic science, etc. I analyze the development of micro-econometric estimations and experiments as a type of "empirical imperialism" – essentially through the replacement of other social scientists by economics experts. Scientific imperialism is not an abstract concept and can be tracked in concrete practices that determine the relative position and status of disciplines in the symbolic and material social order (Fourcade et al. 2015). The role played by the quantitative aspects to define rigorous empirical work is central in the evolution of the relative status of social sciences after World War II. By looking at the social and political environment of sciences, especially within a specific *régime* of expertise, one can put new types of imperialism in the agenda, especially in terms of "standing." Imperialism of standing is defined as the "academic and non-academic prestige, power, and resources as well as the acknowledged technological and political relevance of one discipline [that] increase at the expense of those of another" (Mäki 2013, 11).

1 "Definitions" of economics imperialism

The first use of the term "economic imperialism" is attributed to Ralph W. Souter in 1933 (Swedberg 1990, 14) to imply the unification of all social sciences around one object, individuals in society, by an "enlightened and democratic" economics imperialism.[3] However, few projects on the frontier exist before the mid-1950s. The first works generally labeled as economics imperialism are Becker's (1957) model of discrimination and Downs' (1957) economic analysis of politics. Kenneth Boulding and Gary S. Becker informally used the term in the 1960s,[4] to designate the extension of the analysis of economics to domains previously assigned or considered to belong to other disciplines – mainly sociology, political sciences, law, or psychology (Becker 1971a, 2; Becker interviewed by Swedberg 1990, 39). However, there are relatively few occurrences of the term before the 1970s.

Between the end of the 1960s and the beginning of the 1970s, some economists started proudly to use the notion of economics imperialism. In a battle for redefining their discipline, they appropriated Robbins' definition and increasingly applied the tools of economics to every human decisions.[5] In this perspective, the "economic approach" (Becker 1976) is defined by three elements:

> First, economists assume that individuals engage in maximizing rational behavior. Second, economics adheres strictly to the importance of equilibrium

> as part of any theory. Third, economists place a heavy emphasis on a clearly defined concept of efficiency.
>
> (Lazear 1999, 100)

Finally, starting in the 1980s, reflexive works on economics imperialism (Bourdieu 1984; Swedberg 1990), as well as criticisms and condemnations of it flourished, by economists (Sen, Akerlof, Solow) as well as sociologists (White, Granovetter) and other social scientists.

Hence, the uses of the concept were ambiguous. It was used: (1) to defend the unification of social sciences by economics as the universal "grammar of social sciences" (Hirshleifer 1985); (2) as a label to condemn this project of unification that assumed the superiority of a certain type of economics (Fine 2000); (3) to describe an historical state of the disciplinary frontiers at a particular moment (Swedberg 1990); (4) to describe one category of relationships between social sciences (Mäki 2013); or (5) to systematize a dynamic process of domination assuming various forms, as I try to use it in this chapter.

This chapter focuses on economics imperialism as a particular type of scientific imperialism.[6] Our definition of economics imperialism is based on two elements. First, my use of the concept refers to the traditional definition of economics imperialism in terms of crossing disciplinary boundaries: importing objects of study traditionally considered as beyond the scope of economics (Mäki 2013, 11).[7] In this chapter, I build on this general definition of economics imperialism, but I extend it to encompass any borders established *between* sciences and *within* sciences. This element calls for a careful historical appraisal of the state of the disciplinary frontiers at a particular period of time. Hence, when talking about economics imperialism, one has to analyze the intellectual struggle to define the discipline,[8] and, if this is conflicting, one has to pay attention to the arguments and the postures of the opponents. Disciplines are not homogeneous bodies of knowledge (Davis 2010; Mäki 2013, 13). Traditions of thought, but also approaches defined by specific methods are eligible as imperialist proponents. To carry the geopolitical metaphor further, regions are important in the study of frontiers, as well as the relation between the center and the periphery. Hence imperialism may be concerned with approaches not merging totally with whole disciplines.

The second part of my definition is concerned with the successes and failures of the diffusion (and imposition) of ideas. The analogy with political imperialism and the Marxist origins of the concept[9] as well as the implicit normative connotation of the term in common language, imply a reflection on *domination* in the production of hierarchy in social sciences. Was the imperialistic element of knowledge successful? What counts as a success? If one goes further than using economics imperialism as merely a geopolitical metaphor and if one were to build a useful concept for the history and philosophy of sciences, one would need to introduce power to describe the imperialistic projects. The objective here is not only to look at the imported objects of study, but also to analyze the exporting part of the exchange: how successful were exported methods, how new markets or new trades developed as a result, etc. I use two criteria here. The first criterion is the replacement of the previous element of knowledge within the conquered discipline – or, at least, the recognition of this element of knowledge. A work is imperialistic if it displaces other approaches that got "there" first, and if its author believes he or she is building a "better" economic explanation of the causes or the consequences of a phenomenon. Consequently, this more restrictive definition of economics imperialism is based on the following epistemic aspects: to be imperialistic a statement has to propose a theoretical explanation of a phenomenon or a class of phenomena so far explained by other disciplines,[10] and this explanation has to be accepted by other disciplines as a better explanation. The implicit argument here aims at introducing the idea of "substitution" of two approaches using the scientific criteria proper to the imperialistic approach. Of course the conception of what counts as an explanation or what is evidence of causality are discipline-related and subject to historical shifts. I focus in this text on tentative rather than on effective imperialistic projects. Actors of scientific imperialism aim at producing or think they produce what is considered *better* knowledge. The fundamental element in this project deals with the capacities of a framework to unify all the social sciences – and the status of universal categories of economic knowledge. Not only has the explanation to be considered better, but also (still) economic, i.e., related to economics as a discipline.

The second criterion is more practical. The question of replacement could also be treated as the idea of a better recognition of the new element of knowledge in terms of external uses, i.e., external to scientific discussion, as for example within the expertise sphere – what Mäki calls imperialism in terms of "standing." This last point lies in the stream of sociology of knowledge by enlarging the social inscription of science, as analytical content is not sufficient to explain historical evolution of the status of sciences.

To sum up, economics imperialism refers to:

1 the extension of economics – or some approaches within economics – to domains previously assigned or considered to belong to other disciplines or domains;
2 the displacement of previous elements of knowledge
   a by offering an economic explanation that is considered a better explanation (*theoretical imperialism*);
   b by replacing other social sciences in practical usages of science (*empirical imperialism*).

This definition is conjunctive (1 and 2, 1 and 2a, 1 and 2b).

Given this definition of economics imperialism, the question becomes not only whether the economics of discrimination is imperialistic, but also, if it is imperialistic, of what type of imperialism it is a result. Discrimination as a frontier object offers a great working example to test the usefulness of economics imperialism as a reflexive concept. In the case of discrimination, the relative weight of economic, social, and cognitive factors has been an ongoing question in social sciences. The epistemological question of the different causes of discrimination is invariably linked to the frontiers of market activities, its social embeddedness, and the role of market as a lever either to change or control the other social spheres. This object of study appears as a good case study for assessing the way different social sciences build their explanations in relation or opposition to other perspectives. What are the economic explanations of discrimination? Have they been accepted outside economics as better explanations? Moreover, discrimination is a subject with important policy-oriented work and social scientists have long been solicited for policy advice or evaluation. What social science provides the greatest number of experts on discrimination? Was there a change in the formation of these experts after the development of the economics of discrimination?

2 Becker's imperialism re-contextualized: imperialism within economics?

In 1955, the University of Chicago imposed a sociologist[11] on Becker's thesis committee, since the subject of his dissertation was considered controversial for a PhD in economics. The committee episode echoes the rejection of the first manuscript of Becker's thesis by the University of Chicago Press, eventually being published through pressure from Milton Friedman. Less than 30 years later, Becker, at that time full Professor of Economics, was jointly appointed Professor of Sociology in the same university. He perceived this appointment as a signal to the sociology profession "that his rational choice approach was an acceptable theoretical paradigm" (Horn 2009, 135). In 1992, Becker was awarded the Nobel Prize "for having extended the domain of microeconomics analysis to a wide range of human behavior and interaction, including non-market behavior."[12] I argue in this section that this recognition of a shift in the legitimate object of study in economics is the result of imperialism within economics rather than a kind of imperialism towards other social sciences.

Becker's work on discrimination is commonly viewed as one of the first expressions of "economics imperialism." The idea of discrimination appeared as an object of study in economics long before Becker's 1955 dissertation. Several classical economists, and among them a strong (albeit diverse) utilitarian tradition, socialist economists and early feminists, produced economic theories to understand the differential treatment that minorities and women received in the economic sphere (Chassonnery-Zaïgouche 2014, Chapter 1). Although discrimination does not constitute the core of economic theory, the concept had never been outside the legitimate territory of economics.

The dramatic change of economics in the postwar period clearly had an impact on the perception of economists who considered the works on discrimination before Becker as not part of economic theory. The shifting perception of Gunnar Myrdal's work and his personal trajectory are clear illustrations of this point.

Myrdal was recognized as an economist for his work on monetary theory.[13] Recruited for being a foreign economist – hence supposedly more objective on American values – he designed the collective study on racial inequalities in the United States that eventually became in 1944 the 1,500-page report *An American Dilemma.*[14] Although gathered in the part entitled "Economics" (Myrdal 19969 [1944], 205–428), chapters 9 to 19 were generally seen as a gigantic set of data on the socio-economic conditions of African-Americans in the United States, and not as economic theory. A masterpiece of sociology for Talcott Parsons and Howard W. Odum (Fleury 2012, 6), or a text of propaganda for the Carnegie Corporation program manager Eli Evans (quoted by Cohen 2004), the book was not noticed for its sharp economic theoretical developments (Dewey 1958, 493). What was at stake in this appraisal of Myrdal's work was the soul of economics.

Myrdal builds a theory of discrimination, among an ocean of stylized facts, but a theory that accurately illustrates his conception of social science. From Sweden to the United States, Myrdal did not move away from economic theorizing but tried to integrate value judgments in a scientific method (Cherrier 2009). For instance, he was one of the first to use the concept of discrimination essentially for its normative content.[15] While the political aura of his work was largely stressed, his theoretical contribution has been largely ignored. In the section "The Theory of Vicious Circle" (75–78) and an appendix,[16] Myrdal developed a theory of cumulative causality based on his critique of the concept of equilibrium, formalized in *Monetary Equilibrium* (1939). There, he presented his arguments against the research of a unique causality, especially of an economic causality, and showed his skepticism of the possibility to track the isolated effect of specific causes. Myrdal did not move away from analytical formalism, but simply refused to produce a model of discrimination.[17] As the canon of the discipline was gradually moving to modeling as *the* rigorous method of economics (Mirowski and Hands 1998), Becker soon came to be considered the first to build a theoretical work on discrimination in economics in a market model. His status was helped by his disregard for the previous literature and his careful building of an image of a pioneer.

The mere fact that other economists had written about discrimination before Becker's PhD thesis does not rule out the possibility of his work being imperialistic. As Davis (2010, 11)[18] points out, economics imperialism is not always about the whole discipline. On the contrary, one needs to specify smaller-scale elements as a unit of analysis in order to get a better understanding of economics imperialism.

The impact of Becker's modeling of discrimination is deeper and much more visible within the field of economics. The reception of his model in other social sciences is ambiguous, since the first reactions resulted in criticisms that led to confrontations rather than to immediate followers, or in positive reviews that considered Becker's work as an attempt to build a general framework on the subject that was flooded with facts and lacked a theory (Fleury 2009, 78). Within economics, his work received more attention at the end of the 1960s, and not immediately after the time it was first published: in the early 1970s, it was praised as a first step in the renewal of applied microeconomics. The argument of this section is that Becker actually rebuilds this crossing of boundaries on the occasion of the second edition of *The Economics of Discrimination* in 1971, overcoming legacies of the past, but participating in a collective reorientation of economics.

When he produced his model, price theory and labor issues were considered as separate fields characterized by an evident lack of "co-operation and consequent cross-fertilization" (Schumpeter 2006 [1954], 913–914, n.6).[19] Becker's model can be seen as imperialism towards other approaches within economics, and it illustrated the rise of "analytical labor economics" (Heckman 2003, 2011). This field has been characterized, in retrospect, as the emergence of "modern labor economics." Analytical labor economics rejects inductive methods and relies on deductive modeling of price theory and empirical validation. Becker's model of discrimination – and later works on human capital – are embedded in this project of micro-foundations of labor relations, started at Chicago with the works of Theodore Shultz, Melvin Reder and, above all, H. Gregg Lewis and his PhD students. At the core of this (re)foundation lies the relaxation of the assumption of labor homogeneity, as one of the fundamental criticisms that neo-classical economists have to face. The "counter-revolution" at Chicago (and at Columbia) aimed at completing the shortcomings of neo-classical distribution theory. This "counter-revolution" paved the way to a renewed analysis of the supply of labor in economics. The whole project was directed against the institutionalists' vision on labor issues. It exhibited a fundamental difference in the way theory should be articulated to empirical analysis in economics.[20] More crucially, the whole project encompasses a very strong ideological core[21] (Nik-Khah and van Horn 2012; Cherrier 2011). Unsurprisingly, the strongest reactions within economics had come from institutionalists such as Northrup,[22] a specialist of trade unions discrimination who participated in the *American Dilemma* project.

The development and diffusion of Chicago price theory is one of the biggest institutional successes after World War II[23] (Emmett 2010; van Horn et al. 2011), even though the death of institutionalism is not absolute. The "'neo-classicalization' of older institutionalist themes was very much a trademark of the Chicago economics of the 1960s and 1970s" (Rutherford 2010, 36), and it was concerned with family economics as well as with labor issues, all perceived as "conquests" afterwards (Rutherford 2010, 27–29). Becker proudly used the terms economics imperialism[24] at the very moment of the change of the leading school within labor economics. The analysis of textbooks made by Boyer and Smith (2001) clearly shows the shift from labor issues as an institutionalist province to a neo-classical applied field. At that time, the early 1970s, the conquest was effective and over. The second edition of *The Economics of Discrimination* was consensual at the University of Chicago Press and discrimination was definitely a legitimate object within economics.[25] Imperialism towards other social sciences (on crime, for example) came later, essentially in the 1980s, with the growing discussion of Rational Choice Theory (RCT) in sociology (essentially at Chicago), once the battle within economics was won. Hence, the new approach first had to conquer its "heartland" (Hirshleifer 1985, 53) within the field of economics, and could then expand outside. The economics of discrimination is thus a true example of imperialism from a particular (neo-classical) towards another (institutionalist) approach.[26] On labour issues, the neo-classical approach had made its way from the periphery to the core of economics in the postwar period.

Becker's approach on discrimination in fact crossed boundaries, but I insist here on the impact that his work produced within the frontiers of economics, with no impact on other social sciences, which is a kind of "domain-only imperialism" (Mäki 2013, 12).[27] For Vromen (2009), this does not qualify as imperialism, because the frontier that was crossed was situated within the discipline. This conception of imperialism implies a view of economics (and of social sciences) as characterized by unified territories. The movement described in this section, however, corresponds to a replacement of one piece of knowledge by another, imposing a new way of doing economics.[28] In the case of Becker's work, economic imperialism is clearly "an expression of economics turning inward and isolating itself" rather than representing "the beginning of a dialogue between economics and other social sciences," contrary to Swedberg's optimistic account of economics imperialism (Swedberg 1990, 18). The question of whether this process may be called imperialism or not is based on a belief in the superiority of (a type of) economics as a tool-based discipline that gains strong support within the discipline, as shown by the rise of respectability illustrated by the Nobel Prize received by proponents of this approach. The rise of this approach within economics constituted a springboard for a kind of imperialism towards other social sciences that occurred afterwards. The rhetoric of economics imperialism was part of the conquest of economics itself and mainly occurred via an ex-post rationalization in the 1970s–1980s of a 1950s–1960s phenomenon. Economics imperialism was for Becker a rhetorical tool that served a broader project: the main impact of his approach (and tradition) was not the aggressive invasion of sociology. It was the radical transformation of large sections of the discipline of economics.

3 The economics of information applied to inequality: making racial discrimination an endogenous market phenomenon

> There is no question that my work came at the edge of social analysis, though I don't think I crossed.
>
> (Arrow, interviewed by Swedberg 1990, 138)

What happened after Becker's study of discrimination within economics? The neo-classical research program on discrimination took a new and crucial direction at the end of the 1960s, influenced by a new analysis of the role of information and strategic behaviors. Arrow's theory of statistical discrimination, a term coined by Phelps (Arrow 1976, 235), fits into this general trend, in which economics moves from "a science of interdependent isolated individuals to a science of emerging effects of social interactions" (Gautié 2007, 930, my translation). Arrow's argument had an important impact outside economics since his aim was to build a theory of market discrimination rather than a market theory of all discriminations. Arrow's work illustrates an example of economics imperialism according to definition 1 and 2a. Yet, Arrow is not identified with economics imperialism,[29] because he contends, for example, that differences in the social patterns of ethnic groups are "very sociological and totally inexplicable by any ordinary economic theory" (Arrow, interviewed by Swedberg 1990, 148). I argue in this section that even though Arrow's work on discrimination represents theoretical imperialism in the sense that it gains influence outside economics, his perspective is tempered by reflections on the limits of economics.

The question of crossing boundaries in terms of importing objects of study (definition 1) is not relevant here as Arrow built on Becker's work: the frontier was already crossed. The first argument of this section is that Arrow built an economic explanation of discrimination that gained influence outside economics.

What then counts as an economic explanation of discrimination? Let us return briefly to Becker. He builds an economic explanation of the *consequences* of market operations on discrimination. Concerning the causes of discrimination, Becker's model offers a tautological – or vacuous, as Dupré (1996) characterizes it – "explanation" of discriminatory behavior: discrimination occurs because individuals have "tastes for discrimination." Indeed, Becker is concerned with the question of whether the market is able to eliminate discriminations, and he does not ask why there are discriminations in the first place. As Arrow rightly points out, tastes for discrimination are an *ad hoc* hypothesis, keeping the causes of discrimination outside economics. The assumption of a heterogeneous distribution of these preferences is a sufficient hypothesis to test the prediction of the model: profit-maximizing agents, who do not bear the cost of discriminatory tastes, eliminate utility-maximizing agents. Hence, market forces rule out discriminatory agents. Becker's perspective is based on a Robbinsian view of economics as defined by a method and by unlimited scope.[30] It also reveals a stable moment of the relationship between economics and psychology (Gautié 2007, 928): the acceptance of Pareto's position illustrated by the use of the arguments of revealed preferences. In this view, economics is concerned with effective choices and not with preferences themselves. What Pareto used as a criterion to separate sociology and economics was "later [turned] into a reason for economists to ignore sociology" (Swedberg 1990, 13). Textbooks in sociology do not even quote Becker … at least not for this specific work. Hence, Becker's type of imperialist project does not correspond to the strong sense we gave to the word replacement in the second aspect of our definition (theoretical imperialism towards other disciplines).

By contrast, in Arrow's work, not all social interactions occur through market exchange or *as if* there were markets for everything.[31] Yet, Arrow builds an endogenous explanation of the persistence of *market* discrimination. Indeed, discrimination is an endogenous result of Arrow's modeling, not an element "to be eliminated" by market interactions. His theory of discrimination, based on the integration of strategic behaviors in uncertain environments, is an application of tools previously developed in his work on medical care insurance and moral hazard.

For Arrow, discrimination implies a market valuation of "personal characteristics of the worker unrelated to productivity" (Arrow 1973, 3). Statistical discrimination occurs when individuals use a "projection" of group productivity (based on statistical reasoning or beliefs) to hire or value an individual in a market. In this model, discrimination is a rational process of decision-making, without any additional hypothesis on preferences. Then, discrimination "can be thought of as reflecting not tastes but perception of reality" (Arrow 1973, 23).

There are two types of theories of statistical discrimination. In the first case, exemplified by Phelps, "differences in the company's beliefs are rooted in a 'technological', exogenous difference between groups" (Moro 2009, 2). The analysis takes as given a difference between two demographic groups; the explanation of discrimination determines how these differences translate (imperfectly) into different outcomes. In Phelps' (1972, 659) exact model, judgments emerge from statistical experience or from the fact that beliefs and stereotypes are common knowledge in a society. The differences in perception are determined outside the model. In the second case, "asymmetric beliefs may feed back into differences in the type of behavior that generates them" (Moro 2009, 2). The differentials in perception are the object of the analysis. Arrow mentions several explanations to these differential judgments on economic variables.

The first one is based on Festinger's psychological work[32] concerning "cognitive dissonance": projections concerning average productivity of particular groups tend to be justified according to beliefs and to previous (or current) experience. This hypothesis is very close to Phelps' argument. The second explanation states that the average productivity of the groups differs in the real world, even if the intrinsic abilities of each individual are identical. Hence, if there is no discrimination at the group level (groups differing in terms of labor experience *in average* are treated differently), there is still discrimination at the individual level (the individual may differ from the average characteristics of his group). The last explanation is a self-fulfilling prophecy: inaccurate beliefs on the average productivity and the distribution of productivity among group members could create the self-sustaining conditions to avoid learning effects of the actual productivities: e.g., an employer will not learn about the real productivities of some individuals because he or she will not employ them in the first place. In the abundant literature that followed, beliefs come from homogamy in labor relations, differences in terms of visibility of competences, risk aversion and auto-selection.[33] The second and third arguments locate the origin of discrimination in the market mechanism itself, and not outside it. What lies beyond this difference is the way one conceptualizes the relation between the labor market and other spheres of social life, hence the relation between market and "pre-market" discrimination. The question becomes, then, how to isolate market discrimination from other types of discrimination and from other spheres of social action. It follows that economists building theories of statistical discrimination within the neo-classical tradition end quasi-systematically with a reflection on the proper sphere of the market, and the proper scope of economics. This reflection is very important in Arrow's work, and is best illustrated in his later work on discrimination.

In an article published in 1998, Arrow asked whether "a phenomenon whose manifestations are everywhere in the social world [could] really be understood, even in only one aspect, by the tools of a single discipline" (Arrow 1998, 91). As in his contributions of the 1970s, his objective was to explore discrimination as a phenomenon *and* to assess the "scope and limits of ordinary economic analysis": essentially because "[studying discrimination] is important not only in itself but as a test of standard theories [in economics]" (Arrow 1998, 91). However, he went even further: "What Has Economics to Say About Racial Discrimination?" (Arrow 1998) is not a piece arguing for proud economics imperialism, but a methodological reflection on the advantages and failures of economics that eventually puts an emphasis on markets as a special case of social networks, quoting abundantly economic sociology (Granovetter 1973; White 1995).[34] Does Arrow, then, suggest that all economists "[s]hould [...] pack up and became sociologists?" (Bowles and Gintis 2000, 20), or does he suggest "that economic theorists should fold up their tents and go home to make room for psychoeconomists?" (Thaler 1992, both quoted by Gautié 2007, 937). I do not argue that Kenneth Arrow does, in practice, follow the strong position where the market is considered as a special case of social networks, but I argue that his explicit "interdisciplinary position" implies a very different standing on the border.

Arrow retrospectively made harsher criticisms of Becker's model and pointed out the problem of a market-based explanation isolated from other factors, and discussed the possibility of quantitatively disentangling the different effects of these factors. He clearly stated that economics is limited by its tools, and that economics should appeal to the results of other social sciences for the analysis of discrimination. Arrow then restores a classical division of labor between the social sciences. However, in offering this classical view, Arrow also calls for the specificity of economics as a formalized field of inquiry, focusing on the question of efficiency and profit-maximizing discrimination. This focus places measurement issues at the core of the discipline.

Is Arrow's theory imperialistic? The economics of information (in the sense described here, and not in Stigler's version) has been considered as "theoretical imperialism" (Gautié 2007; Fine 2000). According to Fine, the economics of information represents a "new economics imperialism" essentially because "[i]nstead of necessarily taking the social as given, the social is now open to explanation, despite continuing dependence upon optimizing individuals" (Fine 2000, 14). Once more, I am not concerned here with the accusation of reductionism, but with the frontiers of the discipline. The reception of Arrow's work in other social sciences seems much more positive and constructive than the reception of Becker's work – who essentially plays the role of a foil, underscoring by contrast the distinctive characteristics of the discipline. The statistical discrimination argument is built within economics and is discussed (and not directly rejected) in sociology and political sciences. Its economics origins are generally acknowledged (Bielby and Baron 1986; Neckerman and Kirschenman 1990; Pager and Karafin 2009). Even if the debate on the rationality and efficiency of the employers' attitudes is not seen through the same lens, psychologists, sociologists, and economists are discussing a theory that was first formalized within economics. Whether statistical discrimination is an efficient screening procedure is central to both literatures (Rothschild and Stiglitz 1982; Arrow 1998; Farmer and Terrell 1996; Pager and Shepherd 2008). As a specific way of conceptualizing the role of stereotypes – a long explored question both in social psychology and sociology – this literature represents a true illustration of theoretical imperialism, where an economic argument has been integrated and discussed outside the frontiers of economics.

Imperialism is closely associated with non-academic aspects and with the actors' posture (Mäki 2013, 12–13): Becker's provoking statements are part of his imperialism. At the antipodes of Becker, Kenneth Arrow advocates for powerful economics that finds its limits in its set of tools. As shown in a late contribution, his imperialism is quite "humble" – an element that could foster its acceptance outside the realm of economics. For 30 years, from his first contribution on racial discrimination in 1968 to the last one in 1998, his work has been an excuse for a reflexive contribution on both the limits of economics and the usefulness of its tools. Hence, qualifying a work as imperialistic is not sufficient to evaluate its scientific value. Indeed, as Harrison White argued, economics imperialism is a label that sometimes "prevents people from seriously evaluating scientific work on its own merits" (White, interviewed by Swedberg 1990, 325).

4 "Reverse imperialism" or "empirical imperialism"?

According to Ashenfelter and Oaxaca, although *The Economics of Discrimination* has left a large scholarly legacy, empirical analyses of discrimination "have had a still larger impact on practical matters" (Ashenfelter and Oaxaca 1987, 321). Hence, this last section is focused on the empirical analysis of discrimination as imperialism of standing.

Since the 1980s, the trend to import tools and concepts from psychology and other social sciences seems to be stronger in economics. This is what some authors (like Glaeser) call "reverse imperialism." "Reverse imperialism" refers to imperialism towards supposedly imperialistic sciences, like economics, from previously colonized disciplines, e.g., psychology or sociology (Davis 2010, 7–10). This corpus would be crossing boundaries between the disciplines not because economics is invading other disciplines, but because it is colonized by methods from other fields. In the case of the economics of discrimination, the "empirical turn" of the field beginning in the early 1970s would be the scene of this new imperialism. I argue here that an external criterion could be used to assess this type of imperialism. The argument is twofold: (1) the importation of methods originating in other social sciences is not a "counter-imperialism" (Glaeser 2003) in the case of discrimination, but a successful incorporation; (2) empirically applied microeconomics is imperialistic in the sense that it undermines the expertise tools of other social sciences, e.g., non-quantitative policy evaluation. The study of economics' "standing" is crucial to build a complex appraisal of the institutional position of economics (Fourcade et al. 2015). The quantitative techniques that are at the core of the "empirical imperialism" I described in this section do not belong necessarily to economics, but are now perceived to be the prerogative of the discipline. This evolution is both due to the history of recent economics and to the increasing availability of data and techniques (Glaeser and Shleifer 2014, 1233). This section develops three arguments: (1) there is no evidence of "reverse imperialism" in the case of discrimination; (2) economics increasingly tends to be identified with rigorous quantitative analysis; and (3), as a consequence of (2), economics expertise is becoming a growing field, because quantitative analysis is considered as better scientific research.[35]

The development of an empirical analysis of discrimination has a double origin. The first consists in the (internal) necessity to test theoretical hypotheses and predictions. The second consists in the necessity to produce an evaluation of the extent and evolution of market discrimination: do market forces eliminate discrimination? Are African-Americans better off after the Civil Rights Act? At the same time that Becker's work was awakening a renewed interest (with the second edition of *The Economics of Discrimination*), the opening of new data sets entailed a big push in the development of this type of measurement.[36] Like in the theoretical discussions, the bones of contention are centered on the relative weight that economic factors should have in relation to other kinds of factors.

In economics, the measurement of discrimination is seen to have started with the so-called Oaxaca-Blinder decomposition method to analyze wage differentials (Oaxaca 1973; Blinder 1973). The statistical apparatus comes from the expected case method used to decompose the differences in the mortality rates in demography (Kitagawa 1955), which sociologists also use to decompose the wage rates (Turner 1952; Duncan 1961). The methods were developed in economics in the wake of the operationalization of human capital theory, developed first at Columbia University and then at Chicago University: the work was essentially done by Jacob Mincer through what was later called "the Mincer Earnings Equation." What is new in the Oaxaca-Blinder method is that the difference of reward of the human capital variables is identified as discrimination.[37] Beyond the classical problems of econometrics (endogeneity, omitted variables, biased estimations), the main criticism addressed to this method is that control variables (human capital variables) may be the result of past or pre-market discrimination – e.g., discrimination in promotion and hiring – or even auto-selection may result in differences in years of experience or type of diploma. In the 1980s, lab and field experiments challenged the micro-econometric measurement of discrimination, using direct methods of observation. Experimental methods thus expanded the scope of the economic analysis of discrimination. Field and lab experiments measure other types of discrimination that could not be captured only by adjusted differences in wages – via the measurement of differential outcomes from written application, paired audit testing, or lab protocols.[38]

These are methods imported from other social sciences. There is room for arguments in favor of a "reverse imperialism": from mathematics and statistics, from behavioral sciences in a broad sense, from experimental psychology, from quantitative demography, etc. In the case of discrimination, the concept of "empirical imperialism," rather than of "reverse imperialism," is better suited. The former is better suited because of the kind of transformation that the element of knowledge (or the method) undertakes when "imported" in economics. First, the methods imported come from specific traditions of other social sciences, largely compatible with the core of economics. As an example, experimental psychology has been influential on laboratory experiments on discrimination,[39] while works in social psychology on stereotypes have not. Second, when imported, the methods are integrated within economics with some transformations – e.g., in contrast to experimentation in psychology, the addition of monetary incentives is systematic in the case of lab experiments in economics. More generally, the type of formalism is bounded by the use of econometrics as a specific way of analyzing statistical causality within the experimental data. The correspondence testing methods gained new luster with the article by Bertrand and Mullainathan (2004), essentially because of their use of randomization. Economics then has been clearly influenced by other social sciences in recent years, and once again the capacity of economics to absorb, transform, and fit exterior pieces of knowledge into standard mainstream economics has proved impressive. In the case of the analysis of discrimination, the influence of other social sciences has been "domesticated" and its empirical side is now largely part of mainstream economics. Contrary to the early controversies, the debate with other social sciences fades away rapidly when transposed into applied fields and the isolation of economics is reinforced (Davis 2013, §2). The lab experiments in economics are based on previous results in standard experimental economics. While previous works do exist in psychology and sociology, the discussions of their results occurred within economics and not within psychology or sociology.

The second argument follows from this specific "domestication": the methods "reframed" in economic terms are perceived as more precise and more accurate quantitative analyses than in the other social sciences.[40] This grip over quantitative methods is illustrated by the question of whether economists and sociologists are substitutes in the case of mathematical and statistical skills. Swedberg asked Becker if the trend of hiring economists in sociology departments would be imaginable if it were the other way around: according to Becker, it would be more difficult because of sociologists' weak ability in mathematics and statistical methods (Becker, interviewed in Swedberg 1990, 36).

This very strong economists' self-identification to rigorous quantitative methods (Fourcade et al. 2015) is mirrored by the "empirical turn" displayed in the 1970s, especially with the development of micro-econometrics. This evolution is also a strong phenomenon related to the more general transformation of the relationship between policy and economics: from a technocratic system to New Public Management, supported by accounting methods and due diligence processes (Desrosières 2003). As evidence supporting market discriminations, the results of the economics of discrimination are widely used in policy evaluation and forensics economics (economic expertise in court cases, essentially in class actions lawsuits). This "empirical imperialism" beyond academia stems from the reconfiguration of disciplinary frontiers and the reinforced quantitative orientation of economics, but is also related to the specific political and juridical context of the United States. As discrimination became liable in courts, jurisprudence became the main instrument of the struggle against racial discrimination. Economics is well suited for forensics[41] and the evolution of the US Federal Rules of Evidence reinforced its role. Some economists explain this "standing" in terms of a market for ideas (Mandel 1999, 116). An empirical analysis would be needed to assess the extent of this superiority and the fact that economists would replace other experts or create a new type of expertise. One illustration of this can be given. In the Supreme Court decision that ended segregation in 1954 (*Brown vs. Board of Education*), Myrdal's *An American Dilemma* and the "doll's study" by the psychologists Clark and Clark (1950) were quoted as the scientific justification of the decision. Nowadays in the United States, forensic economics based on micro-econometrics analysis is the main source of testimonies used in discrimination lawsuits.

5 Conclusion

Discrimination is an adequate case study to observe the shifting boundaries of economics and, while looking at an outlying object, to explore the evolution of the core of the discipline. Looking at the history of the economics of discrimination provides a case study that brings out another perspective on economics imperialism and on the history of recent economics. As an object conquered from the institutionalist tradition, the analysis of discrimination mirrors the shift of the discipline "from interwar pluralism to Postwar neoclassicism" (Morgan and Rutherford 1998), and more recently, from the neoclassical dominance to mainstream economics (Davis 2006), characterized by a (relatively) new pluralism of methods.

The three main conclusions of the chapter are as follows. First, Becker's economics of discrimination is (re)contextualized as a kind of "within-economics" imperialism: this work is not an extension of price theory to non-market domains, but an extension of neo-classical price theory to institutionalist themes. Hence, from a methodological standpoint, the chapter argues in favor of the use of smaller units of analysis – e.g., the tradition of thought – rather than the bigger unit of the whole discipline to better determine the relation at the frontiers of disciplines.

Second, Arrow's statistical discrimination is characterized as a humble theoretical imperialism, producing an explanation that gains a acceptance outside economics and which (re)opens a discussion on the complementarity of social sciences. This result emphasizes the fact that imperialist work is not always "bad science." One needs further epistemic criteria beyond the fact that disciplinary boundaries were crossed to judge an element of knowledge as "bad" or "good" science. There is no *a priori* reason to anticipate a negative impact – as Dupré (1996) seems to think – or a positive impact of scientific imperialism. This fact does not preclude any evaluative *a posteriori* analysis that shows a negative impact, if, for example, imperialism "prevents colonized discipline from progressing in ways in which it might otherwise progress" (Clarke and Walsh 2009, 201). Here I argue that the normative criterion to judge science is not to be found in its imperialistic nature.

Third, the empirical analysis of discrimination can be understood as an "empirical imperialism" in terms of standing, essentially through the replacement of the expertise of other social sciences by economics expertise. Even if economics has been widely influenced by other social sciences, the discipline has transformed the imported methods in a way to ensure a dominant position, based on an appropriation of quantitative analysis.

Apart from the analysis of the changes and the necessity to assess the successes and failures of imperialist incursions, the normative characterization of imperialism depends ultimately on the status that one assigns to economics as the "grammar of social sciences" (Hirshleifer 1985), or as a social science with its proper limits. *In fine*, the study of the frontier sheds light on the changing specificities of a discipline in terms of methodology, analytical tools, and self-identity.

Acknowledgments

This paper was presented at the Workshop on Scientific Imperialism at the University of Helsinki and at the 12th STOREP Conference at the University of Torino. It received the Young Scholar STOREP Prize 2015. I would like to thank Manuela Fernández Pinto, Uskali Mäki, Adrian Walsh, Steven Medema, Jérôme Gautié, Jean-Baptiste Fleury, Pedro Teixeira, Alain Marciano, Erich Pinzón-Fuchs, Nicolas Chachereau, Maxime Desmarais-Tremblay, Annie L. Cot, and Judith Favereau. All errors remain mine.

Notes

1 Centre Walras-Pareto (University of Lausanne), contact: cleo.chassonnery-zaigouche@unil.ch.

2 For criticisms of economics imperialism from economists as well as sociologists, see Swedberg (1990). For accounts from philosophers of sciences, see Dupré (1996, 2001), Clarke and Walsh (2009, 2013), Kidd (2013) and Mäki (2013).

3 This "enlightened and democratic" imperialism is supposed to invade the territories of its neighbors, "not to enslave them" but "to aid and enrich them and promote

their autonomous growth" (Souter 1933, 94–95, quoted by Swedberg 1990, 14). This perspective is close to what Mäki labels the Kipling Principle: "the superior discipline's burden to bring scientific enlightenment to other disciplines" (Mäki 2013, 6).

4 In a speech before the American Economic Association, Boulding commented: "The economic theory of democracy indeed as developed by Anthony Downs and others is a very good example of what I have sometimes called 'economics imperialism', which is an attempt on the part of economics to take over all the other social sciences" (Boulding 1969, quoted by Fleury 2009, 8).

5 The "celebration" of the "economic approach" has a long history (Nik-Khah and van Horn 2012, 260). For major contributions, see Tullock (1972), Stigler (1984) and Radnitzky and Bernholz (1987). For a parody, see Blinder (1974). For an inspired description of the "tribe" of the promoters of the "economic approach," see Leijonhufvud (1973).

6 Scientific imperialism is here considered as a synonym of academic or disciplinary imperialism.

7 I use the pagination of the 2012 version of the paper, available at www.helsinki.fi/tint/maki, visited on September 14, 2015.

8 Economics is defined as the historical process to build its object of study and its method as a separate science. Hence, economics is neither entirely defined by its object (a substantive definition *à la* Polanyi), nor by a universal method, nor only by the institutional component ("economics is what economists do," attributed to Viner). Hence, "the institution of a rift between that which comes under the heading of economics and that which does not, has evolved over time" (Cot and Lallement 1994, 55).

9 Imperialism in a general sense is defined as the annexation of territory without a will of integration, annexation to the benefit of a minority of investors, aristocrats, and missionaries (Hobson 1938 [1902]). A fundamentally political term, imperialism is the "the highest stage of capitalism" in Lenin's theory. See also the works of Rosa Luxemburg.

10 This first criterion is close to that of "explanatory expansionism" (Mäki 2009).

11 The sociologist is Everett C. Hughes (Becker, in Swedberg 1990, 30). This appointment did not overcome Becker's disinterest for sociology (Chassonnery-Zaïgouche 2014, 94–96) and Becker and Hughes had very little interaction.

12 "The Sveriges Riksbank Prize in Economic Sciences in Memory of Alfred Nobel 1992," Nobelprize.org. Nobel Media AB 2014. Web. September 1, 2015. www.nobelprize.org/nobel_prizes/economic-sciences/laureates/1992/

13 For which he received the Nobel Prize in 1974, the same year as Friedrich von Hayek.

14 For a detailed history of the project, see Cohen (2004).

15 "[Discrimination] is defined in relation to the norm of equality of opportunity in the American Creed. In this sense it is, naturally, a 'value-loaded' term, and rightly so. But it lacks nothing in scientific preciseness and definiteness" (Myrdal 1996 [1944], 215).

16 Appendix 3, "A Methodological Note on the Principle of Cumulation" (Myrdal 1944, 1065–1070).

17 He nevertheless pointed the direction of further research towards a kind of dynamic general equilibrium model of discrimination (Myrdal 1996 [1944], 1068).

18 I use the pagination of the working paper version.

19 See also Dunlop (1957), Kaufman (2010), Boyer and Smith (2001).

20 "Without a theory [the institutionalists] had nothing to pass on except a mass of descriptive material waiting for a theory, or a fire" (Coase, quoted by Boyer and Smith 2001, 201).

21 Friedman traces ethical consequences to Becker's model (Friedman 1962, 108–119), and strongly advocates against anti-discrimination legislation. The core of the Chicago view is Marshallian price theory with a "belief that free markets promote democracy" (Heckman 2011, 142). See also Emmett (2010) for various definitions of the "Chicago School" in relation to (neo)liberalism.

22 "Mr. Becker's book is actually an interesting exercise in logic which is limited in usefulness by insufficient knowledge, by failure to consult empirical evidence, and by some fairly naive assumptions" (Northrup 1958, 298).

23 "After 1985, Chicago-style labor economics became so widely accepted and imitated that the approach became largely synonymous with labor economics itself [...] Production of doctoral students accelerated at Chicago after 1940 and in the next three decades Chicago produced more PhDs in labor than any other university" (Kaufman 2010, 139).

24 "It is my belief that economic analysis is essential in understanding much of the behavior traditionally studied by sociologists, anthropologists, and other social scientists. *This is a true example of economic imperialism!* In other words, I argue that the broad definition of economics in terms of scarce means and competing ends should be taken seriously and should be a source of pride rather than embarrassment to economists since it provides insights into a wide variety of problems" (Becker 1971a, 2, my emphasis).

25 "It was ten years after the publication of my book before discrimination, or minority, economics started to become a big field in economics. In the preceding period there was little of this subject" (Becker, interviewed by Swedberg 1990, 32).

26 "This might be a case of an imperialism that is internal to the field of economics operating at the level of research programs" (Davis 2010, 9).

27 "The weakest and most innocent version of imperialism of scope is one in which the expansionist discipline simply sets out to explain phenomena in domains belonging to other disciplines while ignoring these disciplines themselves [...] There is no intrusion into other disciplinary cultures and practices. Much of perceived scientific imperialism is of this kind" (Mäki 2013, 12).

28 It is also clear that Becker makes no ontological distinctions between disciplines and their object of study (Swedberg 1990, 42).

29 "Some of the works from the period 1950–1980 that are important to the dialogue between economists and sociologists were produced by economists who are *not* identified with economic imperialism. Albert O. Hirschman, Kenneth Arrow, and Thomas Schelling belong to this category. They have all raised questions in economics that are relevant for sociology – but in a way that is much less hostile to traditional sociology than economic imperialism" (Swedberg 1990, 17).

30 For Lionel Robbins, "any kind of human behavior falls within the scope of economic generalizations [...] there are no limitations on the subject matter of economic science" (Robbins 1932, 15, quoted by Gautié 2007, 929).

31 "I think that the way I would like to see the relationship between economics and social sciences is rather different from what I take to be Becker's program [...] [It] seems to me that the attempt to explain all social interaction as economic interaction in a generalized sense of the word 'economic', only represents one side of the coin. The other side is that a lot of the environment in which economic transactions take place is social and historical in nature" (Arrow, interviewed by Swedberg 1990, 136).

32 Beyond psychology, Arrow acknowledges the influence of anthropology, especially Franz Boas' work on "race" (Arrow, interviewed by Swedberg 1990, 134).

33 For a review, see Chassonnery-Zaïgouche (2014, chapter 3).

34 This field is sometimes described as a response to economics imperialism (Swedberg 1990, 17), but it could also be analyzed as imperialism from sociology.

35 On this rhetoric aspect of formalism and quantitative analysis, see Dupré (1996) and Porter (1996).

36 Data sets are one of the results of Johnson's "War on Poverty." This set of social reforms culminates with the *Civil Rights Act* of 1964, which creates the Office of Economic Opportunity (EEOC), which gathered the *Survey of Economic Opportunity* to the Census data, later transferred at the University of Michigan (Panel Study of Income Dynamics).

37 Before the Oaxaca-Blinder method, discrimination was associated with unexplained differences in wages. The new methods measure the causal effect of human capital variables (years of schooling, diplomas, experiences, etc.) on wage, *and* the difference of reward of these productive characteristics, which are interpreted as discrimination.

38 For a survey of audit studies in economics, see Riach and Rich (2002). For a recent contribution, see Bertrand and Mullainathan (2004). For a survey of lab experiments in economics, see Anderson et al. (2006).

39 See Davis for an analysis of competing theories within psychology, which would have influenced behavioral economics (Davis 2013).

40 "Empirical imperialism sanctions the fact that, due to its comparative advantage in terms of quantitative technics and the extension of its object to all social phenomena, the economist becomes the ultimate (social) expert" (Gautié 2007, 932).

41 Or perceived and promoted as such for various reasons: because of its precise definition of discrimination (Ashenfelter and Oaxaca 1987), because of the use of counterfactual in both economics and legal theory (Posner 1999, 91), or because more and more legal theorists were trained in economic theory.

References

Anderson, Lisa R., Roland G. Fryer and Charles A. Holt. 2006. "Discrimination: Experimental Evidence from Psychology and Economics". In William M. Rodgers III (ed.). *Handbooks on the Economics of Discrimination*. Cheltenham: Edward Elgar, pp. 97–115.

Arrow, Kenneth J. 1968. *Measurement of Racial Discrimination in the Labor Market*. Santa Monica, CA: RAND Corporation.

Arrow, Kenneth J. 1971. *Some Models of Racial Discrimination in the Labor Market*. Santa Monica, CA: RAND Corporation.

Arrow, Kenneth J. 1972. "Models of Job Discrimination". In A.H. Pascal (ed.). *Racial Discrimination in Economic Life*. New York: Lexington Books, pp. 83–102.

Arrow, Kenneth J. 1973. "The Theory of Discrimination". In Albert Ashenfelter, O.C. Rees and A. Rees (eds). *Discrimination in Labor Markets*. Princeton, NJ: Princeton University Press, pp. 1–33.

Arrow, Kenneth J. 1976. "Comment I". *Signs* 1(3): 233–237.

Arrow, Kenneth J. 1998. "What has Economics to Say About Racial Discrimination?" *The Journal of Economic Perspectives* 12(2): 91–100.

Ashenfelter, Orley C. and Ronald L. Oaxaca. 1987. "The Economics of Discrimination: Economists Enter the Courtroom". *The American Economic Review* 77(2): 321–325.

Becker, Gary S. 1971a. *Economic Theory*. Chicago, IL: University of Chicago Press.

Becker, Gary S. 1971b [1957]. *The Economics of Discrimination*. Chicago, IL: University of Chicago Press.

Becker, Gary S. 1976. *The Economic Approach to Human Behavior*. Chicago, IL: University of Chicago Press.

Bertrand, Marianne and Sendhil Mullainathan. 2004. "Are Emily and Greg More Employable than Lakisha and Jamal?" *The American Economic Review* 94(4): 991–1013.

Bielby, William T. and James N. Baron. 1986. "Men and Women at Work: Sex Segregation and Statistical Discrimination". *The American Journal of Sociology* 91(4): 759–799.

Blinder, Alan S. 1973. "Wage Discrimination: Reduced Form and Structural Estimates". *Journal of Human Resources* 8(4): 436–455.

Blinder, Alan S. 1974. "The Economics of Brushing Teeth". *Journal of Political Economy* 82(4): 887–891.

Bourdieu, Pierre. 1984. "Réponses aux Economistes". *Économies et Sociétés, Série «Histoire de la pensée économique»* 18(10): 23–32.

Boyer, George R. and Robert S. Smith. 2001. "The Development of the Neoclassical Tradition in Labor Economics". *Industrial and Labor Relations Review* 54(2): 199–223.

Chassonnery-Zaïgouche, Cléo. 2014. *Expliquer, Quantifier, Prouver: Une Histoire de l'économie des Discriminations (1957–2010)*. PhD Thesis, University Paris 1 Panthéon-Sorbonne.

Cherrier, Béatrice. 2009. "Gunnar Myrdal and the Scientific Way to Social Democracy, 1914–1968". *Journal of the History of Economic Thought* 31(1): 33–55.

Cherrier, Béatrice. 2011. "The Lucky Consistency of Milton Friedman's Science and Politics, 1933–1963". In R. Van Horn, P. Mirowski and T. Stapleford (eds). *Building Chicago Economics: New Perspectives on the History of America's Most Powerful Economics Program*. Cambridge, MA: Cambridge University Press, pp. 335–367.

Clark, Kenneth B. and Mamie P. Clark. 1950. "Emotional Factors in Racial Identification and Preference in Negro Children". *The Journal of Negro Education* 19(3): 341–350.

Clarke, Steve and Adrian Walsh. 2009. "Scientific Imperialism and the Proper Relations Between the Sciences". *International Studies in the Philosophy of Science* 23(2): 195–207.

Clarke, Steve and Adrian Walsh. 2013. "Imperialism, Progress, Developmental Teleology, and Interdisciplinary Unification". *International Studies in the Philosophy of Science* 27(3): 341–351.

Cohen, Shari. 2004. *The Lasting Legacy of an American Dilemma*. New York: Carnegie Corporation Reports.

Cot, Annie L. and Jérôme Lallement. 1994. "The Historiography of Economics, A Methodological Approach". In L.S. Ross (ed.). *Joseph A. Schumpeter, Historian of Economics. Perspectives on the History of Economic Thought. Selected Papers from the History of Economics Society Conference*. London: Routledge, pp. 44–59.

Davis, John B. 2006. "The Turn in Economics: Neoclassical Dominance to Mainstream Pluralism?" *Journal of Institutional Economics* 2(1): 1–20.

Davis, John B. 2010. "Mäki on Economics Imperialism". *Marquette University Working Paper*. Later published in A. Lehtinen, J. Kuorikoski and P. Ylikoski (eds). *Economics for Real*. 2012. London: Routledge, pp. 203–219.

Davis, John B. 2013. "Economics Imperialism Under the Impact of Psychology: The Case of Behavioral Development Economics". *Œconomia* 2013(01): 119–138.

Desrosières, Alain 2003. "Managing the Economy: the State, the Market, and Statistics". In T. Porter and D. Ross (eds). *The Cambridge History of Science*, vol. 7, *Modern Social and Behavioral Sciences*. Cambridge: Cambridge University Press, pp. 553–564.

Dewey, Donald. 1958. "Book Review: *The Economics of Discrimination*, by Gary S. Becker". *Southern Economic Journal* 24(4): 494–496.

Downs, Anthony. 1957. *An Economic Theory of Democracy*. New York: Harper and Row.

Duncan, Otis D. 1961. "Occupational Components of Educational Differences in Income". *Journal of the American Statistical Association* 56(296): 783–792.

Dunlop, John T. 1957. "The Task of Contemporary Wage Theory". In J.T. Dunlop (ed.). *The Theory of Wage Determination*. New York: Macmillan, pp. 3–27.

Dupré, John. 1996. "Against Scientific Imperialism". *Proceedings of the 1994 Biennial Meetings of the Philosophy of Science Association* 2: 374–381.

Dupré, John. 2001. *Human Nature and the Limits of Science*. Oxford: Clarendon Press.

Emmett, Ross B. 2010. *The Elgar Companion to the Chicago School of Economics*. Cheltenham: Edward Elgar Publishing.

Farmer, Amy and Dek Terrell. 1996. "Discrimination, Bayesian Updating of Employer Beliefs, and Human Capital Accumulation". *Economic Inquiry* 34(2): 204–219.

Fine, Ben. 2000. "Economics Imperialism and Intellectual Progress: The Present as History of Economic Thought". *History of Economics Review* 32: 10–36.

Fleury, Jean-Baptiste. 2009. *L'Extension de la Science Économique hors de ses Frontières Traditionnelles: Le Cas Américain (1949–1992)*. PhD Thesis, University Lumière Lyon 2.

Fleury, Jean-Baptiste. 2012. "Wandering Through the Borderlands of the Social Sciences: Gary Becker's Economics of Discrimination". *History of Political Economy* 44(1): 1–40.

Fourcade, Marion, Etienne Ollion and Yann Algan. 2015. "The Superiority of Economists". *Journal of Economic Perspectives* 29(1): 89–114.

Friedman, Milton. 1962. *Capitalism and Freedom*. Chicago, IL: University of Chicago Press.

Gautié, Jérôme. 2007. "L'économie à ses frontières (sociologie, psychologie)". *Revue économique* 58(4): 927–939.

Glaeser, Edward L. 2003. "Psychology and the Market". *NBER Working paper* n°10203.

Glaeser, Edward L. and A. Shleifer. 2014. "Gary Becker (1930–2014)". *Science* 344(6189): 1233.

Granovetter, Mark. 1973. "The Strength of Weak Ties". *American Journal of Sociology* 78(6): 1360–1380.

Heckman, James J. 2003. "Some Brief Remarks on the Life and Work of Jacob Mincer". *Review of Economics of the Household* 1(4): 245–247.

Heckman, James J. 2011. "Private Notes on Gary Becker". IZA Discussion Paper.

Hirshleifer, Jack. 1985. "The Expanding Domain of Economics". *The American Economic Review* 75(6): 53–68.

Hobson, John A. 1938 [1902]. *Imperialism: A Study*. Nottingham: Spokesman Books.

Horn, Karen Ilse. 2009. *Roads to Wisdom, Conversations with Ten Nobel Laureates in Economics*. Cheltenham: Edward Elgar Publishing.

Kaufman, Bruce E. 2010. "Chicago and the Development of the Twentieth-Century Labor Economics". In R.B. Emmett (ed.). *The Elgar Companion to the Chicago School of Economics*. Cheltenham: Edward Elgar Publishing, pp. 128–151.

Kidd, Ian J. 2013. "Historical Contingency and the Impact of Scientific Imperialism". *International Studies in the Philosophy of Science* 27(3): 315–324.

Kitagawa, Evelyn M. 1955. "Components of a Difference Between Two Rates". *Journal of the American Statistical Association* 50(272): 1168–1194.

Lazear, Edward P. 1999. *Economic Imperialism*. Cambridge, MA: National Bureau of Economic Research.

Leijonhufvud, Axel. 1973. "Life Among the Econ". *Western Economic Journal* 11(2): 327–337.

Mäki, Uskali. 2009. "Economics Imperialism: Concept and Constraints". *Philosophy of the Social Sciences* 39(3): 351–380.

Mäki, Uskali. 2013. "Scientific Imperialism: Difficulties in Definition, Identification, and Assessment". *International Studies in the Philosophy of Science* 27(3): 325–339.

Mandel, Michael J. 1999. "Going for the Gold: Economists as Expert Witnesses". *The Journal of Economic Perspectives* 13(2): 113–120.

Medema, Steve G. 1997. "The Trial of Homo Economicus: What Law and Economics Tells Us about the Development of Economic Imperialism". *History of Political Economy* 29(1): 122–142.

Mirowski, Philip and Wade Hands. 1998. "A Paradox of Budgets: The Postwar Stabilization of American Neoclassical Demand Theory". *History of Political Economy* 30: 260–292.

Morgan, Mary S. and Malcolm Rutherford. 1998. "From Interwar Pluralism to Postwar Neoclassicism". *History of Political Economy* 30(Supplement).

Moro, Andrea. 2009. "Statistical Discrimination". In Steven N. Durlauf and Lawrence E. Blume (eds). *The New Palgrave Dictionary of Economics*. Online Edition. Palgrave Macmillan. February 24, 2017. www.dictionaryofeconomics.com/article?id=pde2009_S000544. doi:10.1057/9780230226203.1905.

Myrdal, Gunnar. 1939. *Monetary Equilibrium*. London: Hodge.

Myrdal, Gunnar. 1996 [1944]. *An American Dilemma: The Negro Problem and Modern Democracy*. New Brunswick, NJ: Transactions Publisher.

Neckerman, Kathryn M. and Joleen Kirschenman. 1990. "Statistical Discrimination and Inner-City Workers: An Investigation of Employers' Hiring Decisions". Paper presented at the annual meeting of the American Sociological Association.

Nik-Khah, Edward and Robert Van Horn. 2012. "Inland Empire: Economics Imperialism as an Imperative of Chicago Neoliberalism". *Journal of Economic Methodology* 19(3): 259–282.

Northrup, Herbert R. 1958. "Review of *The Economics of Discrimination*: An Economic View of Racial Discrimination by Gary S. Becker". *Industrial and Labor Relations Review* 11(2): 298.

Oaxaca, Ronald L. 1973. "Male–Female Wage Differentials in Urban Labor Markets". *International Economic Review* 14(3): 693–709.

Pager, Devah and Diana Karafin. 2009. "Bayesian Bigot? Statistical Discrimination, Stereotypes, and Employer Decision Making". *The Annals of the American Academy of Political and Social Science* 621(1): 70–93.

Pager, Devah and Hana Shepherd. 2008. "The Sociology of Discrimination: Racial Discrimination in Employment, Housing, Credit, and Consumer Markets". *Annual Review of Sociology* 34: 181–209.

Phelps, Edmund S. 1972. "The Statistical Theory of Racism and Sexism". *The American Economic Review* 62(4): 659–661.

Porter, Theodore M. 1996. *Trust in Numbers: The Pursuit of Objectivity in Science and Public Life*. Princeton, NJ: Princeton University Press.

Posner, Richard A. 1999. "The Law and Economics of the Economic Expert Witness". *Journal of Economic Perspective* 13(2): 91–99.

Prasch, Robert E. 2004. "One Hundred Years of American Economists on Race and Discrimination, 1881–1981". In D.C. Colander, R.E. Prasch and F.A. Sheth (eds). *Race, Liberalism, and Economics*. Ann Arbor: University of Michigan Press, pp. 145–181.

Radnitzky, Gerard and Peter Bernholz (eds). 1987. *Economic Imperialism: The Economic Method Applied Outside the Field of Economics*. New York: Paragon House.

Riach, Peter A. and Judith Rich. 2002. "Field Experiments of Discrimination in the Market Place". *The Economic Journal* 112(483): F480–F518.

Rothschild, Michael and Joseph E. Stiglitz. 1982. "A Model of Employment Outcomes Illustrating the Effect of the Structure of Information on the Level and Distribution of Income". *Economics Letters* 10(3): 231–236.

Rutherford, Malcolm. 2010. "Chicago Economics and Institutionalism". In R. Emmett (ed.). *The Elgar Companion to the Chicago School of Economics*. Cheltenham: Edward Elgar Publishing, pp. 25–39.

Schumpeter, Joseph A. 2006 [1954]. *History of Economic Analysis*. Taylor & Francis e-Library.

Souter, Ralph W. 1933. *Prolegomena to Relativity Economics: An Elementary Study in the Mechanics of an Expanding Economic Universe*. New York: Columbia University Press.

Stigler, George J. 1984. "Economics: The Imperial Science?" *The Scandinavian Journal of Economics* 86(3): 301–313.

Swedberg, Richard. 1990. *Economics and Sociology: Redefining their Boundaries: Conversations with Economists and Sociologists*. Princeton, NJ: Princeton University Press.

Tullock, Gordon. 1972. "Economics Imperialism". In James Buchanan and Robert Tollison (eds). *Theory of Public Choice*. Ann Arbor: Michigan University Press, pp. 317–329.

Turner, Ralph H. 1952. "Foci of Discrimination in the Employment of Nonwhites". *American Journal of Sociology* 58(3): 247–256.

Van Horn, Robert, Philip Mirowski and Thomas Stapleford. 2011. *Building Chicago Economics: New Perspectives on the History of America's Most Powerful Economics Program*. Cambridge, MA: Cambridge University Press.

Vromen, Jack J. 2009. "The Booming Economics-Made-Fun Genre". *Erasmus Journal for Philosophy and Economics* 2(1): 70–99.

White, Harrison C. 1995. "Social Networks Can Resolve Actor Paradoxes in Economics and in Psychology". *Journal of Institutional and Theoretical Economics* 151(1): 58–74.

Part III

A selection of case studies on scientific imperialism

9 Scientific subordination, molecular biology, and systems biology

Miles MacLeod

Introduction

One of the important roles that philosophical discussion about the concept of scientific imperialism might play is providing resources and analysis that help clarify and assess different types of interdisciplinary relationships, and the commitments they involve. This is particularly relevant for disciplines and fields not yet clear on how to interpret or understand the nature of their relationships with fields from other disciplines, particularly ones that might be moving into their territory. It is important for the groups involved to be clear on their options, as well as the ramifications of each option for their disciplines, and how desirable each option might be on philosophical, particularly epistemological and methodological, grounds. Such advice, when given, needs to take into account not just theoretical arguments for interaction between fields but the practicalities of research on the ground given the social, technical, and other constraints that govern scientific practice and scientific interactions. We look for practical, not just abstract, solutions.

In the case of molecular biology and systems biology, many molecular biologists are unsure what the advent of systems biology means for them, and their attitudes towards it are often uninformed and ambivalent (Calvert and Fujimura 2011). Systems biology seeks to build large-scale computational models of biological systems, particularly biochemical networks, which have been the traditional subject of molecular biology. To do so requires experimental data and experimental testing from molecular biologists. In this chapter we will see that it is possible to identify one set of goals shared by some systems biologists that necessitate some degree of "scientific subordination," a specific and perhaps novel type of *disciplinary imperialism* (see Mäki 2009, 2013). We define *scientific subordination* as the case in which a discipline or field co-opts the explanatory and empirical practices of another discipline (and its researchers) to service its own, on its view, more explanatory or predictive theories. At the same time the subordinating field seeks to take control of the research agenda of the subordinated field and direct it towards these more productive ends. A loss of independence has potential disciplinary consequences on many levels beyond just the loss of explanatory priority, affecting how research projects are designed and selected, how researchers are educated, who has policy influence and so on … (Mäki 2013). We will see that particular goals with respect to modeling in systems biology and claims about the necessity of computational and mathematical methods for understanding and predicting biological systems, when rigorously asserted, demand specific changes in experimental practices to suit modeling endeavors. We will delve into the implications of subordination for both fields and its potential risks in this case, and explore some of the criteria that can be used in this case, and cases like it, to assess whether subordination is warranted. Such analysis applied in this case suggests that subordination to any significant degree will be difficult to justify. There are good methodological and practical reasons to favor more collaborative-type interactions.

As such, this chapter will contribute most directly to two of this volume's projects: expanding our typologies of scientific imperialism; and formulating means of evaluating imperialistic moves. Some of the insights raised in this chapter will be drawn from a four-year ethnographic study of model-building practices in two systems biology labs led by Professor Nancy Nersessian. This study took the form of laboratory observation and interviews with lab participants, some over the course of their graduate research projects.

1 Studying scientific imperialism

As the Introduction to this volume indicates, we already have a number of good papers analyzing at a conceptual level kinds of imperialistic relationships between fields, and whether "imperialism," given its pejorative connotations, is appropriate at all for describing such relationships. I follow Mäki (2009, 2013), however, in defining "scientific imperialism" neutrally, using descriptive rather than normative terms. Scientific imperialism encompasses moves by one field to intrude on the domain of another field. These moves may turn out to be warranted upon careful evaluation on methodological, conceptual, or other grounds. Such moves are thus not to be taken as *prima facie* unjustifiable, but only after some methodological, epistemological, or other evaluative criteria are applied (see the Introduction to this volume).

Two potential areas of fruitfulness for this neutral approach motivate this particular chapter. First, discussions over scientific imperialism fit within the emerging agenda of philosophers of science to clarify the nature of interdisciplinary relationships in science. Philosophers can play an important role delineating the potential forms of interdisciplinary interactions particular fields can engage in or accept (imperialistic, collaborative, trading-based, or otherwise), and the kinds of epistemological commitments and methodological practices they require. Second, philosophers can play a role evaluating the desirability on philosophical grounds (see Introduction) – given what we know from modern and historical case studies of "effective scientific investigation" – of how the various options perform with respect to one another, by weighing up their risks, their costs, their inconveniences, and so on against the strength of their epistemological and methodological claims. In many interdisciplinary contexts researchers are often not aware of what is happening, what the options are or what is at stake. This is particularly the case when relationships are uncertain or potentially threatening.

The context I will discuss today is ripe for this kind of investigative analysis. In this context we have two fields or domains (namely molecular biology and systems biology) currently coming into contact. Both accept that there should be or needs to be interaction between them, but are unsure about how that should happen. At the same time there is large uncertainty about the methods and claims of the potentially imperializing field, and uncertainty over what it can really achieve. This is a context for which philosophical analysis, and the luxury of being able to provide a broad and relatively neutral perspective on what might be happening, stands to be particularly helpful to the scientific participants who are bound up with their day-to-day activities.

In this chapter I want to examine closely one possible type of emerging relationship between these two fields, which I call scientific subordination. Elaborating this notion should serve to enrich our conceptions of scientific imperialism and interdisciplinarity in general. The particular normative or evaluative issue arising from scientific subordination is the following question: on what grounds is subordination desirable, and when might preserving the independence of two fields be preferable? In this chapter I will show that there are grounds for preferring the independence of these fields and a more collaborative form of interaction between them which respects their boundaries.

2 Systems biology and molecular biology

Modern systems biology, often called computational systems biology to distinguish it from older attempts to model biological systems before the advent of widespread computing, aims to build mathematical simulation models of biological systems. Most modelers come from engineering or other quantitative fields and few at present have formal training in biology or biological experimentation. The field's participants share a range of goals, such as modeling large-scale biological systems in order to understand how to control and manipulate them to, for instance, inhibit diseases (Hood et al. 2004). Others aim to discover the design or organization principles that structure biological networks, by identifying structures through computational and mathematical analysis that are optimized for their particular functions (Alon 2006). Although biological systems are present at all levels of biological organization, most systems biologists build models of gene regulatory, metabolic or cell signaling systems which track regulatory states or biochemical concentrations of genes and metabolites over time. The majority of models produced in the field are coupled ordinary differential equation models. Systems biologists rely on biological information assembled by experimenters (molecular biologists) to build their models, and obtain experimental validation of models once they are built, as well as biological expertise to correct and improve their models. Although in theory much of the data they require could be collected through automated high-throughput technology, which can perform thousands of time-series experiments automatically, these are not yet in systematic use, and in any case would not necessarily free systems biologists from the need to rely on molecular biologists for biological expertise.

Molecular biology is an almost purely experimental field that traces its origins back to the late 1930s and the first investigations of the functional relationships between genes and proteins. The field and its close relative biochemistry[1] investigate the structure and synthesis of bio-molecular agents, and how these agents interact and combine to produce particular biological functions, as well as how these interactions are regulated. In practice, many molecular biologists work on particular interactions and try to build up an understanding of what is happening around these locally by expanding their experimental investigations. The work is highly non-quantitative. Molecular biologists rarely have any mathematical training, and do not as a rule build mathematical models.

3 Scientific subordination

Interactions between these fields can be characterized and structured in various ways. In this chapter we examine one such potential manner of structuring relations between these fields which can classed as an instance of scientific subordination. The classical case of scientific imperialism involves attempts by one field to squeeze out another from a domain of study and promote its own approaches as those that have the best chances of success, and thus should be predominantly funded and used to inform policy decisions. Historically, disciplines or fields may not have simply sought to exclude each other but rather have attempted to take over another discipline's agenda and determine where that discipline should best direct its energy. Such cases occur in situations in which there are relations of dependency between fields that cannot be escaped, thus putting the fields into potential conflict over where best to direct resources. With the advent of quantum theory and nuclear physics, theoretical physics in the 1920s and onwards began to drive the agenda of experimental nuclear physics, which had previously operated almost completely independently (MacLeod 2010).

An historical argument can be made that at least for a time this was at the cost of the independence of experimental physics. Certainly some physicists at the time saw the situation this way. For instance, in 1929 Italian physicist and colleague of Fermi, Orso Corbino, gave a speech to a meeting of the Italian Association for the Advancement of Science in Florence entitled "The New Goals of Experimental Physics." In this he stressed that:

> [t]o participate in the general movement, however, either in the present directions or in the direction of a more remote future, it is indispensable that experimentalists acquire a ready and sure grasp of the results that theoretical physics is reaching and that, at the same time, they acquire ever greater experimental means.

Corbino anticipated that experimenters would need quantum theory to guide their research, calling the attack upon the nucleus the true field for the physics of tomorrow (cited in Amaldi 1984, 171). This shifting attitude was the product of an awareness by physicists of the limitation of traditional experimentalism when it came to the nucleus, and the need for theory to inform the most productive and important potential paths of experimentation. If quantum theory applied to the nucleus then, according to Corbino, in a speech from 1929 to the Italian Association for the Advancement of Science in Florence, trying to experiment on the nucleus using traditional techniques would be "the same as trying to win a modern battle without airplanes and without artillery" (cited in Amaldi 1984, 121). In particular, with Chadwick's proof of the neutron, the dynamics of the relation between theory and experiment began to change profoundly. The neutron concept in particular brought theory into the context of investigation of the nucleus. Experimenters had no choice but to rely on it in order to know what the relevant areas of investigation were (MacLeod 2010). Philosophers today readily debate the virtues or otherwise of having independent experimental investigations in fields like physics, and indeed the extent to which experimental physics really is, or always has been, driven by theory (Steinle 1997; Hacking 1983). However, at least in this historical area one can single out a definite change in the nature of relationships between two dependent, but up until that point very independent, fields, in which one took on a more subordinate role.[2]

Cases, then, in which the independence of another field is challenged by a field operating in the same domain or one moving into its domain we are going to call instances of scientific subordination. Identifying cases of scientific subordination should help advance our knowledge about scientific imperialism and the forms it can take.[3] Mäki (2009, 2013) contrasts *domain-only imperialism* with *disciplinary imperialism*. In the case of the former there is no interaction between domains. The imperializing discipline seeks only to explain phenomena within another domain without interfering in the practices of the other discipline. This may of course in the end displace the importance of the original domain, but not directly. For the latter the expansionist discipline does intrude on the other's practices. "Disciplines are institutional structures that involve conventions and convictions, standards and strategies, ideals and identities, principles and practices, styles and statuses, rules and resources, ordinances and organizations. Disciplinary imperialism challenges the disciplinary institutions of the imperialized disciplines" (Mäki 2013, 335). Scientific subordination should be considered, then, as a type of disciplinary imperialism, since challenging the independence of another field and applying it to one's own ends is likely a significant intrusion. Of course, doing so requires subverting the institutions of the imperialized disciplines.

I propose here, then, a set of potential conditions important, if not necessary, for identifying instances of scientific subordination where they occur and have occurred.

A field **A** seeks to subordinate a field **B** if,

1 **A** and **B** have distinct epistemic standards and practices for producing explanations and predictions.
2 **A**'s epistemic practices, however, depend on the practices of **B** for producing explanations and predictions, or other epistemic goals (such as producing well-functioning designs).
3 **A** seeks to exclude the legitimacy of **B**'s epistemic standards.
4 **A** seeks to co-opt the epistemic practices of **B** to produce explanations and predictions that meet its own standards.

These conditions capture the idea that fields in potential subordinating relationships have relations of dependence, but in order for **A** to take a preeminent position, epistemological arguments (or attitudes) are needed that seek to subvert or constrain **B**'s legitimacy and help justify subordinating **B**'s practices to **A**'s goals. Subordination comes naturally in degrees, depending on the extent to which **A** wishes to dominate **B**'s time and exclude any independent work by **B**.

I do not want to claim that these conditions necessarily exhaust instances that might be called subordination but I think they do help capture the essence of what it would take to subordinate another field. They form a good starting point for exploring subordination when and wherever it might have occurred. I will show, then, that attitudes amongst some systems biologists do satisfy these criteria with respect to molecular biology, demanding various degrees of subordination with consequences for both fields. The bases on which they believe this are philosophically interesting and certainly open to critical examination. Let us consider each criterion in turn.

3.1 Distinct epistemic standards and practices

Both systems biologists and molecular biologists seek to investigate the causes of biological functions at particularly the intra-cellular level. However, their approaches are distinct. As we saw, systems biology asserts that understanding, explanation and prediction of biological systems requires computational and mathematical representations of systems. This belief is underwritten by a "systems-view" of biological organization. Systems biologists argue that the role of biological parts in creating organization is highly nonlinear, quantitatively sensitive and dependent on large networks of interactions. The qualitative causal investigations of traditional experimentation are simply insufficient to capture and understand how the coordination and interaction of those parts produce specific biological functions. For instance, given the quantitative sensitivity, nonlinearity and large scale of many systems, effective prediction, particularly at the level of individual organisms, requires sophisticated and high-powered models that can be analyzed mathematically through computationally driven techniques. Explanations require abstract techniques that can draw out the relevant elements, patterns, and processes responsible for a system-level phenomenon. Higher-level patterns and processes, themselves discovered through mathematical analysis, can play their own explanatory roles in grouping and understanding systems. Further, only through the aid of mathematical models and the opportunities for mathematical analysis they bring is it possible to find underlying principles of design or organization in biological networks. To discover a design principle one must show, for instance, that a particular network design is optimal with respect to some function compared to all other possible designs (see Alon 2006). Computation is required to run through and compare these possibilities. On the systems-view, explanations and predictions of biological functions thus depend on computational models that can capture the network dynamics and underlying design principles that give rise to these functions.

The practices and explanatory standards of systems biologists can certainly be distinguished from those of molecular biologists. Molecular biology is a mostly qualitative field. Its approach to investigating and understanding biological function centers around two roles: an exploratory role; and an explanatory role. For instance, in the first case experimenters use their specific knowledge of sets of interactions to hypothesize and infer other potential influences on molecular activity, and then check these through experimental manipulation. Researchers identify molecular agents relevant for a particular function, and then build up knowledge of how those agents operate to achieve their part through controlled testing. The specific nature of the biochemical interactions agents engage in are relevant to understanding how these functional roles happen, when they can be predicted, when they break down and when additional elements might be needed to account for them. This role amounts in effect to building out pathways of interactions and exploring their variability across different samples and under different conditions for clues about what else might be going on. Second, molecular biology work plays an explanatory role. By assembling this information molecular biologists give causal-mechanistic explanations of molecular function.

Molecular biologists and systems biologists do differ specifically, then, over the kinds of information they think important for establishing and understanding the mechanistic causes of biological function. These pull along different directions. Molecular biologists, who are very much schooled in the problems of biological variability and experimental sensitivity to background conditions, think much detail is required about biochemical interactions and mechanisms in order to understand precisely how particular molecular agents participate. Molecular biologists are often considered to be, in philosophical terms, reductionist or at least highly detail driven (Fagan 2016). Systems biologists, on the other hand, tend to think that mathematical abstraction techniques can filter out exigencies to reveal robust network behaviors. They proceed when building models by simplifying interactions and leaving out elements that they judge not to affect the dynamics of a network to a large extent. In this way they hope to put themselves in a position to find robust relationships without being overwhelmed by the complexity of their systems. While molecular biologists tend to evaluate results by assessing the quality of the data underlying them, systems biologists value predictive testing of their models as the best gauge. This can generate a significant point of contention, as molecular biologists are not necessarily convinced that models are really anything more than data-fitting devices, thus highly sensitive in their reliability to the quality and variability of the data. Systems biologists, on the other hand, trust abstractive methods to wash out extraneous detail and noise (MacLeod and Nersessian 2014). It can thus be very difficult for both groups to find a common basis of agreement on the important question of how much detail matters to assembling accounts of biological systems. Without this it is hard to build common trust in modeling results.

3.2 Dependence

At the same time systems biology is dependent on molecular biology for biological expertise relevant to assembling biological networks for modeling, for parameter data for these models, and for experimental validation of models. As mentioned, most systems biologists are trained engineers, computer scientists, or mathematicians, and usually do not have experimental skills. Indeed, most systems biological research takes place at this time in a computer lab, not a wet lab. Hence there is no particular way that systems biologists could produce the explanations and predictions they aim for without having access to experimental results. One should not underestimate just how much experimental work is required to build an effective model. Most modelers complain that they do not get nearly as much cooperation from biologists, and particularly not enough data, as they require (see MacLeod and Nersessian 2014). This leaves them having to fit many parameters, generating often considerable uncertainty about their model results. It also leaves them having to make their own inferences ("guesses") about system structure without expert guidance. The intensity of interactions between molecular biologists and systems biologists required to meet modeling requirements was put succinctly by one lab principal investigator in systems biology: "I've said it before and I'll say it again: we need 10 experimenters for every modeler." Modeling in the ideal requires an enormous redirection of time and work by molecular biologists.

3.3 Exclusion

A systems-view is usually considered an anti-reductionist view of explanation, and most systems biologists are committed to the idea that a mathematical model of a system gives much better explanations and predictions of system function than the more qualitative causal explanations molecular biologists might give. A systems-view tends to undercut the legitimacy of attempts by molecular biology to understand the regulative and other roles that molecular agents play in biological systems through experimental methods alone. For instance, as many systems biologists argue, a fundamental flaw in molecular biology is the tradition of *in vitro* experimentation. In vitro experimentation tries to isolate the properties of molecular agents in test-tube environments in which other potential interactions from the normal cellular environments have been filtered out. On a systems-view the properties of molecular agents and the parameters that govern their interactions are partially determined by these environments and cannot be well calculated or understood through in vitro methods. Further, simply churning out more details about individual interactions will not get us closer to solving basic problems in medicine like cancer without techniques that connect those details together and filter out the relevant from the irrelevant ones. This cannot be done independently of a mathematical and computational approach. As such, simply carrying on traditional practices of exploration and explanation in molecular biology alone will not lead us closer to resolving pressing problems like these.

Such attitudes do not necessarily exclude the value of traditional practices in molecular biology, although they do imply that these are strongly insufficient. Hence molecular biology needs to turn more of its attention to modeling efforts and spending their time providing the resources and expertise modelers need. However, more substantive arguments are sometimes made by systems biologists, which deny the efficacy of molecular biology for building even small-scale accounts of biological organization and function.

Westerhoff and Kell (2007) make seven such arguments. Lack of quantization, for instance, means that molecular biology cannot predict or explain emergent features of biological networks through sequences of qualitative interactions. Without any ability to account for systems-level phenomena molecular biology cannot be integrated "upwards" with organismal biology and physiology, but neither can it be integrated with physics and chemistry in any meaningful way since interactions are too nonlinear and context dependent to be put in physical and chemical terms, without at least much more powerful tools of analysis. More essentially molecular biology has never been in a good position to develop testable hypotheses, since without a knowledge of all the elements and dependencies that contribute to the stability of one interaction, it has been impossible to erect truly controlled experiments that could pick out the role that a particular molecular element might be playing. "The biology of entire living systems was observed to be too complex and ill-defined for the hypotheses to be strict, testable and falsifiable. To some extent this made molecular biology and biochemistry appear to remain stamp collecting" (Westerhoff and Kell 2007, 37). Certainly traditional experimental practices are inadequate on this account. Some of these issues are being resolved with the advent of functional genomics and other -omics, however, which aim, using computational large-scale experimental techniques, to chronicle all the elements that participate in a given function. Nonetheless, molecular biology even with the aid of these computational methods still lacks the analytical tools to calculate and comprehend the nonlinear nature of these interactions, or develop useful insights from them. Informative experimentation and interpretation of experiment is, according to this view, largely blind on its own independent of a systems model. These various positions serve to discount the possibility that molecular biology can really build accurate accounts of biological function and organization at all, in any substantial way on its own, independently of systems models.

3.4 Co-option

Exclusionary arguments in turn help justify arguments that molecular biologists should divert more of their attention and energy to participating in model-building projects. It is quite evident that researchers in systems biology do believe that experimenters should be more devoted to assisting modeling tasks. The practicalities and data requirements of modeling more or less necessitate it. Arguments like those above provide systems biologists with ammunition for taking the position that computational and mathematical approaches are much more informative and often essential for the investigation and understanding of biological systems. Models provide not only better explanations and predictions than are possible with traditional methods, but also the important questions that need to be experimentally answered in order to produce these. Practices in molecular biology should thus be reorganized around model building. When strongly put, the implication is that model building should set the agenda of work in molecular biology, replacing both the independent explanatory and exploratory roles molecular biology currently has. In our ethnographic studies researchers sometimes imply that they see the function of experimenters to produce data relevant for model building, more or less on request or as they require it, as though it were something to be "ordered," to use the language of one of the modelers we interviewed. The tension between what modelers expect from experimenters, given their own preconceptions about the necessity of systems biology, and the effort experimenters are actually willing to provide, leads to much hostile positioning between members in the two fields (MacLeod and Nersessian 2014).

However, producing the experimental work required by systems biologists is often a major inconvenience for molecular biologists, taking time away from their usual patterns of research. The data being asked for are often not data that molecular biologists would produce for their work, especially insofar as they are data specific to estimating the dynamic behavior network elements. For instance, the kinds of kinetic parameters that systems biologists need for modeling metabolic systems are often not priorities for molecular biologists. Such data do not necessarily advance the disciplinary goals of molecular biologists insofar as they are not necessary for identifying and establishing functional relationships and the chemical nature of their operations. Thus, often modelers are forced to try to estimate these themselves based on available data. Westerhoff and Kell, pointing to an earlier time in the field, promote the need to transform experimentation on this basis to suit model-building:

> it is seen that there has therefore been a great dearth of systematic measurements of the properties that we actually wish to measure, viz. the binding and kinetic constants of individual proteins (and other molecules). Such measurements were commonplace in the 1960s and early 1970s … and we need these times to return to biology, with concomitant modernization of the way in which and the scale at which experiments are done.
> (Westerhoff and Kell 2007, 59)

Westerhoff and Kell take an even stronger or more systematic philosophical stance, by perceiving systems biology as a way to build in biology, finally, a traditional methodologically rigorous hypothetico-deductive model of scientific practice. The function of systems biology on this view is to construct large classes of robust models that will facilitate not just the understanding and control of biological systems but also allow the formulation of a general mathematical theory of biological systems based on mathematical laws of structure and organization that tie together biological, biochemical, and physical principles. According to Westerhoff and Kell:

> [s]ystems biology should discover new scientific laws, which may relate as much to physical-chemical, organizational and fitness aspects as to biochemical principles. With respect to this aim, mathematics should not take the form of modeling but rather constitute a way of codifying proposed or verified laws or principles.
>
> (Westerhoff and Kell 2007, 48)

. Kitano (2002) presents a similar viewpoint of the goals of systems biology. In his terms, systems biology will discover constraints on the structure and operation of biological systems. He takes BCS theory of superconductivity from physics as an analogy for what such laws will look like. Such laws or theories will not be fundamental, they will be "structural." They will define the specific constraints that allow specific phenomena to manifest themselves given background fundamental laws and theories. Such structural theories and laws will be developed principally from accurate computational simulation models, and mathematical analysis of these models. They will facilitate basic understanding of network structure and organization, but also the ability to synthetically construct networks which function more efficiently than actual systems or which can perform novel functions.

This strong theoretical view implies a high level of integration of experimentation with systems biology. Systems biology takes on the character, like physics, of a theory-driven hypothetico-deductive science. From mathematical theoretical frameworks, hypotheses can be derived and tested. As Westerhoff and Kell put it, functional genomics might well have collected enormous amounts of data, but "without the systems biology dimension" they would "remain in a cycle of data collection, pattern recognition and the generation of ad hoc empirical laws and hypotheses describing those data phenomenologically." As such, functional genomics is in need of "a principled hypothesis-generating arc that leads from data to knowledge" (Westerhoff and Kell 2007, 52). The implication is that molecular biology and modern genomics have both lacked a substantial dimension of what constitutes canonical scientific practice. Systems biology, through its integration of experiment with a theoretical framework, achieves "an interactive and spiraling interplay between deductive and inductive reasoning …" (Westerhoff and Kell 2007, 54).

In a theory-led system like this the exploratory and the detail-driven causal-mechanistic explanatory roles of molecular biology become less important (see also Fagan 2016). Molecular biology may take on principally a confirmatory role, testing the hypotheses systems biologists produce and correcting models. Although this ideal picture of the role mathematical and computational modeling can play in biology is likely far off, it has a strong resonance with the model of scientific organization we often have in mind when thinking about physics, and perhaps other fields like economics. In the case of systems biology the notion of the importance of theory-driven science embodied powerfully by physics gives philosophical force to the claim that experimental practices should follow theoretical ones.

This kind of very strong position is far from shared by all systems biologists, who possess a diversity of attitudes towards the aims and goals of systems biology (see O'Malley and Dupré 2005). However, there is nonetheless a widely shared belief that the agenda of molecular biology should be diverted better towards the goals and requirements of modeling based on the beliefs that systems biology is a more productive approach. Systems biologists thus see the need to co-opt experimentation to varying degrees for their purposes.

4 Evaluating subordination

As mentioned earlier, I follow Mäki's neutral descriptive approach when defining moves in science as "imperialistic." Imperialism can be warranted in scientific contexts if there are good epistemological or methodological reasons for it that override the difficulties and obvious tensions such situations create. Although originally some experimenters, such as the Curies, had been dismissive of theoretical physics, the quality and strength of the theoretical discoveries, and the complexity they revealed, provided compelling arguments against experimentation proceeding alone in the study of the atomic nucleus. By the late 1930s both groups accepted the need to subordinate experimentation for at least this domain of study. The force of the case is not so strong or evident when it comes to systems biology, however, as we will see. Philosophers can play constructive roles, given what we know in particular about what good and effective scientific practice looks like, informing practitioners involved in these relationships about the potential costs, risks, and benefits of subordination, as opposed to other types of interdisciplinary or non-interdisciplinary alternatives.

That said, what kind of philosophically informed evaluative criteria should play a role assessing subordination in particular? I do not pretend to have an exhaustive answer to this question, but I raise a few factors here that are at least relevant to the systems biology context. First, and most obviously, subordination should be assessed according to how strongly the exclusion claims about best practices for investigation in a given domain hold up to philosophical analysis and what empirical evidence we have that these claims are borne out in research results. These claims should be weighed against the value that independence might have as a general virtue in scientific investigation but also the benefits it might have in the context at hand. In general, leaving certain channels of independent investigation open can well be a virtue if it is unclear just how strong the epistemological claims made by a subordinating field are.

Claims about the best ways to solve problems or advance knowledge in a given domain have to be discounted by the fact that we are always relying on the incomplete state of knowledge at any time. Scientific prognostication is a fraught business and restricting a domain to a limited set of problem-solving strategies (for instance, restricting maverick ones (see Kitcher 1990; Weisberg and Muldoon 2009)) is not a good risk-management strategy. Second, one must factor in the disruption that subordination would cause to the subordinated field by essentially requiring it to refit itself for its new roles, resulting in structural and educational costs, and loss of expertise, that would result from changing its practices to those most wanted by the subordinating field. The stakes in such cases are always likely to be large. Research funding and careers are very much in play, as well as the authority and status a field has and its ability to play a role in policy settings. Finally, given these issues and assuming we accept that the subordinating field has good epistemic claims, one must weigh up the presumed advantages of subordination against what could be achieved through other possible modes of interdisciplinary interaction that would preserve the independence of the fields at much lower costs.

With these in mind we can turn to consider the case at hand, albeit briefly, to motivate how an analysis in this case might look. I think there are plausible grounds to suggest that any moves to subordinate molecular biology are not yet compelling enough, and that a more collaborative model of interaction that respects the independence of molecular biology is the better option. Even if the claims about systemic dependency and quantitative sensitivity are true for predicting and controlling biological networks, this cannot be done without independent detailed experimental investigation of system structure. Molecular biologists produce rigorous, reliable data sets and painstakingly constructed accounts of biological networks, interactions, and their variabilities through processes that cannot be driven by the current mathematical models that exist. Most current models, for instance, abstract out information about variability in order to try to identify robust relationships. Hence the extent of variability cannot be known independently of explorations in molecular biology. One can be skeptical that mathematical reasoning can produce robust high-validity models given biovariability and parameter-sensitivity problems. As such, mathematical models may well not be effective enough to replace their inferential processes and tacit knowledge that come with the expertise molecular biologists build up over careers examining specific interactions.

There are, in fact, genuine worries to be had about the reliability and validity of current models, which cast doubt on the use of models to effectively structure investigation towards explanatory and predictive goals. The most significant parameter-related problem is that of parameter compensation, which leads to model underdetermination (MacLeod 2016). The more complex a model, the more likely it is that parameters or sets of parameters can compensate one another. This facilitates multiple solutions that fit the data. Such models behave the same with respect to the available data but may vary their behavior otherwise, making it difficult to find models of a system that can be relied upon to give correct predictions. The more uncertain the data are, the more of a problem this becomes. The computational problem of fitting unknown parameters and the cognitive challenges of trying to debug models of very complex large-scale systems restricts the scale at which models can be built. In the words of Voit et al.:

> [i]f one would survey all computational systems models in biology, published during the past decade, one would find that the vast majority are neither small enough to permit elegant mathematical analyses of organizing principles nor large enough to approach the reality of cell or disease processes with high fidelity.
>
> (Voit et al. 2012, 23)

Finally the attempt to build models runs up against the persistent problem of biovariability. Biovariability limits the representativeness of a model, and the conclusions or inferences that can be drawn from it with respect particularly to individuals. Parameters formulated with respect to one cell-line or cells within one individual organism may not function well for others, whereas parameter sets derived using parameters estimated from data from different organisms or cell-lines (using average or median values, for instance) may not represent any of these populations well (Marder and Taylor 2011). Unless the range of variability is well understood, it can be difficult to generalize a successful model as representative of the population at large.

These problems put limits on the validity and scale of the models produced currently in systems biology and create uncertainty about just how effective systems biology will be, certainly in the near future. Some of these problems are due to a lack of good data and could be improved if molecular biologists assisted the model-building process better, but problems like biovariability and complexity are not strictly problems of a lack of data, but problems intrinsic to modeling biological systems.

The limits on current practices and concerns about models like these do not lend confidence that computational and mathematically led approaches are in practice always more superior and more effective than traditional ones for directing investigation and building up information about biological systems. However, there is an even deeper worry to be had here about subordination. Independent information from molecular biology is likely necessary for making any objective assessments about the effectiveness of modeling and of models. Without molecular biology playing an independent exploratory role at least, we would have no strong basis for assessing claims about the effectiveness of models. Experimenters often claim to have a much more acute sense of the likely fragility of model results due to the quality and limitations of the experimental information underlying it. While models might well be able to facilitate inferences and explanations impossible for molecular biologists to make about systems and system structure, molecular biology can provide, through its roles, a context of results and information that help understand the constraints on models and understand their overall ability to represent systems. Simply constraining experiments to produce information relevant to the models alone cuts down the opportunity to produce this contextual information independently of models about bio-variability, for instance, that ultimately help evaluate the effectiveness of model-building processes and model performance. This is where a good claim can be made for the continued independence of molecular biological investigation.

Finally, there exists a serious non-disruptive alternative to subordination. This option is a collaborative option. A collaborative option as I take it here is one in which different fields coordinate their practices in ways that serve the disciplinary goals of both fields, without either field taking a stance that would seek to exclude the goals of the other. In practice this kind of attitude does exist. Modelers and experimenters work to coordinate their practices and expertise to produce results that serve the goals of both fields. Indeed, just as it is possible to frame experimentation as an extension of the modeling process, it is possible to reframe many modeling practices as an extension of experimental biology. Models on this view assist experimenters by raising and assessing hypotheses about the structure of the networks they are studying, and the accuracy and consistency of the data they are using, which is difficult or impossible for experimenters to recognize alone (MacLeod 2016). In fact, modelers themselves often give this interpretation in published papers of what the models they have produced achieve, de-emphasizing theory-building, prediction, or control objectives. These practices employ computation to survey different structural and parameter possibilities and draw inferences about which might be plausible and which not. Elsewhere I have called these heuristic uses (MacLeod 2016). Their function can be construed as contributions to experimental investigation. Indeed, when experimenters are asked about what they would want from models, it is more or less guidance on where and how to direct their experiments.

> So it would have to be a very broad question that we couldn't answer with sort of a shotgun approach that would help direct us to finding you know … in other words narrow down the experiments we had to do. Then I would be very interested in testing it afterwards. It's not a good time investment for me to test something we have already shown or to test a system. Making a small perturbation that is not going to advance our knowledge of how this enzyme system works. Just understanding one more agonist well may or may not help us move it forward big enough. It could predict that under certain conditions you change the binding partners which is pretty difficult to study in cells. And we knew that we wanted to test one particular binding partner under specific set of conditions it would be *very* useful.
>
> (experimenter)

The goals of both fields are respected in collaborative enterprises that exploit these functions of models. Building these kinds of models can enrich experimental exploration but also promote the development of better, more sophisticated models which might support the objectives of systems biology later down the line. Some systems biologists are invested quite deeply in the treatment of models as extensions of experimental investigation, by running combined wet and computational labs, and training their modelers to do their own experiments (MacLeod and Nersessian 2014). The goal of one such lab we specifically investigated was to contribute to molecular biology literature directly, and in turn show the benefits of models to investigation in molecular biology. In none of these cases is any attempt made to direct the roles and aims of either field. Rather, the relationship is one of mutual gain.

If one, then, is to weigh this information together in order to evaluate subordination as an option, there are good reasons to caution against the idea that systems biologists should assume the role of directing molecular biology. There are other interdisciplinary options that might not get systems biologists to their goals as quickly as they might have, but do so with much less disruption and potential conflict, and much more caution about what models can achieve and what might still be there for experimenters to discover.

5 Conclusion

In this chapter I have tried to interpret certain attitudes and statements in systems biology as expressing or implying a form of scientific imperialism that would reduce another field, in this case molecular biology, to a less or non-independent role. The concept of subordination should help develop the notion of scientific imperialism, and the different forms it may take. I have shown how one might assess or evaluate such an attitude and weigh-up such moves against other interdisciplinary options. Such analysis, I believe, can help practitioners in the fields involved since communication channels across disciplinary boundaries are often very inefficient, and researchers may not be aware of the motivations or epistemological commitment of the other side, or the diversity amongst the views of researchers on both sides. This is certainly the case in interactions between molecular and systems biology. Molecular biologists are not generally aware of what systems biology might mean for them and how computational and mathematical methods might shape research practices and their roles in research in the coming years. Nor are they aware that different systems biologists have different views and attitudes about the role of systems biologists, some of which manifest themselves in different interdisciplinary practices, some of which are orientated towards productive collaborative gain or indeed towards operating as an arm of molecular biology. What analysis such as that above can help do is provide the tools for a normative assessment deriving both from philosophical argument and scientific practice, which researchers can use for their own considerations of how to combine and integrate different fields in the most productive ways. This seems a very useful role for philosophers of science whenever imperialism surfaces.

Having said that, I do not pretend to have given a complete analysis here of the case of systems biology – only a kind of sketch of what would be useful for such an analysis. There is much more to be said and understood about the potential relationships between molecular and systems biology. There are other issues to be wary of here too. In practical terms molecular biology is far bigger and far more established than systems biology and thus it is hard to imagine that systems biology could really ever have an authoritative position to dictate the roles molecular biology should have in research. On the other hand, the compelling arguments systems biologists have used for computational and mathematical methods are changing the research structure of science. Funding bodies increasingly require biomedical projects to include modeling components in order for research to get funding. With these power relations in mind, perhaps the more likely scenario is one in which in fact modelers become principal resources for experimenters, while it is experimenters who run the show and determine when and where systems biology is applied. Indeed, the notion of subordination I outlined above in any interaction between fields may run both ways, and in fact it seems very plausible, given the funding pressures to produce models, that experimenters themselves might seek to co-opt modeling to fit their practices. This would raise another independence question but from the other direction.

The notion of subordination could thus be construed in some sense as an attempt to capture the attitudes that develop whenever two disciplines or fields begin to take on relations of dependency, for which it is natural to seek to be in the controlling position. As such, collaborative options lie somewhere in the middle between the two poles of subordination. In the context of practice, arguably a collaborative option has good natural advantages, the extent to which in fact allows fields to explore what they can achieve through interdisciplinary interactions without disruption and thus in the long run, guided by empirical results, explore what would be the most efficient or effective ways of organizing the study of their particular domain.

Notes

1 I will not make a big distinction between molecular biology and biochemistry in this chapter, since they overlap substantially. Most of the arguments used by systems biologists apply to both, whichever way their relationship is construed.

2 However, another instance of this may be emerging in relations between engineers and business scientists in certain contexts. Current results of an Academy of Finland-funded investigation of interdisciplinary research collaborations at a technological university in Finland, many of which have been somewhat compelled by new internal funding requirements, suggest that business scientists are often forced into the role of data collectors for engineers. The engineers themselves show no real interest in any theoretical agenda the business scientists might have, and instead direct the business scientists to collect the psychological or other data they want to optimize their designs. (Project: "Interdisciplining the University: Prospects for Sustainable Knowledge Production," Academy of Finland Research Project, 2015–2019.)

3 My case below is also another case of theory and experiment, which has some resonance with historical cases like this in physics. Nonetheless, I do not want to imagine all such cases are in some way replays of this old dynamic, so I keep my definition of subordination open to these.

References

Alon, Uri. 2006. *An Introduction to Systems Biology: Design Principles of Biological Circuits*. Boca Raton, FL: Chapman & Hall/CRC Press.

Amaldi, Edoardo. 1984. "From the Discovery of the Neutron to the Discovery of Nuclear Fission". *Physics Report* 111(1–4): 1–331.

Calvert, Jane and Joan H. Fujimura. 2011. "Calculating Life? Duelling Discourses in Interdisciplinary Systems Biology". *Studies in History and Philosophy of Science Part C: Studies in History and Philosophy of Biological and Biomedical Sciences* 42(2): 155–163.

Fagan, Melinda B. 2016. "Stem Cells and Systems Models: Clashing Views of Explanation". *Synthese* 193: 873–907.

Hacking, Ian. 1983. *Representing and Intervening.* Cambridge: Cambridge University Press.

Hood, Leroy, James R. Heath, Michael E. Phelps and Biaoyang Lin. 2004. "Systems Biology and New Technologies Enable Predictive and Preventative Medicine". *Science* 306: 640–643.

Kitano, Hiraoki. 2002. "Looking Beyond the Details: A Rise in System-Oriented Approaches in Genetics and Molecular Biology". *Current Genetics* 41(1): 1–10.

Kitcher, Philip. 1990. "The Division of Cognitive Labor". *Journal of Philosophy* 87: 5–22.

MacLeod, Miles. 2010. *What is there to be Realist About?* PhD diss., University of Vienna. http://othes.univie.ac.at/8583/

MacLeod, Miles. 2016. "Heuristic Approaches to Models and Modeling in Systems Biology". *Biology and Philosophy* 31(3): 353–372.

MacLeod, Miles and Nancy J. Nersessian. 2014a. "Strategies for Coordinating Experimentation and Modeling in Integrative Systems Biology". *J. Exp. Zool. (Mol. Dev. Evol.)* 9999: 1–10.

MacLeod, Miles and Nancy J. Nersessian. 2014b. "Modeling Systems-Level Dynamics: Understanding without Mechanistic Explanation in Integrative Systems Biology". *Studies in History and Philosophy of Science Part C – Biological and Biomedical Science* 49: 1–11.

Mäki, Uskali. 2009. "Economics Imperialism Concept and Constraints". *Philosophy of the Social Sciences* 39(3): 351–380.

Mäki, Uskali. 2013. "Scientific Imperialism: Difficulties in Definition, Identification, and Assessment". *International Studies in the Philosophy of Science* 27(3): 325–339.

Marder, Eve and Adam L. Taylor. 2011. "Multiple Models to Capture the Variability in Biological Neurons and Networks". *Nature Neuroscience* 14(2): 133–138.

O'Malley, Maureen A. and John Dupré. 2005. "Fundamental Issues in Systems Biology". *BioEssays* 27(12): 1270–1276.

Steinle, Friederich. 1997. "Entering New Fields: Exploratory Uses of Experimentation". *Philosophy of Science* 64: S65–S74.

Voit, E.O., Zhen Qi and Shinichi Kikuchi. 2012. "Mesoscopic Models of Neurotransmission as Intermediates between Disease Simulators and Tools for Discovering Design Principles". *Pharmacopsychiatry* 45(1): 22.

Weisberg, Michael and Ryan Muldoon. 2009. "Epistemic Landscapes and the Division of Cognitive Labor". *Philosophy of Science* 76(2): 225–252.

Westerhoff, Hans V. and Douglas B. Kell. 2007. "The Methodologies of Systems Biology". In Fred C. Boogerd, Frank J. Bruggeman, Jan-Hendrik S. Hofmeyr and Hans V. Westerhoff (eds). *Systems Biology: Philosophical Foundations.* Amsterdam: Elsevier, pp. 23–70.

10 Against neuroscience imperialism

Roberto Fumagalli

Introduction

Over the last few decades, many authors have debated about the justifiability of so-called scientific imperialism (henceforth, SI), the systematic application of a discipline's methods and findings to model and explain phenomena investigated by other disciplines (see e.g. Cartwright 1999; and Dupré 1995, 2001). As part of this debate, increasingly sophisticated conceptualizations of SI have been provided and extended discussions have taken place about the merits of specific instances of SI (see e.g. Clarke and Walsh 2009, 2013; and Mäki 2009, 2013; see also various chapters in this volume). Consider, for example, economics imperialism. On the one hand, some maintained that economists' "analytic categories [...] are truly universal in applicability" (Hirshleifer 1985, 53), and attempted to model a wide range of phenomena using economists' constrained optimization techniques (see e.g. Becker 1981, on family relations and racial discrimination). On the other hand, others complained that economists "invade the traditional territory" of other disciplines (see e.g. Fine 1999, on sociology, and Lewin 1996, on psychology) and export "an entire methodology that is in many cases almost entirely inappropriate" (Dupré 2001, 128).[1]

In recent years, several authors advocated what we may call "neuroscience imperialism" (henceforth, NI), an instance of SI whereby neuroscience methods and findings are systematically applied to model and explain phenomena investigated by other disciplines. Calls in favour of NI target a vast range of disciplines, including psychology (see e.g. Bickle 2003; and Churchland 1986), economics (see e.g. Camerer et al. 2005; and Glimcher 2010), and philosophy (see e.g. Greene 2007; and Greene and Cohen 2004). Some go as far as to maintain that "any endeavor that depends on being able to understand, assess, predict, control, or improve human behavior is, in principle, a potential application area for neuroscience" (Farah 2012, 573; see also Churchland 2007). There are at least two reasons why NI deserves detailed philosophical scrutiny. First, various instances of NI exemplify the features of SI in especially clear terms, and thus make NI an interesting test case for assessing the merits of SI contributions. And second, NI contributions have potentially widespread implications for modelling and theorizing across the vast range of natural and social disciplines they target. To date, however, NI has not received detailed attention by philosophers, and the debate concerning its identification and normative assessment is relatively underdeveloped. In this chapter, I aim to remedy this situation by making some conceptual distinctions about NI and by providing a normative assessment of prominent calls for NI.[2]

The contents are organized as follows. In section one, I identify and appraise several criteria that have been proposed to ground a normative critique of SI. I shall argue that such criteria provide a valuable basis for assessing various instances of SI, but do not enable us to explicate the reasons (if any) why recent calls for NI are objectionable or unjustified. In sections two and three, I articulate and defend two interrelated claims that, I argue, undermine prominent calls for NI. First, the proponents of NI significantly overstate the evidential and explanatory relevance of neuroscience methods and findings for the disciplines they target. And second, prominent calls for NI point to an untenable reductionist position, which rests on unsupported empirical and normative presuppositions. These two claims do not imply that all attempts to apply neuroscience methods and findings to model and explain phenomena investigated by other disciplines are objectionable. Still, they challenge the proponents of NI to qualify their imperialistic calls and ground such calls on more plausible empirical and normative presuppositions.

Before proceeding, three preliminary remarks about the focus of this chapter are worth making. First, debates about SI typically presuppose that we can reliably demarcate disciplinary boundaries. However, speaking of SI does not commit one to hold that disciplines possess precise and unalterable boundaries (see e.g. Dupré 1983). In particular, one may meaningfully speak of SI, and yet allow that the proposed demarcations of disciplinary boundaries depend on several factors, ranging from researchers' pragmatic interests to specific technological developments (e.g. think of how progress in brain-imaging technology expanded the set of phenomena amenable to neuroscientific enquiry). This variability can make it difficult to demarcate the domain of specific disciplines and establish whether the examined interdisciplinary contributions qualify as instances of SI (see e.g. Downes, this volume). However, as I illustrate in sections two and three, it does not prevent us from identifying clear paradigmatic instances of SI.

Second, I shall speak of SI broadly to encompass a wide range of systematic interdisciplinary applications of methods and findings that directly intrude in the modelling and explanatory practices of the targeted disciplines (see e.g. Mäki 2013, on what he calls disciplinary imperialism). In doing so, I use the term "methods" to include both theoretical (e.g. classificatory) tools and pragmatic (e.g. problem-solving) procedures, with the expression "modelling and explanatory practices" covering different types of models (e.g. causal and mathematical models) and distinct accounts of explanation (e.g. unificationist and mechanistic accounts).

Finally, I shall focus on two sets of calls for NI, which respectively target the economic modelling of choice (see e.g. Camerer 2008; and Glimcher 2010) and entrenched philosophical conceptions of free agency (see e.g. Haggard 2008; and Wegner 2002). I focus on these (rather than other) calls for NI because such calls have recently gained widespread support among scientific practitioners and raise pressing pragmatic and normative concerns (see e.g. Morse 2005; and Vincent 2013, on the implications of neuroscience findings for widely held conceptions of moral and legal responsibility). However, I shall comment in various places on how my critical evaluation bears on other calls for NI.

1 Assessing scientific imperialism

In this section, I identify and appraise four criteria that have been proposed to ground a normative critique of SI. I consider in turn: (i) the objection from the disunity of science (see e.g. Cartwright 1999; and Dupré 1983); (ii) the objection from disciplinary autonomy (see e.g. Aizawa and Gillett 2011; and Fodor 1974); (iii) the objection from counterfactual scientific progress (see e.g. Clarke and Walsh 2009, 2013); and (iv) the objection from cumulative constraints (see e.g. Mäki 2009, 2013). I shall argue that these objections provide a valuable basis for assessing various instances of SI, but do not enable us to explicate the reasons (if any) why recent calls for NI are objectionable or unjustified.[3]

(i) The *objection from the disunity of science* (see e.g. Cartwright 1999; and Dupré 1983) can be explicated as follows. Modelling and explanatory unifications are frequently regarded as significant achievements both in general philosophy of science (see e.g. Friedman 1974; and Kitcher 1981) and in specific disciplines (see e.g. Bechtel and Hamilton 2007, on biology; and Ferejohn and Satz 1995, on economics). Both proponents and critics of SI praise attempts to increase the variety of phenomena that can be modelled and explained within specific disciplines (see e.g. Rolin, this volume). However, SI contributions often target disciplines that sharply differ from the imperializing ones in terms of methods, evidential standards and categorizations (see e.g. Haggard 2008; and Wegner 2002, on NI contributions targeting philosophical conceptions of free agency). These interdisciplinary differences, in turn, severely constrain the modelling and explanatory relevance of SI contributions for the targeted disciplines. As Dupré puts it, there is "no common essence to the diverse [...] projects generally collected under the heading of science. The contrary assumption is not only mistaken, but can be seen to justify bad, and even sometimes harmful, science" (2001, 71).

There are at least two reasons to doubt that this objection grounds a convincing normative critique of NI. First, various NI contributions target disciplines whose practitioners rely on methods, evidential standards and categorizations that resemble those entrenched in neuroscience (see e.g. Piccinini and Craver 2011, on cognitive neuroscientists' attempts to systematically integrate cognitive psychologists' functional models of specific cognitive capacities with neuroscience mechanistic models of the same capacities). And second, NI contributions may yield major modelling and explanatory benefits to neuroscience and/or the targeted disciplines even if the targeted disciplines adopt methods, evidential standards, and categorizations that sharply differ from those entrenched in neuroscience. Indeed, several calls for NI are premised on the assumption that neuroscience methods, evidential standards and categorizations differ from – and are superior to – those adopted by the targeted disciplines (see e.g. section two on calls for NI targeting the economic modelling of choice). In this respect, emphasizing that some NI contributions target disciplines whose methods, evidential standards and categorizations sharply differ from those entrenched in neuroscience does not *per se* cast doubt on the modelling and explanatory relevance of such contributions. Doing so, in fact, would require one to supplement observations pertaining to the differences between the involved disciplines with a detailed specification of how exactly such differences hinder the modelling and explanatory relevance of the examined NI contributions.

To be sure, some authors defend the objection from the disunity of science by claiming that the modelling and explanatory relevance of SI contributions reliably decreases when these contributions target disciplines that sharply differ from the imperializing disciplines (see e.g. Dupré 1995, against various instances of economics imperialism). Still, these claims do not license wide-ranging opposition to NI. For the history of science abounds with cases where specific disciplines' methods and findings have been successfully applied to model and explain phenomena investigated by several different disciplines (see e.g. Kitcher 1981, on various cases of explanatory unification that span several different disciplines). Similarly, it would be of little import to hold that SI contributions frequently fail to provide major modelling and explanatory benefits when the imperializing discipline is in its early stages of development. For the rapid advances made in neuroscience observational tools and experimental protocols have already enabled this discipline's practitioners to model and explain a range of phenomena besides those traditionally investigated by neuroscientists (see e.g. Ross et al. 2008, on pathological gambling). That is to say, neither appealing to selected cases where interdisciplinary differences hampered SI contributions nor pointing to the putative developmental immaturity of neuroscience grounds a convincing normative critique of NI.

(ii) The *objection from disciplinary autonomy* (see e.g. Aizawa and Gillett 2011; and Fodor 1974) opposes SI contributions on the alleged ground that these contributions reduce or threaten the autonomy of the targeted disciplines from the imperializing ones in terms of various epistemic and non-epistemic criteria (e.g. availability of separate methods and modelling frameworks, existence of independent journals and research groups). I am not concerned here with assessing this characterization of disciplinary autonomy or with articulating an alternative characterization of such a notion. For present purposes, it suffices to note that the proposed attempts to ground a normative critique of NI on considerations of disciplinary autonomy fall prey to the following dilemma.

On the one hand, appealing to the alleged fact that some discipline is *in fact* autonomous from neuroscience falls short of implying that NI contributions targeting this discipline are unjustified. For example, a proponent of NI may consistently allow that the targeted discipline as it has been practised is autonomous from neuroscience, yet insist that NI contributions enable neuroscientists to build models that are more predictive and explanatory than the models built by the practitioners of the targeted discipline (see e.g. Churchland 1986, part II, for NI contributions targeting folk psychological models of intentional action). On the other hand, appealing to the putative fact that some discipline *should be* autonomous from neuroscience challenges one to explicate why exactly the mere fact that some NI contributions reduce or threaten the autonomy of this discipline would make such contributions unjustified. To date, the opponents of NI have provided a few pragmatic and epistemic reasons to think that specific disciplines should retain some degree of autonomy from neuroscience (see e.g. Aizawa and Gillett 2011, on the autonomy of psychology). However, these authors have not explicated why exactly the mere fact that some NI contributions reduce or threaten the autonomy of the targeted disciplines would make such contributions unjustified. Moreover, there are reasons to doubt that the mere fact that some NI contributions reduce or threaten disciplinary autonomy makes such contributions unjustified. For instance, NI contributions may yield to neuroscience and/or the targeted disciplines major modelling and explanatory benefits that compensate for the loss of disciplinary autonomy involved in such contributions (see e.g. Piccinini and Craver 2011, on the predictive and explanatory benefits yielded by the systematic integration of various functional models in cognitive psychology with neuroscience mechanistic models).[4]

(iii) The *objection from counterfactual scientific progress* (see e.g. Clarke and Walsh 2009, 2013) opposes SI contributions insofar as they preclude the targeted disciplines from progressing in ways these disciplines would have progressed in the absence of such contributions. Whether or not a discipline is plausibly taken to make progress may critically depend on what account of progress one adopts (see e.g. Lakatos 1970; and Popper 1963, ch. 10–11). Suppose, for the sake of argument, that both proponents and critics of NI agreed to evaluate NI contributions in terms of some shared account of progress. Even so, there are various reasons to doubt that the objection from counterfactual scientific progress grounds a convincing normative critique of NI. Below I explicate three such reasons, which hold on most entrenched accounts of scientific progress.

First, neuroscience often makes considerable progress thanks to its imperialistic contributions to other disciplines. This progress, in turn, may vindicate NI contributions even if they preclude the targeted disciplines from progressing in ways these disciplines would have progressed in the absence of such contributions (see e.g. Churchland 2007, ch. 8–10, on cognitive neuroscientists' attempts to supplant folk psychological models of individuals' processing of sensory experiences). Second, there seems to be no general reason to think that NI contributions prevalently (or more likely) hamper – as opposed to promote – progress in the disciplines they target. Indeed, as noted in points *(i)–(ii)*, NI contributions may prompt significant modelling and explanatory advances in the disciplines they target. This does not exclude that considerations of counterfactual scientific progress may bear against the justifiability of NI contributions (e.g. think of cases where the targeted disciplines would have made even more progress in the absence of NI contributions). However, it seriously constrains the range of situations where appealing to disciplines' counterfactual developments can ground a convincing normative critique of NI. And third, scientific theorists and practitioners are rarely in the epistemic position to establish that NI contributions preclude the targeted disciplines from progressing in ways these disciplines would have progressed in the absence of such contributions. To see this, consider the following example.

Suppose we wanted to ascertain whether NI contributions have detrimental (as opposed to beneficial) effects on the progress of psychology. Doing so would require us to estimate and assess the developmental trajectories of both psychology and neuroscience in a number of actual and counterfactual scenarios. Regrettably, the limitations affecting our epistemic access to counterfactual disciplinary developments would severely hinder our ability to make reliable estimates and assessments. Perhaps NI contributions would lead psychology to collapse into a cluster of fragmented degenerating research programs. Or perhaps psychology would make more rapid modelling and explanatory progress thanks to NI contributions. In short, the vast range of involved scenarios and the difficulty of reliably estimating these scenarios' probability of occurrence would make it quite difficult (not to say arbitrary) to single out one particular scenario (see Fumagalli 2015, for analogous remarks about the influence of neuroscience methods and findings on the development of economics).

(iv) The *objection from cumulative constraints* (see e.g. Mäki 2009, 2013) subordinates the justifiability of SI contributions to the satisfaction of a series of evaluative constraints. Four such constraints have been put forward in the recent literature. The *ontological* constraint subordinates the justifiability of SI contributions to their ability to foster so-called ontological (as opposed to mere derivational) unification. The idea is that such contributions should "successfully represent how things are related in the causal structure of the world", rather than merely derive "large classes of explanandum sentences from a parsimonious set of theoretical sentences" (Mäki 2009, 363–364). The *epistemological* constraint "recommends a strong sense of fallibility and openness to critical conversation across disciplinary boundaries" (ibid., 373). The *axiological* constraint states that "theories that unify insignificant phenomena [by] ignoring or marginalizing significant ones are much less supportable than those that unify significant phenomena at the expense of less significant ones" (Mäki 2013, 337). Finally, the *institutional* constraint proscribes "against dismissive hegemonic arrogance in the relations between disciplines [and against] intellectual monopolies protected by non-argumentative means of exclusion" (ibid., 337).

These four constraints provide a valuable basis for assessing various instances of SI. For example, as noted by Mäki (2009), one may build on the divide between ontological and derivational unification to criticize various instances of economics imperialism (see also Fernández Pinto, this volume, for an application of the epistemological constraint to appraise social epistemologists' rational choice models of scientists' division of cognitive labour). However, as they stand, the four aforementioned constraints are articulated at an exceedingly high level of abstraction to enable us to assess the justifiability of NI contributions. This does not preclude the proponents of these constraints from refining the formulation of those constraints so as to license informative verdicts regarding specific instances of NI. Still, the proposed refinements of such constraints face non-trivial challenges, which hamper these constraints' potential to ground a convincing normative critique of NI. To illustrate this, let us briefly consider the ontological constraint (for a detailed evaluation of the four constraints, see Fumagalli 2017a).

Two interpretations of the ontological constraint have been put forward in the literature (Kuorikoski and Lehtinen 2010, 349–350). On the one hand, the *causal interpretation* holds that the ontological constraint is met when the phenomena targeted by the imperializing and the targeted disciplines have a shared causal basis (e.g. think of the micro-physical substrates of the behavioural patterns studied by distinct decision sciences). On the other hand, the *structural interpretation* holds that the ontological constraint is met whenever the phenomena targeted by the imperializing and the targeted disciplines display similar structural characteristics (e.g. equilibria with equivalent mathematical descriptions). Neither of these interpretations seems to provide an informative basis for assessing the justifiability of NI. More specifically, the causal interpretation seemingly overlooks that imperialistic contributions may be justifiable even when the phenomena targeted by the imperializing and the targeted disciplines lack a shared causal basis (see e.g. Fumagalli 2017b; and Ylikoski and Aydinonat 2014, on the explanatory insights yielded by systematic applications of economists' constrained optimization techniques across disciplinary boundaries). Conversely, the structural interpretation targets properties that are instantiated by too many and overly diverse systems to be plausibly regarded as an informative evaluative criterion (see e.g. Kuorikoski and Lehtinen 2010, on the vast variety of systems that display equilibria with equivalent mathematical descriptions). The causal and structural interpretations do not exhaust the set of interpretations one could give to the ontological constraint. Still, the above remarks challenge the proponents of this constraint to provide more precise and plausible interpretations of such constraint. In the absence of such interpretations, those remarks can be provisionally taken to cast doubt on the informativeness of the ontological constraint.

To recapitulate, several criteria have been proposed to ground a normative critique of SI. These criteria provide a valuable basis for assessing various instances of SI, yet do not enable us to explicate the reasons (if any) why recent calls for NI are objectionable or unjustified. In the next two sections, I articulate and defend two interrelated claims that, I argue, undermine prominent calls for NI. First, the proponents of NI significantly overstate the evidential and explanatory relevance of neuroscience methods and findings for the disciplines they target. And second, prominent calls for NI point to an untenable reductionist position, which rests on unsupported empirical and normative presuppositions. To substantiate these two claims, I shall draw on two sets of influential calls for NI, which respectively target the economic modelling of choice (section two) and entrenched philosophical conceptions of free agency (section three).

2 Neuroscience imperialism and economic choice

As anticipated in the Introduction, several leading economists have advocated imperialistic contributions targeting other disciplines. In recent years, however, economics has been increasingly targeted by other disciplines' imperialistic contributions. NI contributions are especially prominent in this context. The idea is that although economic and neural findings can be combined to build more predictive and explanatory models of choice, neuroscience will ultimately reduce (see e.g. Glimcher 2010, Ch. 6–7; and Kable and Glimcher 2009) or eliminate (see e.g. Camerer et al. 2005; and Camerer 2008) several economic constructs. This section examines these NI contributions, focusing on Glimcher's (2009, 2010, Ch. 6–7) influential calls for a "partial reduction" of economics to neuroscience.[5]

Glimcher's calls for a partial reduction of economics to neuroscience aim to identify and formally model the workings of the neural areas that putatively determine reward valuation and choice at the whole-individual level. These calls build on an often-made distinction between the so-called "soft" and "hard" interpretations of expected utility theory (henceforth, EUT). On the one hand, *Soft-EU* holds that individuals behave "as if" they first compute the expected utilities of the available options and then choose the option with highest expected utility (Glimcher 2010, Ch. 6). The idea is that EUT provides a purely formal representation of consistent behavioural patterns and makes no explicit assumption concerning the neural substrates of decisions. On the other hand, *Hard-EU* holds that individuals behave the way they do because "a neural activation [pattern in their brains] encodes outcomes' desirability in a continuous monotonic fashion" (Glimcher 2010, 133–134).

After contrasting these two interpretations of EUT, Glimcher (2010, Ch. 6–8) argues that Soft-EU fails to provide predictive and explanatory models of choice. He then advocates developing increasingly predictive and explanatory versions of Hard-EU to effect a partial reduction of economics to neuroscience. In his view, effecting this partial reduction involves reducing economic constructs such as decision utility (an index of observed choices that lies at the core of EUT) to neuroscience constructs such as neural indices of the activations observed in neural populations that putatively contribute to reward valuation. To be sure, Glimcher doubts that "all of economics can be reduced to neuroscience" (2010, 25), and emphasizes that "there will [likely] be emergent concepts in economics that resist reduction" (ibid., 31). Still, he contends that "regularities that homomorphically map some economic kinds to neurobiological kinds [...] have already been discovered" and that "the data [support partial reduction] incontrovertibly" (ibid., 31; see also Vromen 2012).

To evaluate these claims, let us focus on so-called subjective values (henceforth, SV), a neural construct that figures prominently in Glimcher's calls for partial reduction. SV are defined as the mean firing rates of specific populations of neurons, encoded cardinally relative to a purportedly unique baseline, and are supposed to be linearly proportional to the BOLD signal measured in these populations (Glimcher 2009). In particular, SV are assumed to be always stochastically consistent with individuals' choices, even when these choices are inconsistent with EUT. For this reason, identifying which neural populations compute and integrate SV is regarded as a crucial step in the development of algorithmic models that predict choices even when EUT fails to do so (Glimcher 2010, Ch. 12–15). Building on previous studies of the neural substrates of reward valuation, Glimcher boldly conjectures that the ventral striatum and the medial prefrontal cortex "contain all the neurons we require to extract [SV] for any object" (2009, 509).

Glimcher's studies have greatly advanced former attempts to inform economic modelling by providing experimentally regimented and precise definitions of the basic constructs of neuroeconomic theorizing (see e.g. Ross 2014, Ch. 4). Moreover, his bold conjecture concerning what neural mechanisms putatively encode and integrate SV satisfies strict falsificationist standards and provides a testable basis for developing reductive mappings between economics and neuroscience (see e.g. Fumagalli 2016). Even so, there are reasons to doubt the prospects of Glimcher's calls for partial reduction. My point is not just that the hitherto proposed versions of Hard-EU are not convincingly supported by the available neural evidence. For Glimcher (2010, Ch. 6) himself emphasizes that current versions of Hard-EU do not fit all the available evidence concerning the human neural architecture. Rather, my main concern is that too many and overly speculative inferential steps are required to reduce decision utility to neural measures of subjective reward value. To give one example, several factors whose influence can be only partly accounted for in terms of neural computations of subjective reward value (e.g. ethical commitments) frequently affect individuals' choices in real-life situations (see e.g. Fumagalli 2013). In this respect, it is telling that various authors who advocate constraining economic models with neural findings doubt the prospects of attempts to reduce decision utility to neural measures of subjective reward value (see e.g. Ross 2011).

A proponent of NI may object that although we have limited knowledge of how factors such as individuals' ethical commitments affect choices, we already have sufficiently detailed information concerning the neural substrates of reward valuation to ground reductive mappings between economics and neuroscience (see e.g. Glimcher 2010, Ch. 11). The idea would be that most of the factors that causally influence decisions have to exert their influence through the neural mechanisms associated with reward valuation, and that we already possess detailed information about these neural mechanisms. This objection invites the following twofold rejoinder. First, it is dubious that our current knowledge of the neural substrates of reward valuation is sufficiently detailed to ground the reductive mappings envisaged by Glimcher (see e.g. Rustichini and Padoa-Schioppa 2015, on recent debates concerning how many kinds of value-related signals are computed in the human brain and how exactly these signals are integrated). And second, the claim that a sharply demarcated stage of subjective reward valuation determines choice at the whole-individual level is contested within neuroscience. Indeed, it is still debated under what circumstances decisions are made on the basis of action-related values as opposed to values computed and compared in the space of goods irrespective of sensorimotor contingencies (see e.g. Padoa-Schioppa and Schoenbaum 2015, for a recent discussion).

A proponent of NI may further object that "so far, every scientific discipline that has ever faced a reductive challenge [...] has demonstrated at least a partial set of linkages to the more reduced disciplines", and that "there is no reason to believe that economics will be different" (Glimcher 2010, 31). However, it remains unclear whether the historical record supports this reductionist inference over non-reductionist alternatives. To be sure, scientific theorists and practitioners made significant modelling and explanatory advances by pursuing reductionist research strategies (see e.g. Oppenheim and Putnam 1958, for some historical examples; see also Bickle 2003, for more recent illustrations from the cognitive and neural sciences). Yet, the history of science also abounds with cases where scientific theorists and practitioners made progress by restraining or even relinquishing former reductionist ambitions (see e.g. Craver 2005, on various episodes from the history of neuroscience). In this context, appealing to past reductive accomplishments seems to constitute a questionable enumerative induction, which presupposes (rather than shows) that the intertheoretic relations between economics and neuroscience are analogous to the intertheoretic relations between other disciplines. That is to say, it is one thing to claim that economists who ignore neuroscience findings reject "powerful tools for the improvement and development of economic theory" (Glimcher 2010, 142). It is quite another thing to claim that "the data [support a partial reduction of economics to neuroscience] incontrovertibly" (ibid., 31).

3 Neuroscience imperialism and free agency

Over the last few decades, many proponents of NI have maintained that neuroscience methods and findings challenge entrenched philosophical conceptions of free agency. In particular, several authors (e.g. Greene and Cohen 2004; Haggard and Libet 2001; Libet 1999; and Wegner 2003) have argued that *contra* a widely held intuition, individuals' intentions to act are initiated by unconscious neural processes well before these individuals become aware of such intentions. The thought is that the commonsense conception of free agency (henceforth, CFA) according to which individuals' conscious will is a non-determined cause of their intentions to act (see e.g. Aristotle, NE, Book III; and Kane 1996) is illusory and must be relinquished (see e.g. Wegner 2002; and Haggard 2008). In this section, I briefly review the main findings of neuroscience studies of free will and explicate the alleged implications of these findings for CFA. I then argue that *contra* leading proponents of NI the available findings do not undermine CFA.[6]

The experimental design of early neuroscience studies of free will (e.g. Libet et al. 1983) can be characterized as follows. The experimenter asks the investigated subjects to make a simple specified movement (e.g. flexing the wrist) when they decide to do so within a predefined period of time. The investigated subjects rely on a clock-like device to report the first moment they become aware of their decision to move. During the trial, the experimenter records electric signals from premotor neural areas and from the muscles involved in the examined movement. In their pioneering studies, Libet et al. (1983) documented that the readiness potential – a gradual build-up of activity in neural areas putatively involved in motor preparation – begins on average 550 milliseconds (ms) before the movement onset and 350 ms before subjects' reported time of their conscious decision to move. These findings, in turn, allegedly demonstrate that *contra* CFA individuals' intentions to act are "initiated by unconscious [neural] processes" (Libet 1985, 529). The thought is that although individuals might exert conscious veto decisions over unconsciously initiated actions, it is individuals' unconscious neural processes – rather than their own conscious will – which initiate their intentions to act (see e.g. Libet 1999; see also Brass and Haggard 2007, on the possibility that veto decisions themselves are initiated by unconscious neural events).

A number of potential confounds have been hypothesized to explain away these findings. For instance, some authors doubted that the investigated subjects reliably reported the timing of their own conscious decisions to move (see e.g. van de Grind 2002). Others alleged that the targeted neural signals may provide information about motor preparation, but are likely influenced by subjects' former intention to perform a predefined type of action (see e.g. Zhu 2003) and partly reflect spontaneous sub-threshold fluctuations in neural activity (see e.g. Schurger et al. 2012). Still others argued that neural activity seems to build up before subjects consciously decide to move because of confounding effects related to clock monitoring (see e.g. Miller et al. 2011). In recent years, however, a series of follow-up studies replicated earlier studies' findings while controlling for the hypothesized confounds (see e.g. Haggard 2008, on the alleged biases in subjects' reports of the timing of their conscious decisions; Filevich et al. 2013, for increasingly detailed dissociations of the neural correlates of motor preparation and higher-level action planning; and Soon et al. 2013, on the putative experimental confounds arising from clock monitoring effects).

These experimental findings have demonstrated that unconscious neural processes exert a more pervasive influence on individuals' intentions to act than many authors had previously supposed. Nonetheless, several concerns remain unaddressed, which severely constrain the implications of neuroscience studies of free will for CFA. Below I identify and explicate three interrelated concerns, which respectively relate to: (i) the *questionable conceptualization of free agency* presupposed by many neuroscience studies of free will; (ii) the *constrained evidential relevance* of current studies' findings for the types of decisions (if any) where it is plausible to think that individuals exercise their free agency in the sense indicated by CFA; and (iii) the *limited potential* of neuroscientists' predictions of individuals' decisions to *confirm or undermine* CFA. These three concerns do not exclude that neuroscience findings may prompt significant changes in entrenched philosophical conceptions of free agency. Still, taken together, they cast serious doubt on several authors' claim that the proposed NI contributions undermine CFA.

(i) Questionable conceptualization of free agency. To date, most neuroscience studies of free will have been premised on the assumption that an action is performed freely only if it is *internally caused* in the sense that it is not performed in direct response to any external stimuli, controls or cues (see e.g. Libet 1985; and Haynes 2011). Suppose we could reliably identify whether individuals' actions are internally caused in this sense. Even so, the set of internally caused actions overlaps only partly with the set of actions (if any) where it is plausible to think that individuals exercise their free agency in the sense indicated by CFA. For clearly, not all internally caused actions are plausibly taken to result from individuals' exercise of free agency (e.g. think of compulsive behaviour). Conversely, several actions that are performed in direct response to external stimuli may be plausibly regarded as free in the sense indicated by CFA (e.g. think of choice settings where experimental incentives provide individuals with reasons to perform specific actions without compelling them to perform those actions). Unfortunately, few neuroscience studies explicitly distinguish between factors whose causal influence on decisions rules out individuals' exercise of free agency and factors whose influence is compatible with this exercise (see e.g. Schlosser 2012).

This is not the only respect in which the conceptualization of free agency presupposed by neuroscience studies of free will appears to be overly simplistic. To give another example, most studies presuppose at the outset that individuals' conscious awareness of their own intentions to act varies in an all-or-nothing (as opposed to gradual) fashion. This, in turn, critically hinders the evidential relevance of the collected findings for CFA in all cases where individuals' conscious awareness of their own intentions to act builds up gradually until it reaches a reporting criterion (see e.g. Miller and Schwarz 2014). For in these cases, observing that unconscious neural activations precede individuals' conscious decisions falls short of indicating that these decisions are initiated by unconscious neural processes rather than by individuals' conscious will (e.g. the observed neural activations may reflect intermediate degrees of awareness rather than complete absence of awareness).

(ii) Constrained evidential relevance of current neuroscience findings for CFA. Philosophers often divide choices into two mutually exclusive sets when discussing neuroscience studies of free will (see e.g. Kane 1996; and van Inwagen 1989). On the one hand, we find cases of *free will proper*, where the involved individuals have reasons to perform one particular action rather than alternative actions, or to perform an action at one particular time rather than at some other time (e.g. conflicts of duty versus inclination, intertemporal trade-offs between short-term and long-term rewards). On the other hand, there are cases of *freedom of indifference*, where the involved individuals lack specific reasons to perform one particular action rather than alternative actions, or to perform an action at one particular time rather than at some other time (e.g. choices between qualitatively identical and equally accessible objects).[7]

So far, most neuroscience studies of free will have focused on cases of freedom of indifference rather than cases of free will proper (see e.g. Schlosser

2014). This severely constrains the evidential relevance of these studies' results for CFA. For in primis, cases of free will proper are those where it is comparatively more plausible to think that individuals exercise their free agency in the sense indicated by CFA (see e.g. O'Connor 2009). And second, cases of free will proper differ *ex hypothesi* from cases of freedom of indifference for engaging the involved individuals in additional consideration and weighing of their reasons for action. This difference, in turn, makes it doubtful that neural findings concerning cases of freedom of indifference provide informative insights regarding the tenability of CFA in cases of free will proper. To illustrate this, suppose we observed analogous neural activation patterns across several pairs of otherwise identical cases of freedom of indifference and cases of free will proper. This experimental result might *prima facie* seem to bear against CFA, but may be given several plausible interpretations that are consistent with CFA. For example, one may take such an experimental result to indicate that the observed neural activations fail to capture the neural substrates of individuals' consideration and weighing of their reasons for action, and thus lack direct evidential relevance for the tenability of CFA.

(iii) Limited potential of neuroscientists' predictions to confirm or undermine CFA. In recent years, neuroscience studies of free will succeeded in predicting simple choices such as basic motor decisions for increasing time spans and with growing reliability. For instance, Soon et al. (2008) predict with 60% reliability freely paced motor-decision tasks (e.g. choosing to press either the left or right button using the corresponding index finger) from fMRI observations of neural activity in the parietal cortex up to a few seconds before subjects report making their decision. Similarly, Fried et al. (2011) build on activations observed in the supplementary motor area to predict with more than 80% reliability individuals' self-initiated finger movements and estimate with a precision of a few hundred ms the time point of these decisions to move (see also Haynes et al. 2007, for comparable results obtained from activations observed in prefrontal and parietal cortex areas).

These predictive advances have led many authors to prefigure the provision of reliable predictions for increasingly complex real-life decisions (see e.g. Soon et al. 2013, on arithmetic calculations). Even so, neuroscience studies of free will face stringent trade-offs between the length of the time spans they cover, the complexity of the decisions they target, and the reliability with which they are able to predict such decisions.[8] Moreover, the ongoing advances in neuroscientists' attempts to predict individuals' decisions on the basis of observed neural activations seem to have a limited potential to confirm or undermine CFA. To see this, suppose that an imaginary *superscan* was invented, which enabled researchers to detect all neural events occurring in the monitored subjects' brains in real time and at any level of detail. Assume further that this superscan allowed researchers to unerringly predict complex real-life decisions before subjects claim to become aware of such decisions. Even this would not *per se* undermine CFA. For CFA does not rest on the assumption that human behaviour is inherently unpredictable, and does not exclude that one might be able to predict individuals' decisions based on observed neural activations (see e.g. Roskies 2010).[9]

In sum, neuroscience studies of free will have demonstrated that unconscious neural processes exert a more pervasive influence on individuals' intentions to act than many authors had previously supposed. These experimental results, in turn, may prompt significant changes in entrenched philosophical conceptions of free agency. Even so, one does not have to posit Cartesian souls or irreducible agent causation to doubt that current NI contributions undermine the commonsense view that individuals' conscious will is a non-determined cause of their intentions to act. In particular, leading NI contributions targeting CFA rest on unsupported reductionist presuppositions, which impede (rather than promote) rigorous reflection on the implications of neuroscience findings for entrenched philosophical conceptions of free agency.

4 Conclusion

Calls for NI have recently gained widespread support among scientific practitioners, making inroads into both natural and social disciplines. In this paper, I articulated and defended two interrelated claims that, I argue, undermine prominent calls for NI. First, the proponents of NI significantly overstate the evidential and explanatory relevance of neuroscience methods and findings for the disciplines they target. And second, many calls for NI point to an untenable reductionist position, which rests on unsupported empirical and normative presuppositions. These two claims do not imply that all attempts to apply neuroscience methods and findings to model and explain phenomena investigated by other disciplines are objectionable. Still, they challenge the proponents of NI to qualify their imperialistic calls and ground such calls on more plausible empirical and normative presuppositions.

Acknowledgments

I thank Stephen M. Downes, Manuela Fernández Pinto, Paul Hoyningen-Huene, Uskali Mäki, Adrian Walsh and Petri Ylikoski for their comments on previous versions of this chapter. I also benefited from the observations of audiences at the University of Mainz and the Finnish Centre of Excellence in the Philosophy of the Social Sciences (Helsinki).

Notes

1 I use the expression "economics imperialism" to differentiate this instance of scientific imperialism from the notion of "economic imperialism" (see Mäki 2002, for a similar distinction). I gloss over various fine-grained distinctions between types of imperialistic contributions (see e.g. Mäki 2009, on imperialism of standing, imperialism of style and imperialism of scope) unless the cogency of my critique rests on such distinctions.

2 I focus prevalently on imperialistic interactions between disciplines rather than imperialistic interactions between fields, research programs or specific theories. This does not exclude that one may gain informative insights by focusing on imperialistic interactions between units of analysis other than disciplines. Also, my critique of NI is not meant to cast doubt on the merits of SI in general. I explore in Fumagalli (2017a)

the merits of SI in general and alleged links between SI in general and specific instances of SI such as NI.

3 This list includes what I regard as the most informative criteria put forward to ground a normative critique of SI. Some of these criteria were proposed to assess other instances of SI than NI, so my discussion should not be seen as a direct critique of those criteria.

4 Several criteria may be employed to assess the justifiability of specific SI contributions (e.g. trade-offs between the explanatory benefits and losses yielded by such contributions, availability of non-imperialistic forms of interdisciplinary interaction, etc.). I am not concerned here with demarcating the set of relevant criteria or establishing what weight one should ascribe to each of these criteria. For present purposes, it suffices to note that both the relevance of distinct criteria and the weight ascribed to each of these criteria may vary remarkably depending on several factors, ranging from one's epistemic interests to the modelling practices entrenched in the disciplines involved in the examined SI contributions.

5 I use the term "reduction" broadly to encompass both cases of *derivational* reduction, which obtains when the constructs and the ontology of the reduced theories are retained (e.g. as limiting cases) by the reducing theories, and cases of *eliminative* reduction, which obtains when the constructs and the ontology of the reduced theories are replaced by the reducing theories (see e.g. Churchland 1986). Derivational and eliminative reduction do not exhaust all cases of intertheoretic reduction, but are plausibly regarded as endpoints of a continuum with several intermediate cases (see e.g. McCauley 1986). I gloss over other distinctions between various notions of reduction (see e.g. Nickles 1973, on synchronic and diachronic reduction) and dissimilar accounts of intertheoretic reduction (see e.g. Bickle 1998, on classical and new wave reduction) since my critique targets not so much the reductionist character of some NI contributions, but rather the limited empirical and normative support for these contributions' presuppositions.

6 Philosophers have developed many distinct conceptions of free agency, which differ both as to what conditions must obtain to license the claim that an agent's will is free and as to whether free will and causal determinism are compatible (see e.g. Pereboom 2009). In this section, I focus on CFA since many influential neuroscience studies of free will target this conception of free agency. Exercising one's free agency in the sense indicated by CFA typically involves identifying one's reasons for action and weighing competing reasons for the available actions. Different views have been advocated as to how reasons for action and the process of weighing competing reasons for action are most aptly conceptualized (see e.g. Frankfurt 1971; Habermas 2007; Parfit 1997; and Williams 1980). I take my critique of neuroscience studies of free will to hold irrespective of what position one advocates about the merits of such views.

7 For a choice to count as a case of freedom of indifference, it suffices that the involved individual finds the available options subjectively indiscernible in the context of choice. To put it differently, cases of freedom of indifference encompass choice situations where an individual could in principle – but does not in fact – identify some reasons for choosing one of the available options rather than the others through additional reflection or through a detailed examination of such options.

8 By way of illustration, the studies that reliably predict subjects' decisions a few seconds before subjects claim to become aware of such decisions almost invariably target simple laboratory choices (e.g. pressing one of two buttons) rather than the complex decisions many people face in non-laboratory settings (e.g. choosing between alternative job offers). This is due not just to the technological limitations of current observational tools (see e.g. Klein 2010, on scanners' limited spatial and temporal resolution), but also to the complexity of the interactions between the neural and non-neural factors that jointly cause individuals' decisions in non-laboratory settings (see e.g. Nachev and Hacker 2014).

9 I mention in passing the debate regarding the alleged existence of inherently stochastic neural events (see e.g. Glimcher 2005) since the existence of such events would not *per se* vindicate CFA (e.g. the existence of stochastic neural events is

compatible with the existence of deterministic causal processes operating at more fine-grained spatio-temporal scales than neural events).

References

Aizawa, Ken and Carl Gillett. 2011. "The Autonomy of Psychology in the Age of Neuroscience". In P.M. Illari, F. Russo and J. Williamson (eds). *Causality in the Sciences*. Oxford: Oxford University Press, pp. 202–223.

Aristotle. 1985. *Nicomachean Ethics*. Transl. by T. Irwin. Indianapolis: Hackett Publishing.

Bechtel, William and Andrew Hamilton. 2007. "Reduction, Integration, and the Unity of Science". In T. Kuipers (ed.). *Philosophy of Science: Focal Issues*. New York: Elsevier, pp. 377–430.

Becker, Gary S. 1981. *A Treatise on the Family*. Cambridge, MA: Harvard University Press.

Bickle, John. 1998. *Psychoneural Reduction: The New Wave*. Cambridge, MA: MIT Press.

Bickle, John. 2003. *Philosophy and Neuroscience: A Ruthlessly Reductive Approach*. Dordrecht: Kluwer.

Brass, Marcel and Patrick Haggard. 2007. "To Do or Not to Do: The Neural Signature of Self-Control". *Journal of Neuroscience* 27: 9141–9145.

Camerer, Colin F. 2008. "The Case for Mindful Economics". In A. Caplin and A. Schotter (eds). *The Foundations of Positive and Normative Economics*. New York: Oxford University Press, pp. 43–69.

Camerer, Colin F., George Loewenstein and Drazen Prelec. 2005. "Neuroeconomics: How Neuroscience Can Inform Economics". *Journal of Economic Literature* 43: 9–64.

Cartwright, Nancy. 1999. *The Dappled World*. Cambridge: Cambridge University Press.

Churchland, Patricia S. 1986. *Neurophilosophy*. Cambridge, MA: MIT Press.

Churchland, Paul M. 2007. *Neurophilosophy at Work*. New York: Cambridge University Press.

Clarke, Steve and Adrian Walsh. 2009. "Scientific Imperialism and the Proper Relations between the Sciences". *International Studies in the Philosophy of Science* 23: 195–207.

Clarke, Steve and Adrian Walsh. 2013. "Imperialism, Progress, Developmental Teleology, and Interdisciplinary Unification". *International Studies in the Philosophy of Science* 27: 341–351.

Craver, Carl. 2005. "Beyond Reduction: Mechanisms, Multifield Integration and the Unity of Neuroscience". *Studies in History and Philosophy of Biological and Biomedical Sciences* 36: 373–395.

Downe, Stephen. 2018. "Scientific Imperialism and Expanding Appeals to Evolution in the Social Sciences". In Mäki, U., Walsh, A. and M. Fernández Pinto (eds) *Scientific Imperialism: Another Facet of Interdisciplinaraity*. Abingdon: Routledge.

Ren´ndez Pinto, M. 2018. "Imperialising Epistemology: Shortcomings". In Mäki, U., Walsh, A. and M. Fernández Pinto (eds) *Scientific Imperialism: Another Facet of Interdisciplinaraity*. Abingdon: Routledge.

Dupré, John. 1983. "The Disunity of Science". *Mind* 92: 321–346.

Dupré, John. 1995. "Against Scientific Imperialism". *Proceedings of the 1994 Biennial Meeting of the Philosophy of Science Association* 2: 374–381.

Dupré, John. 2001. *Human Nature and the Limits of Science*. Oxford: Oxford University Press.

Farah, Martha J. 2012. "Neuroethics: The Ethical, Legal, and Societal Impact of Neuroscience". *Annual Review Psychology* 63: 571–591.

Ferejohn, John and Debra Satz. 1995. "Unification, Universalism, and Rational Choice Theory". *Critical Review* 9: 71–84.

Filevich, Elisa, Patricia Vanneste, Marcel Brass, Wim Fias, Patrick Haggard and Simone Kühn. 2013. "Brain Correlates of Subjective Freedom of Choice". *Consciousness and Cognition* 22: 1271–1284.

Fine, Ben. 1999. "The Question of Economics: Is it Colonizing the Social Sciences?" *Economy and Society* 28: 403–425.

Fodor, Jerry. 1974. "Special Sciences (or: The Disunity of Science as a Working Hypothesis)". *Synthese* 28: 97–115.

Frankfurt, Harry. 1971. "Freedom of the Will and the Concept of a Person". *Journal of Philosophy* 68: 5–20.

Fried, Itzhak, Roy Mukamel and Gabriel Kreiman. 2011. "Internally Generated Preactivation of Single Neurons in Human Medial Frontal Cortex Predicts Volition". *Neuron* 69: 548–562.

Friedman, Michael. 1974. "Explanation and Scientific Understanding". *Journal of Philosophy* 71: 5–19.

Fumagalli, Roberto. 2013. "The Futile Search for True Utility". *Economics and Philosophy* 29: 325–347.

Fumagalli, Roberto. 2015. "Five Theses on Neuroeconomics". *Journal of Economic Methodology* 23: 77–96.

Fumagalli, Roberto. 2016. "Choice Models and Realistic Ontologies: Three Challenges to Neuro-Psychological Modellers". *European Journal for Philosophy of Science* 6: 145–164.

Fumagalli, Roberto. 2017a. "Who is Afraid of Scientific Imperialism?" *Synthese*. In press.

Fumagalli, Roberto. 2017b. "How 'Thin' Rational Choice Theory Explains Choices". Manuscript under review.

Glimcher, Paul W. 2005. "Indeterminacy in Brain and Behavior". *Annual Review of Psychology* 56: 25–56.

Glimcher, Paul W. 2009. "Choice: Towards a Standard Back-pocket Model". In P.W. Glimcher, C.F. Camerer, E. Fehr and R.A. Poldrack (eds). *Neuroeconomics: Decision Making and the Brain*. New York: Academic Press, pp. 503–521.

Glimcher, Paul W. 2010. *Foundations of Neuroeconomic Analysis*. Oxford: Oxford University Press.

Greene, Joshua D. 2007. "The Secret Joke of Kant's Soul". In W. Sinnott-Armstrong (ed.). *Moral Psychology, Vol. 3: The Neuroscience of Morality: Emotion, Disease, and Development*. Cambridge, MA: MIT Press, pp. 35–79.

Greene, Joshua D. and Jonathan Cohen. 2004. "For the Law, Neuroscience Changes Nothing and Everything". *Philosophical Transactions of the Royal Society London B* 359: 1775–1785.

Habermas, Jürgen. 2007. "The Language Game of Responsible Agency and the Problem of Free Will". *Philosophical Explorations* 10(1): 13–50.

Haggard, Patrick. 2008. "Human Volition: Towards a Neuroscience of Will". *Nature Reviews Neuroscience* 9: 934–946.

Haggard, Patrick and Benjamin Libet. 2001. "Conscious Intention and Brain Activity". *Journal of Consciousness Studies* 8: 47–63.

Haynes, John-Dylan. 2011. "Beyond Libet: Long-Term Prediction of Free Choices from Neuroimaging Signals". In W. Sinnott-Armstrong and L. Nadel (eds). *Conscious Will and Responsibility*. Oxford: Oxford University Press, pp. 85–96.

Haynes, John-Dylan, Katsuyuki Sakai, Geraint Rees, Sam Gilbert, Chris Frith and Richard Passingham. 2007. "Reading Hidden Intentions in the Human Brain". *Current Biology* 17: 323–328.

Hirshleifer, Jack. 1985. "The Expanding Domain of Economics". *American Economic Review* 75: 53–68.

Kable, Joseph W. and Paul W. Glimcher. 2009. "The Neurobiology of Decision: Consensus and Controversy". *Neuron* 63: 733–745.

Kane, Robert. 1996. *The Significance of Free Will*. New York: Oxford University Press.

Kitcher, Philip. 1981. "Explanatory Unification". *Philosophy of Science* 48: 507–531.

Klein, Colin. 2010. "Images are Not the Evidence in Neuroimaging". *British Journal for Philosophy of Science* 61: 265–278.

Kuorikoski, Jaakko and Aki Lehtinen. 2010. "Economics Imperialism and Solution Concepts in Political Science". *Philosophy of the Social Sciences* 40: 347–374.

Kuorikoski, Jaakko and Petri Ylikoski. 2010. "Explanatory Relevance across Disciplinary Boundaries: The Case of Neuroeconomics". *Journal of Economic Methodology* 17: 219–228.

Lakatos, Imre. 1970. "Falsification and the Methodology of Scientific Research Programmes". In I. Lakatos and A. Musgrave (eds). *Criticism and the Growth of Knowledge*. Cambridge: Cambridge University Press, pp. 191–195.

Lewin, Shira. 1996. "Economics and Psychology: Lessons for Our Own Day from the Early Twentieth Century". *Journal of Economic Literature* 34: 1293–1323.

Libet, Benjamin. 1985. "Unconscious Cerebral Initiative and the Role of Conscious Will in Voluntary Action". *Behavioral and Brain Sciences* 8: 529–566.

Libet, Benjamin. 1999. "Do We Have Free Will?" *Journal of Consciousness Studies* 6: 47–57.

Libet, Benjamin, Curtis Gleason, Elwood Wright and Dennis Pearl. 1983. "Time of Conscious Intention to Act in Relation to Onset of Cerebral Activity (Readiness-Potential). The Unconscious Initiation of a Freely Voluntary Act". *Brain* 106: 623–642.

Mäki, Uskali. 2002. "Explanatory Ecumenism and Economics Impérialism". *Economics and Philosophy* 18: 235–257.

Mäki, Uskali. 2009. "Economics Imperialism: Concept and Constraints". *Philosophy of the Social Sciences* 9: 351–380.

Mäki, Uskali. 2013. "Scientific Imperialism: Difficulties in Definition, Identification, and Assessment". *International Studies in the Philosophy of Science* 27: 325–339.

McCauley, Robert N. 1986. "Intertheoretic Relations and the Future of Psychology". *Philosophy of Science* 53: 179–199.

Miller, Jeff and Wolf Schwarz. 2014. "Brain Signals Do Not Demonstrate Unconscious Decision Making: An Interpretation Based on Graded Conscious Awareness". *Consciousness and Cognition* 24: 12–21.

Miller, Jeff, Peter Shepherdson and Judy Trevena. 2011. "Effects of Clock Monitoring on Electroencephalographic Activity: Is Unconscious Movement Initiation an Artifact of the Clock?" *Psychological Science* 22: 103–109.

Morse, Stephen J. 2005. "Brain Overclaim Syndrome and Criminal Responsibility: A Diagnostic Note". *Ohio State J. Criminal Law* 3: 397–412.

Nachev, Parashkev and Peter Hacker. 2014. "The Neural Antecedents to Voluntary Action: A Conceptual Analysis". *Cognitive Neuroscience* 5: 193–208.

Nickles, Thomas. 1973. "Two Concepts of Inter-Theoretic Reduction". *Journal of Philosophy* 70: 181–201.

O'Connor, Timothy. 2009. "Conscious Willing and the Emerging Sciences of Brain and Behavior". In E. Murphy and T. O'Connor (eds). *Downward Causation and the Neurobiology of Free Will*. Heidelberg: Springer, pp. 173–186.

Oppenheim, Paul and Hilary Putnam. 1958. "The Unity of Science as a Working Hypothesis". In H. Feigl et al. (eds). *Minnesota Studies in the Philosophy of Science*, vol. 2. Minneapolis: Minnesota University Press, pp. 3–36.

Padoa-Schioppa, Camillo and Geoffrey Schoenbaum. 2015. "Dialogue on Economic Choice, Learning Theory, and Neuronal Representations". *Current Opinion in Behavioral Sciences* 5: 16–23.

Parfit, Derek. 1997. "Reason and Motivation". *Proceedings of the Aristotelian Society* Suppl. 71: 99–130.

Pereboom, Derk. 2009. *Free Will*. Indianapolis, IN: Hackett Publishing.

Piccinini, Gualtiero and Carl Craver. 2011. "Integrating Psychology and Neuroscience: Functional Analyses as Mechanism Sketches". *Synthese* 183: 283–311.

Popper, Karl R. 1963. *Conjectures and Refutations*. New York: Harper and Row.

Rolin, Kristina. 2018. "Scientific Imperialism and Epistemic Injustice". In Mäki, U., Walsh, A. and M. Fernández Pinto (eds) *Scientific Imperialism: Another Facet of Interdisciplinaraity*. Abingdon: Routledge.

Roskies, Adina L. 2010. "How Does Neuroscience Affect Our Conception of Volition?" *Annual Review of Neuroscience* 33: 109–130.

Ross, Don. 2011. "Estranged Parents and a Schizophrenic Child: Choice in Economics, Psychology and Neuroeconomics". *Journal of Economic Methodology* 18: 217–231.

Ross, Don. 2014. *Philosophy of Economics*. New York: Palgrave Macmillan.

Ross, Don, Carla Sharp, Rudy Vuchinich and David Spurrett. 2008. *Midbrain Mutiny: The Picoeconomics and Neuroeconomics of Disordered Gambling*. Cambridge, MA: MIT Press.

Rustichini, Aldo and Camillo Padoa-Schioppa. 2015. "A Neuro-Computational Model of Economic Decisions". *Journal of Neurophysiology* 114: 1382–1398.

Schlosser, Markus E. 2012. "Free Will and the Unconscious Precursors of Choice". *Philosophical Psychology* 25: 365–384.

Schlosser, Markus E. 2014. "The Neuroscientific Study of Free Will: A Diagnosis of the Controversy". *Synthese* 191: 245–262.

Schurger, Aaron, Jacobo Sitt and Stanislas Dehaene. 2012. "An Accumulator Model for Spontaneous Neural Activity Prior to Self-initiated Movement". *Proceedings of the National Academy of Sciences* 109: E2904–E2913.

Soon, Chun, Marcel Brass, Hans-Jochen Heinze and John-Dylan Haynes. 2008. "Unconscious Determinants of Free Decisions in the Human Brain". *Nature Neuroscience* 11: 543–545.

Soon, Chun, Hanxi He, Stefan Bode and John-Dylan Haynes. 2013. "Predicting Free Choices for Abstract Intentions". *Proceedings of the National Academy of Sciences* 110: 6217–6222.

Van de Grind, Wim. 2002. "Physical, Neural, and Mental Timing". *Consciousness and Cognition* 11: 241–264.

Van Inwagen, Peter. 1989. "When is the Will Free?" *Philosophical Perspectives* 3: 399–422.

Vincent, Nicole A. 2013. *Neuroscience and Legal Responsibility*. New York: Oxford University Press.

Vromen, Jack. 2012. "Review of *Foundations of Neuroeconomic Analysis*, P.W. Glimcher. Oxford University Press, 2010". *Economics and Philosophy* 28: 108–113.

Wegner, Daniel M. 2002. *The Illusion of Conscious Will*. Cambridge, MA: MIT Press.

Wegner, Daniel M. 2003. "The Mind's Best Trick: How We Experience Conscious Will". *Trends Cognitive Science* 7: 65–69.

Williams, Bernard. 1980. "Internal and External Reasons". Reprinted in *Moral Luck*. 1981. Cambridge University Press.

Ylikoski, Petri and Emrah Aydinonat. 2014. "Understanding with Theoretical Models". *Journal of Economic Methodology* 21: 19–36.

Zhu, Jing. 2003. "Reclaiming Volition: An Alternative Interpretation of Libet's Experiment". *Journal of Consciousness Studies* 10: 61–77.

11 Scientific imperialism and explanatory appeals to evolution in the social sciences

Stephen M. Downes

Introduction

In this chapter I focus on what the charge of scientific imperialism amounts to and whether the charge is (always) appropriate in the case of the relation between evolutionary biology and the social and behavioral sciences. I will then briefly introduce two very different examples of this relationship: evolutionary psychology and evolutionary linguistics. Here I hope to support the claim that relations between evolution and the study of human behavior are not all created equal. Next I move to a discussion of scientific imperialism and the way in which the concept has been proposed, developed, and defended. Here I focus on John Dupré's account of scientific imperialism, as this account was developed to critically assess evolutionary psychology (along with economics). I go on to propose an empirically based approach to scientific change that I hope can help in identifying the strengths and limitations of appeals to scientific imperialism. Here I will draw on an analogy between the ways in which empirical approaches to scientific change helped refine the notion of scientific revolutions and the ways in which related empirical approaches may help refine our notion of scientific imperialism.

My interest in scientific imperialism stems from claims like this made by John Dupré: "Evolutionary psychology can be seen as a failed imperialist adventure from evolutionary biology" (Dupré 2001, 16). I have critically appraised evolutionary psychology from a number of different perspectives (see e.g. Downes 2015), and when I first encountered Dupré's charge of imperialism I assumed that he was invoking an independent critical perspective on evolutionary psychology. Here I will reassess this initial assumption. There are several issues that are relevant to this reassessment. First, evolutionary psychology is a quite distinct "adventure" from evolutionary biology. There are other applications of aspects of biological theory in the social and behavioral sciences that raise very few eyebrows and do not seem to warrant the charge of imperialism, where "imperialism" is understood as a pejorative term. Second, there are many critics of evolutionary psychology who criticize it on the grounds that its proponents do not pay enough attention to evolutionary biology and many base their critical attacks on evolutionary psychology in evolutionary theory. Third, the very notion of scientific imperialism has undergone some serious critical appraisal. There are many promising analyses of scientific imperialism but there are still plenty of issues unresolved (many of these issues are confronted in other chapters in this volume). For example, if scientific imperialism is used in a pejorative sense, is the charge an epistemic one or a moral one? Further, when we discuss change in science in terms of scientific imperialism, are we discussing an issue internal or external to science?

This latter issue invokes discussion of another political metaphor in philosophy of science, Thomas Kuhn's scientific revolutions (Kuhn 1962). I propose that discussion of the metaphor of scientific imperialism in relation to empirical work on scientific change can be as productive as discussion of Kuhn's famous metaphor of revolutions in relation to relevant empirical work. Finally, I touch on the broader connections between empirical work, in the sociology of science, for example, and the epistemic appraisal of scientific work. Can we say that a given piece of scientific work is bad on the basis of an assessment of the institutional practices scientists engage in without any detailed examination of the relevant hypotheses and the testing of those hypotheses? Is it possible to completely distinguish the assessment of institutional practice from epistemic appraisal?

1 Evolution and the social sciences

In the conclusion of *The Origin of Species*, Charles Darwin says:

> In the distant future I see open fields for far more important researches. Psychology will be based on a new foundation, that of the necessary acquirement of each mental power and capacity by gradation. Light will be thrown on the origin of man and his history.
>
> (see e.g. Downes 2015)

Darwin went on to write the *Descent of Man* and *The Expression of the Emotions in Man and Animals*, delivering on his promise rather than waiting for researchers in a distant future. Now, almost every avenue of social and behavioral science has an evolutionary branch or shows the influence of some kind of evolutionary thinking. Evolutionary anthropology is a well-established part of anthropology with several in field journals and one or two entire departments dedicated only to this mode of anthropological research. In cognitive psychology, a large percentage of hypotheses are accompanied by some evolutionary speculation, such as to postulate that the psychological mechanism of interest is part of the evolutionarily older part of the brain. Dual process theory in cognitive psychology is presented against the background of loosely evolutionary considerations. In the last ten years, consumer studies researchers have proposed a reworking of their field that draws upon evolutionary ideas.

The kind of evolutionary thinking invoked in the wide range of social sciences varies greatly. Also the ways in which the relevant evolutionary thinking is brought into the different social sciences varies greatly, as does the source of the relevant evolutionary thought. This variation should be of great interest to all of us interested in scientific imperialism and especially those of us who aim to understand the introduction of evolutionary ideas into the social sciences as a case of scientific imperialism. I only lay out two brief examples of evolutionary social sciences here, but I hope that the contrast between the two adequately illustrates these points. I will briefly discuss the well-known field of evolutionary psychology and then turn to a likely less familiar sub-discipline of the larger field of evolutionary linguistics.

2 Evolutionary psychology

Evolutionary psychology was proposed and defended by a number of different researchers in the late 1980s. Proponents of evolutionary psychology included anthropologists and psychologists along with human sociobiologists and others who studied human behavior. Some key experiments were championed by evolutionary psychologists as providing strong support for evolutionary hypotheses about the way the human mind works and leading thinkers in the field produced theoretical papers, and subsequently books, to lay out the key theoretical tenets and methodologies of the new field. Examples of this early experimental work include David Buss (1989) on mate selection, Leda Cosmides (1989) on cheater detection and Devendra Singh (1993) on women's waist/hip ratio preferences in men. Theoretical work included John Tooby and Cosmides' (1992) long position paper on evolutionary psychology, which they shortened into the still highly cited and much more accessible primer on *Evolutionary Psychology* (1997), and papers by Buss. Buss also wrote the first evolutionary psychology textbook (1999) and edited *The Handbook of Evolutionary Psychology* (2005b).

Evolutionary psychology has always been highly visible, thanks in large part to the success of popular works championing the field. Robert Wright's *The Moral Animal: Why We Are the Way We Are: The New Science of Evolutionary Psychology* (1994) and Steven Pinker's *How the Mind Works* (1997) are hugely popular works. Each author provides nice, accessible introductions to evolutionary psychology and also vehemently attacks all work in the social and behavioral sciences that does not appeal to evolution or, worse still, that is in direct opposition to evolutionary approaches. There are many other popular evolutionary psychology books and several of these are highly controversial and provocative. Books such as Randy Thornhill and Craig Palmer's *A Natural History of Rape: Biological Bases of Sexual Coercion* (2000) and Buss's *The Murderer Next Door: Why The Mind is Designed to Kill* (2005a) proved incredibly controversial and are lightning rods for much of the backlash against evolutionary psychology.

Taking a look at Buss's (1995) theoretical paper and Clark Barrett's (2015) recent book helps us understand what evolutionary psychologists proposed at the outset and reveals the central tenets of the now somewhat more mature field. Buss challenges his colleagues in psychology, saying: "Psychological science is currently in conceptual disarray, characterized by unconnected mini-theories and isolated empirical findings. [...] Evolutionary psychology provides the conceptual tools for emerging from this fragmented state" (Buss 1995, 1). The relevant conceptual tools include treating the human mind as a collection of adaptations that were selected to solve the various adaptive problems our ancestors faced.

Understanding organs as adaptations is a key component of evolutionary biology. Darwin countered design theorists such as Paley by defending the view that an organ as complex as the human eye could arise as the result of gradual modification, guided by selection, over huge stretches of time. Adapting Darwin, evolutionary psychologists propose that we understand the mind as a collection of highly adapted mental organs and we account for the existence of these mental organs by appealing to adaptation or natural selection. Evolutionary psychologists argue that this account of the human mind better explains the results of experiments on human preferences and on human reasoning. Singh's work on waist/hip ratio preferences in men involves first showing a convergence on a particular preferred waist/hip ratio and then explaining this convergence by appealing to an evolved mental mechanism that helps men achieve success in mate selection (see e.g. Singh 1993). Buss provides this assessment of the promise of evolutionary psychology:

> From the perspective of evolutionary psychology, many traditional disciplinary boundaries are not merely arbitrary but are misleading and detrimental to progress. They imply boundaries that cleave mechanisms in arbitrary and unnatural ways. Studying human psychology via adaptive problems and their solutions provides a natural means of "cleaving nature at its joints" and hence crossing current disciplinary boundaries.
>
> (Buss 1995, 1)

Assessing the current state of play in evolutionary psychology 20 years after Buss's paper, Barrett says: "Unfortunately, the rise of evolutionary psychology has led to rampant generation of evolution-flavored hypotheses throughout the social sciences, not all of which pass basic tests of evolutionary plausibility" (Barrett 2015, 12). Barrett claims: "When done properly, evolutionary psychology is just evolutionary biology applied to the mind" (ibid., 12). Barrett's aim is to realign evolutionary psychology with evolutionary thinking in general. He argues that "not all evolutionary hypotheses are created equal, and that careful thinking about how evolution actually works [...] can get you a long way" (ibid., 12), and aims to provide "an evolutionary psychology grounded as rigorously as possible in the logic of biology – the logic of evolution – and no other" (ibid., 12). Barrett is convinced that "the fact that human minds are products of evolution is something that we can be as sure of as any other fact in biology" (ibid., 12). Echoing Buss, Barrett claims that evolutionary models are far preferable "than the currently rather impoverished set of models in psychology" (ibid., 12). Evolutionary psychologists claim to offer their colleagues in psychology and other social sciences a stronger explanatory repertoire that can account for a wide range of phenomena.

3 Evolutionary linguistics

I now turn to a contrasting example of work in the social sciences influenced by evolution: Russell Gray and his colleagues' work on human migration in the Pacific (Gray et al. 2009; Gray and Jordan 2000). In the 1990s there were two main hypotheses about the original human settlement of the Micronesian and Polynesian Pacific islands and neighboring areas. One hypothesis is that this settlement process was quite rapid. This is referred to as the "express train" hypothesis. The other hypothesis is of a slower and less direct migration pattern with looping and back-tracking, sometimes referred to as the "entangled bank" model. Linguists have observed that the inhabitants of the Pacific all share related languages. Gray proposed that quantitative phylogenetic techniques could be applied to the analysis of languages and tree structures relating relevant languages could be proposed and tested. In phylogenetics, trees are produced to assess hypotheses about relations between segments of DNA. Gray observes that languages have a syntactic structure that is amenable to similar analysis. Gray notes that despite the parallels between biological and linguistic evolution, before him, no linguists had appealed to the "quantitative phylogenetic methods that have revolutionized evolutionary biology in the last 20 years" (Gray and Jordan 2000, 1052; Gray et al. 2009).

Gray and his colleagues produced several trees and assessed the best trees for the language data sets using statistical methods from quantitative phylogenetics. In their first work of this kind, they found that the topology of their language tree "was highly compatible with the express train model" (Gray and Jordan 2000, 1052). In more recent work, they developed language trees that were compatible with the "pause-pulse" model of human settlement as opposed to its main competitor, the "slow-boat diffusion" model. Gray and his colleagues exploit the similarity between lexical items and sections of DNA and use the techniques from evolutionary biology used to discover relations between sections of DNA. At this stage of the work, there are no grand claims about the applicability of evolution to all of linguistics or other areas of social science. Rather, Gray and his colleagues have introduced a technique from evolutionary biology into one area of work in linguistics and the study of early human migration. There is potential for the application of phylogenetic analysis to other aspects of human cultural evolution, and Fiona Jordan (see e.g. Jordan 2013) has made some cautious forays into this area.

4 Scientific imperialism

I now turn to scientific imperialism. Recall that my interest in scientific imperialism was sparked by Dupré's claim that "[e]volutionary psychology can be seen as a failed imperialist adventure from evolutionary biology" (Dupré 2001, 16). My focus therefore will be on Dupré's notion of scientific imperialism. My understanding of Dupré's notion of scientific imperialism has been greatly helped by work by Steve Clarke and Adrian Walsh (2009, 2013), and Uskali Mäki (2009, 2013), and their influence will be seen in what follows here.

Dupré says that scientific imperialism is "the tendency for a successful scientific idea to be applied far beyond its original home, and generally with decreasing success the more its application is expanded" (Dupré 2001, 16). I agree with Clarke and Walsh (2009) that Dupré uses imperialism in a pejorative sense and also agree with them that this is clearly a political metaphor used in the service of understanding science. The most familiar political metaphor in the philosophy of science is Kuhn's (1962) scientific revolution. I return to scientific revolutions below. Clarke and Walsh (2009) emphasize the normative dimension of Dupré's scientific imperialism. They express this normative dimension as follows:

> In the human sciences, scientific imperialism raises an additional concern, which is that important human values could fail to be expressed, and we might become the sorts of people that, on reflection, we would not wish to have become.
>
> (Clarke and Walsh 2009, 205)

I agree that Dupré is concerned that scientizing the study of human behavior can lead to a lack of appreciation for values but I also think that there is a strongly epistemic normative component to Dupré's charge of scientific imperialism. He thinks that scientific imperialism produces bad science not just science that leads us to devalue ourselves. Mäki asks an important question in this context too: is evolutionary psychology bad science independent of any considerations about scientific imperialism (Mäki 2013, 238)? Dupré certainly has argued on independent grounds that evolutionary psychology is bad science but I take him also to be arguing that it is bad science because of its imperialism. Let us look at Dupré's version of scientific imperialism in a little more detail to see if this holds up.

Dupré says "imperialistic economics is a deliberate attempt to import the perspective of economics into many diverse facets of human behavior" (Dupré 2001, 16). I think he would be happy with the following reworking of this claim: imperialistic evolutionary psychology is a deliberate attempt to import the perspective of evolutionary biology into many diverse facets of human behavior. According to Dupré, the problem here is that the imperialist imposes a general structural, or scientific theoretical, perspective irrespective of the nuances of context. He goes on to spell this out by discussing structure. The idea is that if you take yourself to have an account of the underlying structure of something, say human beings, you apply that structural account whenever you come across a human being. Dupré says that taking note of the relevant context will undermine such structural accounts and they will be "doomed to failure" (ibid., 16). The positive side to Dupré's proposal is pluralism, which he contrasts with imperialism. He says that we cannot properly understand any central features of human life from "any unique and homogeneous perspective" and "[t]he moral of pluralism, therefore, is not to point to better ways of doing science [...] but to show the limits of science, at least in its application to the complexities of human life" (ibid., 17). Dupré admits that what he takes to be a problem with imperialist ventures in science is for others a mark of their success and what makes them attractive to their adherents. For example, he thinks that what attracts people to evolutionary psychology is its "claims to breadth of application" (ibid., 81). He also says that "[o]bvious rewards and attractions accrue to the pursuit of scientific ideas with the greatest claims to generality of application" (ibid., 84). Dupré is not just concerned with generality understood in terms of the wide scope of application of the imperialist science but also in the sense that imperialist sciences stay above the level of what matters empirically. Evolutionary psychologists' "most imperialistic moments," he says, "are also the moments at which they are most inclined to substitute purely formal demonstrations of possibility for the hard work of demonstrating actuality" (ibid., 138). Perhaps the following characterization, which I have adapted from Mäki's analysis of imperialist economics, best captures Dupré's scientific imperialism:

> Evolutionary psychology is an example of scientific imperialism as it is the result of scientific expansionism of evolutionary biology. Evolutionary psychology locates new types of explanandum phenomena in psychology and other social sciences, and evolutionary psychology presents itself hegemonically as being in possession of superior theories and methods, thereby excluding rival theories and approaches from consideration.
>
> (adapted from Mäki 2009, 374)

Further, Dupré takes the newly imported theories to be mistaken in various ways and hence detrimental to psychology and other social sciences. This assessment is made on independent grounds from the attribution of imperialism but bringing mistaken theories to bear renders the imperialism more pernicious.

Dupré presents what I take to be a somewhat different view of scientific imperialism from the one outlined above when he says the following:

> whether it is from commendable epistemological enthusiasm, or from the desire to become rich and famous through writing best selling books, the tendency to exaggerate the scope of the theory with which one is professionally engaged is a familiar aspect of scientific life. *I referred to this phenomenon above as scientific imperialism.*
>
> (Dupré 2001, 82, emphasis added)

This seems to be a distinct notion from the one we have been focusing on so far. Clearly this notion is related but it pertains to popularizers of imperialistic science rather than practitioners. He gives examples: "physicists envisage final theories of everything, and not surprisingly these turn out to be physical theories" and "Darwinians, or their camp-followers, make almost equally ambitious claims. Recall the remark from Dennett [...] 'Darwin's dangerous idea is reductionism incarnate, promising to unite and explain almost everything in one magnificent vision'" (ibid., 82). I agree that popularizers of various sciences tend to exaggerate their scope of application but I take the charge of scientific imperialism to be part of an epistemic criticism of the relevant science rather than part of a critical appraisal of popular science writing. It is certainly annoying when popular science writers exaggerate and misrepresent the claims of the scientists they promote but the relevant scientists are not directly responsible for these exaggerated claims. There are many popular works that focus on sciences that promote easily digestible and widely applicable general ideas and so the approach to science may influence its broad appeal. It should be noted, though, that there are also bestsellers, for example, Stephen Hawking's *A Brief History of Time* (1988), whose content is completely obscure to most lay readers and yet sales indicate that this work has wide appeal. A detailed sociological examination of evolutionary psychology will doubtless reveal various connections between popular works on progress in the field and actual practice in the lab. For now, though, I want to focus just on Dupré's narrower version of scientific imperialism, characterized in Mäki's terms.

5 Empirical challenges for claims of imperialism

Kuhn (1962) characterizes scientific change in terms of scientific revolutions. We all grasp the outlines of Kuhn's view. A period of normal science guided by a paradigm starts to break down and a period of revolutionary science takes over. Many philosophers of science argue that Kuhn's approach to scientific change is irrationalist or arational, but many in history and sociology of science are not as worried about this issue and focus more on the facts on the ground in times of scientific change. The issues here are the extent to which Kuhn's characterization fits with the way in which actual theory change occurs in science or the way in which theory change has occurred in specific instances in the history of science. Mäki asks "what does it take to empirically identify instances of scientific imperialism?" (Mäki 2013, 338). As he points out, though, there are currently a number of different notions of scientific imperialism. Rather than trying to identify instances of scientific imperialism empirically, I propose comparing particular versions of scientific imperialism to the relevant empirical facts in their specific domain of application. I focus on some of Dupré's claims about evolutionary psychology as an imperialist venture and contrast them with some empirical observations about the field. First, I briefly outline an analogous move directed at Kuhn.

One question about scientific change is how the new theory/paradigm/research program, etc. can take hold when the folks who held the old view are still around. Kuhn held that Planck's principle was explanatorily relevant here. Planck said that new theories do not come to the fore because opponents to the theory come to accept it, but rather the opponents die off. A related idea of Planck's, popular among many scientists and historians and philosophers of science, is that younger scientists take up newer ideas more readily. David Hull and his collaborators (Hull et al. 1978) point out that these claims are open to empirical test, as the birth and death records of scientists throughout history are readily available. Hull also points out that it is more difficult to assess when scientists come to adopt "various positions on scientific issues," but not impossible. Hull focuses on the incredibly rapid adoption of Darwin's ideas in Britain shortly after the publication of *The Origin of Species*. In a mere ten years after the publication of *The Origin of Species*, Darwin's ideas had become dominant. Not all relevant scientists accepted Darwin's ideas after ten years, but Hull's finding is that "age explains less than 10 percent of the variation in acceptance" (ibid., 202). The Darwinian revolution may have come about due to extra-scientific factors but Hull reveals that the age of scientists was not one of the most important factors.

Dupré makes a number of claims about the imperialism of evolutionary psychology that are open to empirical investigation. One is that evolutionary psychology is an "imperialist adventure from evolutionary biology" (Dupré 2001, 16). Evolutionary psychologists certainly couch their claims in evolutionary language but the most consistent criticism of evolutionary psychology since its inception is one or other version of the claim that evolutionary psychologists do not have the biology right. A related criticism is that evolutionary psychologists do not carry out experiments that confront evolutionary hypotheses. Evolutionary biologists have been some of the harshest critics of evolutionary psychology. John Maynard Smith claimed that evolutionary psychologists had not produced any models that were ready for appropriate testing. Richard Lewontin has argued that evolutionary psychologists employ an impoverished notion of adaption, and Jerry Coyne (see e.g. Coyne 2009) attacks evolutionary psychology on a number of evolutionary grounds. Clark Barrett's (2015) recent assessment of evolutionary psychology includes voicing concerns over how far from evolutionary theory his evolutionary psychology colleagues have strayed. Perhaps evolutionary psychology is an imperialist venture but it is a homegrown one. The appropriate political metaphor here is perhaps a coup, a movement cooked up by insiders using what they take to be the conceptual tools of a field that they admire. If the empire is not backing your adventure, it is perhaps not an imperialist adventure.

As we have seen, Dupré sees a connection between ideas with general applicability and imperialist moves in science. He also says "[o]bvious rewards and attractions accrue to the pursuit of scientific ideas with the greatest claims to generality of application" (Dupré 2001, 84). This sounds right, but like Planck's claims about science, it is also amenable to empirical investigation. One way in which we can measure both the rewards an idea accrues and its attractiveness is via citation analysis. If we assume for now that this is a reasonable measure, let us use it to test Dupré's claim. Recent research on the most-cited work of all time reveals a rather surprising result with respect to the generality of scientific work (Van Noorden et al. 2014). The most-cited paper of all time is a paper describing an assay for determining the amount of protein in a solution. No groundbreaking theoretical papers in any field, such as theoretical work leading to Nobel Prizes, come anywhere close to the top 100 most-cited papers. The top 100 papers are overwhelmingly those characterizing biological techniques (see Figure 11.1, for more data such as this). Certainly some of this work has led to Nobel Prizes but none of this work is paradigmatically general. The work certainly has wide applicability in experimental contexts and is tremendously useful but it is not work of the scope of the General Theory of Relativity. Citation analysis is only one way to measure the success and attractiveness of scientific ideas but if we use citations as an index of success, generality is not the route to success in science. It should be noted that citation analysis does not provide clear access to epistemic success. For example, increased citation count could simply mean increased popularity. However, there are good reasons to believe that scientific works are cited, for the most part, for their scientific content. This indicates that we should have some confidence in the conclusion here that generality is not a sure route to success in science.

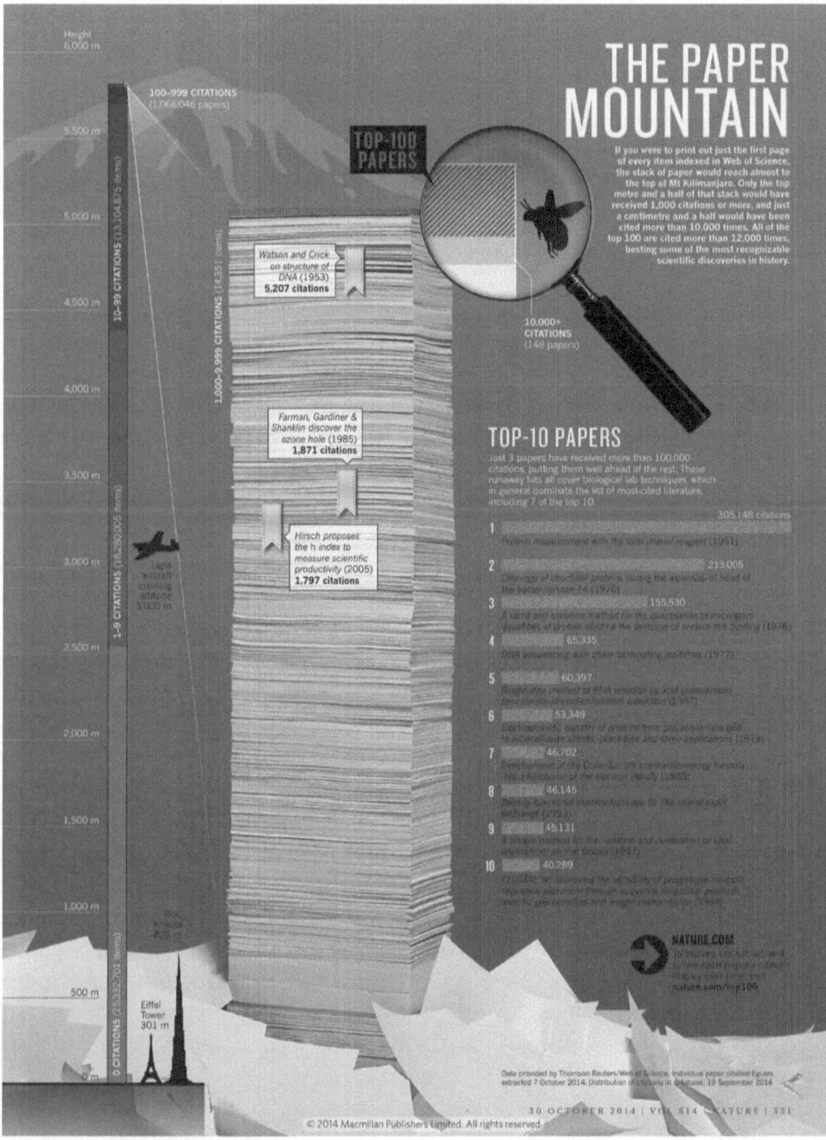

Figure 11.1 "The Paper Mountain" in "The Top 100 Papers"
Source: *Nature* 514: 550–553

I have no doubt that various non-epistemic factors play an important role in sustaining evolutionary psychology but I am skeptical that evolutionary psychology is sustained by evolutionary biologists and I am also skeptical that it is successful because its ideas have general applicability. It may be true that evolutionary psychology has popular appeal because it appears to be based in evolutionary biology and, when sloganized, its central ideas seem widely applicable. As I mentioned above, though, I think that the popular uptake of a given scientific field is neither a measure of its success nor a mark of the extent to which it is imperialist.

6 Conclusion

If the evidence supports my claim that evolutionary psychology is not supported by evolutionary biology and is therefore not best understood as an "adventure from evolutionary biology," perhaps that component can be dropped from our characterization of Dupré's scientific imperialism. Recall, I borrowed from Mäki to give this characterization:

> Evolutionary psychology is an example of scientific imperialism as it is the result of scientific expansionism of evolutionary biology. Evolutionary psychology locates new types of explanandum phenomena in psychology and other social sciences, and evolutionary psychology presents itself hegemonically as being in possession of superior theories and methods, thereby excluding rival theories and approaches from consideration.
>
> (adapted from Mäki 2009, 374)

Here is a characterization without the clause about expansionism from evolutionary biology:

> Evolutionary psychology is an example of scientific imperialism as it locates new types of explanandum phenomena in psychology and other social sciences, and evolutionary psychology presents itself hegemonically as being in possession of superior theories and methods, thereby excluding rival theories and approaches from consideration.

This perhaps better captures the activity of evolutionary psychologists on the ground, but without the expansionist component, does this adequately capture scientific imperialism? Is a hegemonic approach coupled with a broadening of the scope of application and excluding rival theories sufficient for imperialism?

Evolutionary psychology is clearly a very different endeavor from the type of evolutionary linguistics I introduced earlier. Evolutionary psychologists are very confrontational in their approach to their fellow social scientists and much of their work has a triumphal air to it. Even Barrett, who is relatively moderate in his approach, champions evolutionary psychology over the "impoverished set of models in psychology" (Barrett 2015, 12). I am not sure if refining our notion of scientific imperialism will help us better understand a field like evolutionary psychology. Perhaps other related political metaphors are more appropriate in this context. I do think that careful empirical analysis of evolutionary psychology, such as citation analysis, may reveal more about the dynamics of the field and help us understand its success in the face of apparently cogent and damaging criticism of its central ideas. In contrast, evolutionary linguistics appears not to be a candidate for scientific imperialism in Dupré's terms. The evolutionary linguists considered above do not present their approach in order to exclude rivals, and display none of the hegemonic tendencies Dupré sees in evolutionary psychology. The case of evolutionary linguistics does reveal problems for Dupré, as if scientific imperialism in this domain simply consisted in transporting evolutionary views into social science, then it too would be imperialist. To rule out evolutionary linguistics as a case of imperialism requires stressing the clause about excluding rival theories and approaches. Finally, the introduction of careful empirical analysis – such as citation analysis – of any proposed imperialist science should help the philosophical task of better characterizing scientific imperialism.

References

Barrett, H. Clark. 2015. *The Shape of Thought: How Mental Adaptations Evolve*. Oxford: Oxford University Press.

Buss, David M. 1989. "Sex Differences in Human Mate Preferences: Evolutionary Hypotheses Tested in 37 Cultures". *Behavioral and Brain Sciences* 12: 1–49.

Buss, David M. 1995. "Evolutionary Psychology: A New Paradigm for Psychological Science". *Psychological Inquiry* 6: 1–30.

Buss, David M. 1999. *Evolutionary Psychology: The New Science of the Mind*. Boston, MA: Allyn and Bacon.

Buss, David M. 2005a. *The Murderer Next Door: Why the Mind is Designed to Kill*. New York: Penguin.

Buss, David M. (ed.). 2005b. *The Handbook of Evolutionary Psychology*. Hoboken, NJ: Wiley.

Clarke, Steve and Adrian Walsh. 2009. "Scientific Imperialism and the Proper Relations between the Sciences". *International Studies in the Philosophy of Science* 23: 195–207.

Clarke, Steve and Adrian Walsh. 2013. "Imperialism, Progress, Developmental Teleology, and Interdisciplinary Unification". *International Studies in the Philosophy of Science* 27: 341–351.

Cosmides, Leda. 1989. "The Logic of Social Exchange: Has Natural Selection Shaped How Humans Reason? Studies with the Wason Selection Task". *Cognition* 31: 187–276.

Coyne, Jerry A. 2009. *Why Evolution is True*. New York: Viking.

Downes, Stephen M. 2015. "Evolutionary Psychology, Adaptation and Design". In Thomas Heams, Philippe Huneman, Guillaume Lecointre and Marc Silberstein (eds). *Handbook of Evolutionary Thinking in the Sciences*. Dordrecht: Springer, pp. 659–673.

Dupré, John. 2001. *Human Nature and the Limits of Science*. Oxford: Oxford University Press.

Gray, Russell D. and Fiona M. Jordan. 2000. "Language Trees Support the Express-train Sequence of Austronesian Expansion". *Nature* 29: 1052–1055.

Gray, Russell D., Alexei J. Drummond and Simon J. Greenhill. 2009. "Language Phylogenies Reveal Expansion Pulses and Pauses in Pacific Settlement". *Science* 323: 479–483.

Hawking, Stephen. 1988. *A Brief History of Time*. New York: Bantam Books.

Hull, David L., Peter D. Tessner and Arthur M. Diamond. 1978. "Do Young Scientists Accept New Ideas with Greater Alacrity than Older Scientists?" *Science* 202: 717–723.

Jordan, Fiona M. 2013. "Comparative Phylogenetic Methods and the Study of Pattern and Process in Kinship". In Patrick McConvell, Ian Keen and Rachel Hendery (eds). *Kinship Systems: Change and Reconstruction*. Salt Lake City: University of Utah Press, pp. 43–58.

Kuhn, Thomas. 1962. *The Structure of Scientific Revolutions*. Chicago, IL: University of Chicago Press.

Mäki, Uskali. 2009. "Economics Imperialism". *Philosophy of the Social Sciences* 39: 351–380.

Mäki, Uskali. 2013. "Scientific Imperialism: Difficulties in Definition, Identification, and Assessment". *International Studies in the Philosophy of Science* 27: 325–339.

Pinker, Steven. 1997. *How the Mind Works*. New York: W.W. Norton.

Singh, Devendra. 1993. "Adaptive Significance of Female Physical Attractiveness: Role of Waist-to-Hip Ratio". *Journal of Personality and Social Psychology* 65: 293–307.

Thornhill, Randy and Craig T. Palmer. 2000. *A Natural History of Rape: Biological Bases of Sexual Coercion*. Cambridge, MA: MIT Press.

Tooby, John and Leda Cosmides. 1992. "The Psychological Foundations of Culture". In Jerome H. Barkow, Leda Cosmides and John Tooby (eds). *The Adapted Mind*. New York: Oxford University Press, pp. 19–136.

Tooby, John and Leda Cosmides. 1997. *Evolutionary Psychology: A Primer*. Santa Barbara: University of California. www.psych.ucsb.edu/research/cep/primer.html.

Van Noorden, Richard, Brendan Maher and Regina Nuzzo. 2014. "The Top 100 Papers". *Nature* 514: 550–553.

Wright, Robert. 1994. *The Moral Animal. Why We Are the Way We Are: The New Science of Evolutionary Psychology*. New York: Vintage Books.

12 Logical form, the first person, and naturalism about psychology

The case against physicalist imperialism

Frederique Janssen-Lauret

1 The rise and fall of the physicalistic empire

Physicalist philosophers and psychologists trust that all facts about human psychology will ultimately be explained in physical terms. The view that physics is such an exemplary science that it will eventually account for the psychological as well as the traditional domain of physics is a classic case of scientific imperialism as first defined by Dupré (1995). The explanatory capacity of physics, both with respect to its methods and to its domain, is taken to extend beyond the traditional realm of physics, and into that of psychology. Is it, as Dupré assumes imperialism must be, a harmful development? In the 20th century, several practicing psychologists modeled their methods and explanations on those of physics, preferring intersubjectively observable descriptions of behavior or brain states to first-personal self-reports. Assuming a naturalistic point of view, the physicalist approach to psychology and its philosophy seemed well-supported in view of these developments. At the time physicalism seemed like a case of neutral disciplinary expansion (see this volume's Introduction) or even beneficial scientific imperialism (of the kind proposed in Mäki 2009). I will argue in this chapter that more recent theories in psychology are moving in the opposite direction, and so naturalists ought to conclude that the project of physicalist imperialism has failed. As a philosopher of logic, I will focus on the under-appreciated problem of logical form: the recent move towards first-personal methods and theoretical statements in psychology, rather than impersonal ones modeled on the logical form of physics.

Twentieth-century philosophical naturalists had some good reasons to endorse physicalism, given the scientific practices of their contemporaries. At the time, psychologists strove to develop theories that shared a logical form with physics. The physicalist imperialism involved was therefore disciplinary, not domain-only (see this volume's Introduction, 4). The logical form of physics is impersonal; it does not distinguish between first, second, or third person. All observations hold independently of who the observer is. Twentieth-century scientists equated this feature of physics to scientific objectivity, and strove to emulate it in psychology. They replaced first-personal self-ascriptions of psychological qualities with impersonal claims about behavior or brain states. From a naturalistic point of view, physicalism would have been vindicated if these efforts had led to improved scientific theories, theories which were simpler, more explanatory, more fecund, more predictive and testable. A sample of the contemporary scientific literature in psychology reveals that practicing scientists did not find the resulting theories fit for purpose. The 21st century saw a return to first-personal methods and theoretical statements, especially in areas like social psychology. As enthusiasm about physicalist imperialism has waned among empirical scientists in the relevant fields, 21st-century philosophical naturalists should call for a re-examination of physicalism. I make the case that the recent resurgence of the first-personal in psychology points towards a new challenge for physicalism: the problem of expressive power. First-personal theories have a richer logical form, and potentially greater expressive power, than impersonal theories.

My case is based on philosophically interesting consequences of the difference in logical form between physics and psychology, and does not rely at all on familiar philosophical concerns about consciousness or the nature of the mind. A previously published argument of mine which uses only general principles belonging to the philosophy of logic, plus the plausible assumption that objects indiscriminable by physical descriptions are to be identified for the purposes of physics, implies that impersonal theories have diminished expressive power compared to theories with first-personal pronouns. Adding the equivalent of "I" to a theory results in a minor but theoretically fruitful increase in expressive strength. Such theories can, where impersonal theories cannot, discern objects that are descriptively indiscriminable.

2 Naturalism, physicalism, and the explanatory scope of physics

Scientific imperialism occurs when the explanatory scope of one discipline is stretched to encompass a domain traditionally assigned to another (called "domain-only" imperialism, see the volume Introduction, 4), or when the methods of one science are forced onto another science which previously had its own distinct methodology. Physicalistic practitioners of psychology and its philosophy have tried, for philosophical reasons, to bring the methods and explanatory power of physics to bear upon psychology. I am concerned here with one previously neglected way in which this physicalistic movement has been unsuccessful: a difference in logical form. Unlike most anti-physicalist arguments, the arguments of this chapter do not flow from *a priori* philosophical opposition to physicalism. My point of departure is one many physicalists claim to share: philosophical naturalism. Before delving into the ramifications of the differences in logical form between impersonal physics and first-personal contemporary psychology, I will show in this section that there is conceptual space for a naturalist philosophy which is not physicalist. Subsequent sections will make the case that non-physicalist naturalism is preferable to physicalist naturalism.

2.1 Naturalism does not imply physicalism

Naturalists need not be physicalists. Naturalism is a methodological thesis, which states that our best philosophical theories derive from fitting philosophy around the deliverances of the special sciences, and rejects first philosophy. Manuela Fernández Pinto, in her chapter in this volume, calls this type of naturalism the "passive cooperation" type. As she puts it, naturalists of this type use the results of scientific research for limiting the scope of epistemological inquiry (this volume, 277). I take a similar line, and have defended, in a joint paper, letting our best scientific theories guide our answers to questions of ontology and ideology too. We can arrive at credible answers to philosophical questions about the most general categories of things there are only by aiming to unify and clarify the theoretical statements of the special sciences into a coherent overall theory (MacBride and Janssen-Lauret 2015, 294). Naturalists admit no higher authority than our current best scientific theories, arrived at by means of what is presently our best scientific methodology. Physicalism, by contrast, is a metaphysical thesis, the philosophical doctrine that everything is physical. (One popular formulation of physicalism, "everything is physical or supervenes on the physical," is dubiously compatible with methodological naturalism. Supervenience requires quantification over possible worlds, purely philosophical entities rejected as posits by most current theories of physics, as we will see in section six, below.) Naturalists should follow the scientific method where it leads, and endorse whatever our current best empirical theories are. They should be physicalists only if those theories talk exclusively about physical things. Whether or not those theories do is a matter to be investigated, not stipulated.

Physicalism and naturalism are two logically independent doctrines. Physicalism does not entail naturalism. Physicalists quite frequently base their physicalism explicitly upon reasons at odds with naturalism, such as philosophical intuitions or appeals to common sense. Appeals to common sense and intuition have at times been in conflict with, and been overthrown by, scientific progress. Cantor and Einstein did not let philosophical intuitions about the absurdity of infinite wholes the same size as some of their proper parts or about non-Euclidean geometry stand in the way of the development of set theory and general relativity (MacBride and Janssen-Lauret 2015, 291). Nor does naturalism entail that everything is physical. Although it is perfectly possible that at some points in human history our best evidence is in favor of physicalism, we must draw conclusions based on our *present* best theory. As I argue below, our present best theories of psychology are not physicalist. It cannot be assumed that the scientific method inexorably pushes towards the discovery that everything is physical. We cannot maintain physicalism based on the hope that future science will confirm it. To assume that physics will find a way to explain all facts about human psychology without actually having a credible way to explain them in such terms is not naturalism, but a kind of first philosophy, where intuitive philosophical convictions are allowed to take precedence over *a posteriori* empirical findings. We must opt instead for careful philosophical interpretation of recent scientific results, weaving our empirical findings into a unified theory without giving precedence to philosophical arguments from first principles. Demarcations of science from non-science ought to be made on methodological grounds. Maddy has argued that the very fecund and successful discipline of mathematics should not be denied the title of "science" merely because its posits are for the most part not quantified over in physics (Maddy 1997). Similarly, naturalists should favor theories which score highly on our current best criteria of theory choice, whether physicalist or not. The correct naturalistic strategy is to investigate whether psychology can be interpreted in physicalist terms.

2.2 The relationship between physics and physicalism

Our best and fullest knowledge of physical objects derives from the science of physics, not from first philosophy. So naturalist philosophers who wish to be physicalists should derive the subject matter of their theories exclusively from the subject matter of physics, and model their methods as closely as possible on the methods of physics. Physics in itself remains neutral on the question of physicalism. Physics makes a variety of existence claims, quantifying over fields, particles, real numbers, space–time points, and other physical posits, but it does not state that its posits are all there is. Although it arguably quantifies over mathematical objects (Quine 1981; Putnam 1979), and assumes the truth of significant parts of mathematics, it does not say anything at all about the truth or falsehood, ontology or methodology of most other sciences. In particular, it says nothing about the sorts of objects discussed in the life sciences or social sciences, such as organisms, ecosystems, societies, or personalities, and therefore does not claim that they either fail to exist or are purely physical objects.

Where the evidence points towards all other sciences conforming to the methods and subject matter of physics, or to their discourse being reducible to that of physics, the methodological naturalist has good reason to embrace physicalism. The physicalistic project might have looked to naturalists like a harmless venture into disciplinary expansion at first, and there were indeed some good arguments for considering physicalist imperialism a viable project over the course of the 20th century. The first is that practicing scientists were making moves in this direction, especially towards explaining psychology in terms that referenced only impersonally observable behavior. Another derives from the search for a unified science (see also the Introduction to this volume, 5). A unified science seems easier to arrive at if we only need to quantify over the domain of a well-established discipline like physics which we already accept as true, and consider the life sciences and social sciences as mere applications of it. As other disciplines, like biology and psychology, do contain much applied physics, physicalist disciplinary expansion was originally a promising idea. Yet, where empirical results defy such treatment, it would be wrong to allow a philosophical program, physicalism, to take precedence over careful weighing of *a posteriori* empirical findings. Unified science should result not from forcibly subsuming one discipline to another, but treating equally all theories that perform well on our best criteria of theory choice.

A unified scheme may just as well result from one scientific discipline adding to another. In this chapter I offer one argument to suggest that this may be the relationship between current physics and current psychology. Twenty-first-century psychology deploys first-personal methods and statements which suggest that psychology assumes the truth of physics, but adds further theoretical statements that are inexpressible in terms of physics' logical form. Several branches of contemporary psychology rely heavily upon statements in the first person, and theoretical statements about what social psychologists call "the self," the referent of the first-person pronoun. Since the theoretical statements of physics are all paradigmatically impersonal, the methods of psychology differ from those of physics down to their very logical form. The canonical language of psychology does, where the language of physics does not, require grammatical differentiation between self and other. The truly naturalist response is to search for the best interpretation of all deliverances of the empirical sciences, and consider the possibility that recent advances in first-personal psychology point towards a movement away from explanations in terms of physics. Of course, it remains possible that physicalists will answer this challenge, and provide an interpretation of these results in physicalistic terms, but the problem of logical form represents a new challenge for physicalism which must be addressed.

3 Impersonal logical form and criteria of identity in physics

3.1 Impersonal logical form and the regimented language of physics

The characteristic logical form of physics is entirely impersonal. Physics has no theoretical statements in the first or second person. Its methods, too, are impersonal. They are such that all evidence and observations must hold regardless of who the observer is. So the theoretical statements of physics correspond exactly to Quine's model of the canonical form of scientific theories: they have the logical form of classical first-order logic without individual constants. (Proponents of second-order logic may substitute "second-order" for "first-order" here. It makes no difference to my argument, which only pertains to the difference in expressive power between canonical languages with and without individual constants.) All physical pronouns are impersonal pronouns, the formal analogue of "it": "x," "y," "z," etc. Theoretical statements of physics have quantified logical forms like $\exists x\varphi(x)$, $\forall x\varphi(x)$, where "φ" may be replaced with any open formula of the language of physics, and truth functions of those quantified statements (Quine 1960, Ch. 5).

A key feature of physical theory, reflected in its logical form, is that all knowledge of physics is descriptive knowledge. The posits of physical theories are never observed directly. They are not known to us by acquaintance, with our minds reaching out to them and grasping them whole, but only as solutions to puzzles that arise about the pattern of our observations. All we know is that if our best physical theories are true, we have reason to believe that there is something or other which answers to their quantified statements. Physical descriptions convey only general thoughts, thoughts about whatever it is that meets the criteria set out by our theory. They never convey singular thoughts about some specific individual we singled out prior to formulating a theory. As we gradually improve our theories, step by step we readjust our understanding of physical objects too, positing, for example, sub-atomic particles instead of atoms, and space–time instead of three-dimensional space. Moving towards better explanations to account for new experimental evidence and observations, we dispense with the need for things that answer to the old descriptions, and start believing in objects that satisfy the open formulae of our new theories instead. So our grasp of physical objects is always theory-laden. We do not have theory-independent cognitive access to physical objects which outstrips the roles they play within physical theories (MacBride and Janssen-Lauret 2015, 297).

3.2 Criteria of identity in physics

Every physical posit is presumed to be in the domain of physics because it satisfies a distinctive explanatory role, φ. If our best physical theory requires for its truth that something or other answers to φ, we have good reason to believe in the φ. The explanatory roles played by our posits within physical theories impose criteria of identity on physical objects. We only have reason to believe in exactly one object per full explanatory role. A fully specified explanatory role is highly specific, including, for example, the object's exact spatio-temporal location and all its physical relations to other objects. For most physical posits, relative indiscernibility provides a full explanatory role. An exception is the case of certain quantum particles, entangled fermions, which are only weakly discernible (Saunders 2006, 60). Here we need an irreflexive sentence in two free variables to tell them apart. Quine attributes this relation of discriminability in two free variables to Fox, so let us call it "Fox-discriminability" for short (Quine 1976, 115). Now we can conclude that all things posited by physics are subject to a generic criterion of identity: any two posits that are indiscriminable must be identified. Here "x and y are indiscriminable" means that either x and y are not quantum particles, and they are indiscernible, or x and y are merely weakly discernible posits like entangled fermions, and are Fox-indiscriminable.

Why do we never have good reason to posit two distinct objects x and y which are indiscriminable with respect to their explanatory role in physics? Because in that case y would be completely theoretically inert. All the explanatory work that y could do is already fully taken care of by x. So we cannot justify believing in y as well as x. Our theory would carry on ticking over just as well if we had only one of them. Physical objects that satisfy exactly the same full physical descriptions (where necessary, in two variables) count as exactly one object. Physical theories regimented into canonical form cannot have two distinct but indiscernible values of variables in their domains (Janssen-Lauret 2015, 156–157).

3.3 Equivalence relations in the language of physics

As a result, the strongest equivalence relation for physics, as for theories regimented according to Quine's strictures generally, is slightly weaker than identity properly so called (Janssen-Lauret 2015, 155–157). Identity, in systems without constants or logically proper names, is equated with indiscriminability by the predicates of the theory. In our case, this means indiscriminability by means of the predicates of physics. Identity proper is the relation that partitions the domain into singleton equivalence classes, or simply the relation of being the same thing. This relation is stronger than any sort of descriptive indiscriminability. After all, it is always coherent to say that indiscriminable objects are distinct. However detailed our physical predicates may be, it remains logically possible that x and y satisfy all and only the same open sentences (in one or two free variables) and remain distinct. We cannot derive a contradiction from conjoining the regimented version of "x and y are indiscriminable" with "$x \neq y$" (Ramsey 1931 [1925], 31). The only exceptions are cases where one of the physical predicates is of the form "$= x$," but that is ruled out in this case. The language of physics identifies physically indiscriminable things, so it does not admit of a predicate which has "=" occurring within it, where "=" connotes old-fashioned sameness-of-thing identity.[1] We are only entitled to believe in objects that do some explanatory work for us, not in theoretically inert objects that play the exact same explanatory role already performed by some other posit (Janssen-Lauret 2016, 595–597).

There is always a logical possibility that two physically indiscriminable objects are distinct, but the above result implies that physicalists can have no good reason to believe in physically indiscriminable but distinct objects. To make the cut for being in the domain of a physicalist theory, an object has to contribute a distinct physical explanatory role, which means that it satisfies some open formula (possibly in two free variables) which no other object in that domain satisfies. Beyond the use of physical predicates, there is no theoretical apparatus available to distinguish x from y. Nor can physicalists help themselves to a stronger language, with an identity relation not subject to the criterion of identity "indiscriminable by the language of physics." Physicalists must accept the best current empirical theories used by practicing scientists as true, including their criteria of identity.

4 Impersonal vs. first-personal logical form in psychology

4.1 Impersonal psychology?

Twentieth-century physicalists working in psychology and philosophy made explicit efforts in a variety of ways to subject psychological theories to impersonal logical form. Traditional behaviorists regarded first-personal methods, and data acquired by means of them, as unscientific because they failed to resemble the impersonal methods of physics, which behaviorists regarded as a paradigm of scientific objectivity, in one crucial way. First-personal observations are not, like the observations and experiments of physics, observable or repeatable independently of who the observer is. Behavior, by contrast, is intersubjectively observable whether the behavior in question is our own or other people's. It is also describable in completely impersonal terms, with no special authority being given to the first-person point of view. As a result, the project of translating the language of psychology into the language of behavior was taken to be naturalistically sound. At the time, it really was promising from a naturalistic point of view, because behavioristic methods and theoretical statements were the order of the day among working psychologists. Psychologists and psychiatrists began to find behaviorism's strictly third-personal methods rather limiting in the 1960s and 1970s (Miller 2003). First-personal methods began to be used again; nowadays they are in full swing, especially in areas like social psychology, as can be seen from my sampling of the psychology literature below. Philosophical naturalists must adjust their theories along with theoretical and methodological shifts made by practicing scientists.

Post-behaviorism, many physicalists had high hopes for an eliminativist approach to psychology. Eliminative materialists predicted that psychology would fall out of use altogether and be replaced with completed neuroscience (Churchland 1981, 67). First-personal claims about the subject's psychological state would form no part of the evidence or theories of human psychology. They would be replaced with impersonally describable claims about brains and brain states, i.e. physical objects and events. More than 30 years on, the eliminativists' predictions have not come to pass. To date there has been no sign of psychology fading away. Instead, the late 20th and early 21st centuries have seen a robust return to first-personal methods and first-personal theoretical claims. It was always a dubious move for naturalists to profess the truth of physicalism based on predictions that successors to our present best theories would be physicalist. As we reflected above, philosophical naturalists cannot rest content with vague promises for successor theories, but must accept the ontology and ideology of our present best theory. Now that 30 or more years have gone by without any clear sign of psychology being replaced by neuroscience, there is all the more reason to embrace naturalism and follow the evidence, rather than cling to physicalism in spite of the evidence.

Second philosophers who had endorsed physicalism for naturalistic reasons ought to have responded by holding on tightly to their naturalism and re-evaluating their physicalism. Philosophers, for the most part, have done exactly the opposite of that, relaxing their definition of physicalism to "everything is physical or supervenes on the physical." Introducing supervenience here loosens the tie between physicalism and the science of physics, since it requires quantification over possible worlds. Possible worlds are philosophical posits, which most working physicists would either not recognize at all, or be highly dubious of. Possible-world explanations are not completely unheard of in physics. The Everett many-worlds interpretation of quantum mechanics, for instance, relies upon possible worlds. Still, such explanations are both highly controversial among physicists and associated with unresolved debates in the philosophy of physics as to exactly which physical entities are invoked within them (Wallace 2002). Philosophers' accounts of possible worlds generally present them as clearly abstract entities, which are not physical, but mathematical or metaphysical posits (Adams 1974; Plantinga 1974). More rarely, possible worlds are thought of, Lewis-style, as concrete objects. Even these are not uncontroversially physical entities, because some of Lewis's possible worlds contain non-physical components (Lewis 1986, x).

4.2 First-personal methods in contemporary psychology

Twenty-first-century psychologists still use some behavioristic methods, but the use of first-personal data and statements is now also widespread. A key source of data in various branches of psychology is composed of self-reports, where subjects ascribe qualities to themselves in the first person – precisely the sort of thing behaviorists denounced as opposed to scientific objectivity. Attitude surveys, which are very common in contemporary psychology, use such first-personal self-reports as their primary kind of evidence (Sirken et al. 1999). Grammatical differentiation between self and other is a necessary feature of this kind of methodology. Recent research into memory has yielded results that are inexpressible without distinguishing first and third person: for instance, a case study that showed first-personal judgments to be more reliably remembered than third-personal ones by a patient with severe Alzheimer's (Klein et al. 2003). Some neuroscientists, too, are investigating the explanatory potential of the first person, and show more interest in furthering the aims of first-personal psychology than in the eliminativist physicalism beloved of several philosophers of neuroscience. In a series of experiments by Heatherton et al., for instance, brain scans revealed distinct medial prefrontal cortex activations for first-personal vs. third-personal character trait ascriptions, and the authors have drawn upon these results to engage in fruitful dialogue with social psychologists on the question of "the self" (Heatherton et al. 2007). Active researchers in psychology have turned away from the impersonal approach of behaviorism, and the brain state-based approach of eliminativism, having found that they make better third-personal predictions and explanations of people's behavior by taking those people's first-personal statements as good, though not infallible, evidence about them.

In social psychology, many prominent figures go further and freely make theoretical statements about "the self," which they take to be the referent of the first-person pronoun (Baumeister et al. 2007; Baumeister 1987, 2010).[2] The self is explicitly distinguished from any "part of the brain," "the animal body" as well as from "what happens inside the individual mind" (Baumeister 2010, 139). Characteristics ascribed to it include typically psychological ones, but also physical, interpersonal, and social ones. Our cognitive access to the self, in some cases, is presumed to be essentially first-personal and introspective, a form of knowledge by acquaintance: "one important part of the self exists mainly inside the individual's own mind … it starts as people pay attention to themselves" (Baumeister 2010, 143). We also know ourselves in more indirect ways, via knowledge by description of our own social role, or via awareness of our own bodies. For instance, self-awareness is heightened in people who can see themselves in a mirror. Our awareness of how our bodies are perceived by others also has major repercussions for self-awareness. Eating disorders, for example, are associated with changes in self-awareness: binge eating decreases self-awareness, and is used by sufferers to escape from acute self-awareness (Heatherton and Baumeister 1991), while self-imposed dietary restraint is associated with heightened self-awareness (Bleau 1996). The logical form of the theoretical statements used by these psychologists is of a radically different kind from the impersonal form of physics: it relies essentially on grammatical distinctions between self and other, and there is good reason to believe that they cannot be translated into the impersonal language of physics. The problem is not merely that we do not have a clear translation scheme for rendering psychological claims into exactly equivalent physical claims. Although this is true, the problem runs deeper than that. First-personal theories have a richer logical form than impersonal ones, and can aspire to levels of expressive strength inaccessible to impersonal theories.

My argument in the next section combines a result of mine which uses only general principles from the philosophy of logic (Janssen-Lauret 2016, 594–599) with the principle of identification of physically indiscriminable objects used by philosophers of physics, which we discussed in section three, above. It will become apparent that the first-personal statements of psychology cannot be translated back into the impersonal logical form of physics without loss of information. Contemporary psychology assumes the truth of contemporary physics, but adds something extra, some expressive strength in its syntax which the language of physics lacks. In particular, the canonical language of psychology, unlike that of physics, is capable of distinguishing physically indiscriminable objects.

5 Logical ramifications: the increased expressive strength of languages with first-personal logical form

5.1 First-personal logical form and knowledge by acquaintance

From a logical point of view, first-personal pronouns like "I" behave differently from impersonal pronouns like "it" or "*x*." Suppose we follow the early Russell in treating "I" as a logically proper name (Russell 1910–11). Syntactically, it is a kind of constant, "*i*." So where the language of physics is first-order classical logic without individual constants, the language of psychology adds to that logical form statements of the form "φi," with "φi_1," "φi_2," "φi_3" for first-person pronouns with different referents. We need not prejudge the question whether the referent of the first-person pronoun is a stable self or a Humean bundle or succession of different person-stages. As the answer to this question is to some extent empirical, the logically proper names "i_1," "i_2," "i_3" may be used either for different selves or persons, or for different person-stages, whichever are posited by the psychological theory at hand.

Physicalism now faces the objection that this logical form is not expressible in the language of physics. Lacking constants in their syntax, regimented physical theories have only one type of pronoun, and no analogue of "I" or logically proper names. A canonical language with constants or logically proper names has more expressive strength than one with only impersonal pronouns like "it," or "x," "y," and the like. Unlike physical theories, such a language is potentially capable of distinguishing objects that are descriptively indiscriminable. As we saw in section three, physical posits that are physically indiscriminable are identified. The full extent of our knowledge of physical objects proceeds via physical descriptions, so where those coincide, objects must coincide.[3] This is not the case for a language that contains the additional logical form of logically proper names in its theoretical statements. Logically proper names are associated with knowledge by acquaintance rather than knowledge by description. Knowledge by acquaintance, or direct cognitive access to an object, is in itself no less naturalistic a way of knowing an object than indirect knowledge by description is. I have argued, for instance, that Ruth Barcan Marcus makes extensive use of knowledge by acquaintance for naturalistic reasons (Janssen-Lauret 2015, 158–164). First-personal knowledge has the potential to be associated with immediate awareness of ourselves or our own current psychological states. The psychology literature bears out that many active researchers in psychology take first-personal statements as evidence of such awareness, which is taken as a basis for self-knowledge, knowledge of the referent of the first-person pronoun. "The first basis for selfhood is consciousness turning around on itself ... Self-knowledge would be impossible without self-awareness, which is the basic process by which attention turns around toward its source" (Baumeister 2010, 142). Contemporary psychological evidence may include knowledge by acquaintance, reported by the subject in the first person, which bears some resemblance to what philosophers call introspective knowledge, yielding immediate knowledge of the subject's own psychological state by means of privileged access. In psychological research this is rarely used as a stand-alone method, but rather in combination with more indirect ways of knowing the self, for instance by means of knowledge of the body. For example, attitude self-reports are more accurate when subjects are placed in front of a mirror (Pryor et al. 1977).

5.2 Identity vs. criteria of identity

Objects known by means of acquaintance or introspection, and denoted by constants or logically proper names, resist being given criteria of identity based purely upon descriptive explanatory roles. My argument for this claim, as I noted before, is based wholly upon the philosophy of logic and some general principles about criteria of identity. It does not deploy any peculiar assumptions from the philosophy of mind or specifically anti-physicalist philosophical arguments, but flows from the difference in logical form between equivalence relation statements that deploy descriptive indiscriminability and those that deploy identity. It is a syntactic feature of directly referential expressions, which here includes variables as well as constants, that they can legitimately be concatenated with the identity predicate. As a result, it can be true or false *tout court* that $i_1 = i_2$, that these are the same object. By contrast, definite descriptions cannot simply be substituted for constants on either side of the identity sign; their deep logical structure is different from their surface form. "The F = the G" must be regimented to reveal its underlying logical form "$\exists x((Fx \wedge \forall y(Fy \rightarrow x = y)) \wedge \exists z(Gz \wedge \forall w(Gw \rightarrow z = w)) \wedge x = z)$," with only variables on either side of the identity predicate. As we saw above, the language of physics does not have "=" in the sense of identity proper, sameness of thing. Its strongest equivalence relation is indiscriminability by physical theory. So this statement would have to be weakened to "the F is physically indiscriminable from the G."

The difference between "i" and "the F," and between "=" and "is physically indiscriminable from," is not merely syntactic. It has profound philosophical consequences. Identity is the relation that relates every object to itself and to no other thing. Identity does not hold by virtue of any special condition that the object fulfills, but simply and primitively because it is the same object as itself. Now the semantic role of descriptions is to specify some condition an object must meet. An object is denoted by this description only if it meets that condition. The semantic role of names and logically proper names is a different one: it is just to denote an object directly, without specifying anything about what the object is like. The object denoted by a proper name is denoted only qua object, not qua satisfier of some condition. Identity, which is exactly the relation of being the same object, can only be expressed with the help of expressions that single out the object only qua object. By contrast, the fact that an object x meets some condition F a description sets out never guarantees that there is only one such object, nor that any F is the same identical x, unless identity is smuggled into the description by the use of "=." Otherwise, no matter how detailed the description, it remains logically possible that two or more things satisfy it. In the case of constants or logically proper names, by contrast, it is a crucial aspect of their semantic role that they denote that thing, and only that thing. They refer without needing to rely on descriptive conditions. If they had a descriptive component, then their denoting their referents would depend upon their meeting the condition set out in that description, as is the case for the theoretical terms of physics. If that were so, just like for physical posits, any objects which satisfied that condition would be considered identical. The strongest equivalence relation expressible by means of such expressions would not be identity proper, quantitative identity, but some form of descriptive indiscriminability, or qualitative identity (Marcus 1961, 12). Constants and logically proper names, by contrast, single out only the object itself, irrespective of its qualities. So it is no coincidence, but a matter of logic, that once they have a referent, they do nothing but refer to that referent, and that they can feature in statements of quantitative identity (Janssen-Lauret 2016, 595–597).

We saw that physical theories have a slightly weaker relation than identity as their strongest equivalence relation, namely physical indiscriminability. The canonical language of physics also does not have logically proper names, but only descriptions. As a result, physical theories cannot make any meaningful distinction between "$i_1 = i_2$" and "the *F* is physically indiscriminable from the *G*," but these have distinct logical forms, and the latter by no means entails the former (Marcus 1990, 197). By the argument of the previous paragraph, the identity claim is stronger than the indiscriminability claim. Although every object that is physically indiscriminable from something is physically indiscriminable from itself, physical indiscriminability is logically compatible with distinctness. That means a claim of physical indiscriminability is weaker than any identity claim, since identity claims express logical facts (Janssen-Lauret 2016, 597–598). A physicalist may object that a difference expressible only by means of expressions like "$i_1 = i_2$" cannot be expressed in the language of physics, and therefore she need not take them seriously, but this just amounts to an admission that the language of psychology genuinely does have more expressive power than the language of physics. Psychologists who have embraced first-personal logical form do not deny the truths of physics, but extend physical theory with additional first-personal logical forms. Since these working scientists derive better empirical results from theories with that richer logical form, naturalists ought to embrace the logical form they introduce.

6 Conclusion: a new challenge to physicalism

In what sense can physicalism be maintained given these results? Physicalism based on first philosophy is dubiously motivated, but physicalists who reject first philosophy must accept the deliverances of our current best empirical theories, including psychological ones which defy formalization in the canonical language of physics. Their theoretical statements include first-personal ones which embody a richer logical form than that of current physics. As contemporary psychology also assumes the truth of current physics, and contains a large number of statements of applied physics and biology, its canonical language provides a genuine, irreducible addition to that of physics. It adds genuinely psychological truths to the collection of physical truths. It is unclear where this leaves physicalism, the thesis that everything is physical. We derive our best and fullest knowledge of physical things from their explanatory roles described by the science of physics. Since psychology adds statements with a demonstrably richer logical form, we have good reason to avoid classifying these statements as physical knowledge. So it is far from obvious how we could have good reason nevertheless to classify them as knowledge of exclusively physical things.

A physicalist might protest that while she accepts that the methods of psychology differ from those of physics, she nevertheless maintains that their underlying subject matters coincide. One potential route to this conclusion is to say that the psychological supervenes on the physical, but we have seen that the quantification over possible worlds that supervenience involves renders it dubious, because possible worlds are not obviously physicalistically acceptable entities. An alternative argument claims that first-personal language merely provides a new perspective upon, or mode of presentation of, familiar old physical truths.[4] It is difficult to see how a naturalist could justify this argument. Once we admit first-personal theoretical statements, which naturalists must do because they are part of our current best theories of empirical psychology, and admit that these statements are not equivalent to physical statements, which we ought to do because of their richer logical form, no good empirical reason remains for saying that their subject matters must coincide. To identify two posits, we need to have reason to believe they share an explanatory role. Physical posits are always introduced by means of impersonal pronouns, while some psychological posits are introduced by means of first-personal pronouns that cannot be translated into impersonal form. Our current best theories do not allow for the role of the latter to be described in the vocabulary associated with the former, so we do not have good reason to ascribe the exact same explanatory roles. My argument is not that the identification of psychological and physical posits is impossible, or that no future theory will achieve this feat. Perhaps it will. Yet, we cannot identify x and y on the grounds that it is not impossible that they are identical. By way of comparison: it is not impossible that Julius Caesar is identical with a number, but we do not identify any number with the conqueror of Gaul.

A final option that is popular among 21st-century physicalists is so-called "non-reductive physicalism." Non-reductive physicalists believe they are entitled to ascribe biological, psychological, mental, and social properties to objects while maintaining that the objects in question are physical. In a sense, this tallies well with the way in which social psychologists now speak about the self. Physical as well as distinctively psychological and social characteristics are freely ascribed to the self, and self-awareness and self-knowledge are taken to include knowledge of the body just as much as introspective knowledge (Baumeister 2010, 143). Still the question arises how we can justify calling such objects physical objects which happen to have some psychological characteristics, rather than psychological objects which have some physical characteristics. After all, we ascribe to these objects various characteristics of which the science of physics makes no mention, so it is unclear in precisely what sense they are physical. Certainly they are not readily identified with the posits of physics. Psychology builds upon the truth of physics but adds distinctively psychological truths. Physics, by contrast, remains entirely silent on the psychological. Since psychology assumes and adds to the truths of physics, but not the other way around, entities that answer to psychological as well as physical descriptions are more appropriately classified as psychological entities which also have physical characteristics than as physical entities which happen to have psychological characteristics.

In summary, recent developments in the science of psychology, especially social psychology, appear to indicate that our best psychological theories contain first-personal theoretical statements, with a radically different logical form from those of physics. This suggests that the forces of physicalist imperialism, which aimed to model the methods and theoretical statements of psychology upon physics, are now in retreat. If the physicalist is to be a good naturalist, she will have to answer the challenge of logical form outlined here, and explain how the thesis that everything is physical can be maintained whilst giving due diligence to first-personal theoretical statements in the science of psychology.

Acknowledgments

I would like to thank Manuela Fernández Pinto, Uskali Mäki, and Adrian Walsh for their efforts in putting together the conference and volume, as well as for their helpful comments on a draft. Thanks are also due to the people who attended my University of Campinas seminar "Logical Form and Ontology" in the fall of 2014, and to the audience at the international Glasgow-Campinas workshop "Quine, Science, and Ontology," where an early draft of part of this chapter was presented. I am especially grateful to George Bealer, Renée Bleau, Gary Kemp, Nathan Kirkwood, Fraser MacBride, and Erasmus Mayr for discussion, and to Alan Weir for providing a response to my draft at the workshop. This research was supported by a Capes Postdoctoral Research Fellowship Grant.

Notes

1 We could also derive such a contradiction in a language where the identity predicate is defined away completely. Quine proposes such a language, retaining the old-fashioned identity predicate "=" but ditching its old-fashioned meaning, "being the same thing," and its old-fashioned extension of all and only the domain's singleton equivalence classes. Instead he defines the identity predicate for each language as having exactly the sense of indiscernibility according to the predicates of that language (Quine 1961, 1970, 63).
2 Special thanks to social psychologist Dr Renée Bleau for discussion of how the term "the self" is used in social psychology, and for drawing my attention to Baumeister's work.
3 There is some debate in the literature regarding the case of bosons, which are alleged to be absolutely indiscernible (Saunders 2006; Hawley 2006), but bosons cannot be known by acquaintance, and so would not help the physicalist establish that physics does have exactly the same expressive power as first-personal psychology.
4 Thanks to Erasmus Mayr for this objection, even though he does not wholeheartedly endorse it.

References

Adams, Robert Merrihew. 1974. "Theories of Actuality". *Noûs* 8: 211–231.
Baumeister, Roy. 1987. "How the Self Became a Problem: A Psychological Review of Historical Research". *Journal of Personality and Social Psychology* 52: 163–176.

Baumeister, Roy. 2010. "The Self". In Roy Baumeister and Eli Finkel (eds). *Advanced Social Psychology: The State of the Science*. Oxford: Oxford University Press, pp. 139–175.

Baumeister, Roy, Kathleen Vohs and Dianne Tice. 2007. "The Strength Model of Self-control". *Current Directions in Psychological Science* 16: 351–355.

Bleau, Renée. 1996. "Dietary Restraint and Anxiety in Adolescent Girls". *British Journal of Clinical Psychology* 35: 573–583.

Churchland, Paul M. 1981. "Eliminative Materialism and the Propositional Attitudes". *The Journal of Philosophy* 78: 67–90.

Dupré, John. 1995. "Against Scientific Imperialism". *PSA: Proceedings of the Biennial Meeting of the Philosophy of Science Association* 2: 374–381.

Duval, Shelley and Robert A. Wicklund. 1972. *A Theory of Objective Self Awareness*. Oxford: Academic Press.

Dupré, John. 1995. "Against Scientific Imperialism". *PSA: Proceedings of the Biennial Meeting of the Philosophy of Science Association* 2: 374–381.

Hawley, Katherine. 2006. "Weak Discernibility". *Analysis* 66: 300–303.

Heatherton, Todd F. and Roy F. Baumeister. 1991. "Binge Eating as Escape from Self-awareness". *Psychological Bulletin* 110: 86–108.

Heatherton, Todd F., Anne C. Krendl, C. Neil Macrae and William M. Kelley. 2007. "A Social Brain Sciences Approach to Understanding Self". In C. Sedikides and S.J. Spencer (eds). *The Self.* New York: Psychology Press, pp. 3–20.

Janssen-Lauret, Frederique. 2015. "Meta-ontology, Naturalism, and the Quine-Barcan Marcus Debate". In Frederique Janssen-Lauret and Gary Kemp (eds). *Quine and his Place in History*. Basingstoke: Palgrave Macmillan, pp. 146–167.

Janssen-Lauret, Frederique. 2016. "Committing to an Individual: Ontological Commitment, Reference and Epistemology". *Synthese* 193(2): 583–604.

Klein, Stanley B., Leda Cosmides and Kristi A. Costabile. 2003. "Preserved Knowledge of Self in a Case of Alzheimer's Dementia". *Social Cognition* 21: 157–165.

Lewis, David. 1986. *Philosophical Papers, volume II*. Oxford: Oxford University Press.

MacBride, Fraser and Frederique Janssen-Lauret. 2015. "Meta-ontology, Epistemology and Essence: On the Empirical Deduction of the Categories". *The Monist* 98: 290–302.

Maddy, Penelope. 1997. *Naturalism in Mathematics*. Oxford: Oxford University Press.

Mäki, Uskali. 2009. "Economics Imperialism: Concepts and Constraints". *Philosophy of the Social Sciences* 39: 351–380.

Marcus, Ruth Barcan. 1961. "Modalities and Intensional Languages". *Synthese* 13: 302–322.

Marcus, Ruth Barcan. 1990. "A Backward Look at Quine's Animadversions on Modalities". In R. Barrett and R. Gibson (eds). *Perspectives on Quine*. Oxford: Blackwell, pp. 230–243.

Miller, George A. 2003. "The Cognitive Revolution: A Historical Perspective". *Trends in Cognitive Sciences* 7: 141–144.

Plantinga, Alvin. 1974. *The Nature of Necessity*. Oxford: Oxford University Press.

Pryor, John B., Frederick X. Gibbons, Robert A. Wicklund, Russell H. Fazio and Ronald Hood. 1977. "Self-focused Attention and Self-report Validity". *Journal of Personality* 45: 513–552.

Putnam, Hilary. 1979. "What is Mathematical Truth?" In *Mathematics, Matter and Method: Philosophical Papers, volume 1*. Cambridge: Cambridge University Press.

Quine, W.V.O. 1960. *Word and Object*. Cambridge, MA: MIT Press.

Quine, W.V.O. 1961. "Reply to Professor Marcus". *Synthese* 13: 343–347.

Quine, W.V.O. 1970. *Philosophy of Logic*. Englewood Cliffs, NJ: Prentice Hall.

Quine, W.V.O. 1976. "Grades of Discriminability". *The Journal of Philosophy* 73: 113–116.

Quine, W.V.O. 1981. "Things and Their Place in Theories". In *Theories and Things*. Cambridge, MA: Harvard University Press, pp. 1–23.

Ramsey, Frank. 1931 [1925]. "The Foundations of Mathematics". In *The Foundations of Mathematics and Other Logical Essays*. London: Kegan Paul, pp. 1–61.

Russell, Bertrand. 1910–11. "Knowledge by Acquaintance and Knowledge by Description". *Proceedings of the Aristotelian Society* 11: 108–128.

Saunders, Simon. 2006. "Are Quantum Particles Objects?" *Analysis* 66: 52–63.

Sirken, Monroe G., Douglas J. Herrmann, Susan Schechter, Norbert Schwarz, Judith N. Tanur and Roger Tourangeau. 1999. *Cognition and Survey Research*. New York: John Wiley and Son.

Wallace, David. 2002. "Worlds in the Everett Interpretation". *Studies in History and Philosophy of Science Part B: Studies in History and Philosophy of Modern Physics* 33: 637–661.

13 Is the behavioral approach a form of scientific imperialism?

An analysis of law and policy

Magdalena Małecka and Robert Lepenies

Introduction

The legal sciences are currently witnessing a sweeping intellectual transformation. Applications of the behavioral sciences to law and policy (henceforth: the behavioral approach) are now a flourishing and expanding field of academic inquiry.[1] Outcomes of this research, even though often still preliminary, are eagerly adopted by policymakers interested in enacting laws and regulations that more effectively impact citizens' behavior.[2] The rapid development and growing popularity of behaviorally informed analyses of law has, however, also been met with skepticism. Are the legal sciences the target of "behavioral imperialism"?

The aim of this chapter is twofold. First, we would like to understand the behavioral approach by using conceptual distinctions made in the philosophy of science in discussions on scientific imperialism. Second, we would like to shed light on scientific imperialism – on its definition and assessment.

Proponents of bringing the behavioral sciences to law (henceforth: behavioralists) share the conviction that it is necessary and important that references to findings of the behavioral sciences should be made in order for law and policy to be informed by more "realistic" models of human behavior in legal and policy settings. The authors of the foundational texts of this movement say: "The task of behavioral law and economics, simply stated, is to explore the implications of *actual (not hypothesized) human behavior* for the law" (Jolls et al. 1998, 1476, our emphasis). Behavioralists advocate a "systematic framework" that aims both to "explain the effects and content of law" and "model and predict behavior relevant to law" (ibid., 1474).[3] They are interested in producing policy-relevant scholarship, which is based on empirical evidence and thereby "empirically informed" (Sunstein 2011; Strassheim and Kettunen 2014).

Behavioralists, so we argue, understand law primarily as an *instrument* that influences human behavior. Their goal is to see how a new understanding of human behavior "bears on the actual operation and possible improvement of the legal system" (Jolls et al. 1998, 1480). Behavioralists hence believe both that there is an accessible body of scientific knowledge about the regularities and patterns of human behavior, and further, that behavioral scientific findings can and should be used in order to inform law and policy, which as a result will lead to "better" (that is, more effective) legal regulations.[4]

Recently, the impact of behavioral analysis for the legal sciences has become a topic that is under scrutiny. Commentators have begun to ask whether and how the behavioral approach changes the legal sciences – and whether a widespread application of behavioral sciences to law is justifiable.[5] It is here that we draw the connection to the debate of scientific imperialism.

Scholars thinking about scientific imperialism have asked themselves a similar question: when (if ever) does a scientific theory, research program, or discipline encroach unduly upon another? Further, why could this be problematic? The recent debate on scientific imperialism in the philosophy of science promises to offer some guidance in answering these questions. Here, the debate has revolved around the question of the *permissibility* of the application of scientific theories and methods outside the discipline in which they were initially introduced (which will also be referred to as "scientific trespassing"). Philosophers of science have attempted to clarify what it means for a theory or a discipline to be applied outside its own field or domain, and whether such an application can be understood as *imperialistic.*[6] Attempts to *identify* instances of scientific imperialism have sometimes been accompanied by formulating criteria for *evaluation.*

The starting point here is the account of scientific imperialism proposed by Uskali Mäki (2013), in which he defines scientific imperialism and situates recent voices in the debate[7] on scientific imperialism. According to Mäki, scientific imperialism is a phenomenon that may occur between two disciplines. Thus, for him, scientific imperialism is a "dynamic interdisciplinary relationship." Mäki then distinguishes between certain types of imperialism: imperialism of scope, imperialism of style, and imperialism of standing. He argues that imperialistic endeavors should be constrained if they do not advance the pursuit of explanatory unification across disciplinary boundaries. He also argues that scientific imperialism can be defined "neutrally in terms of scope, style and standing in interdisciplinary relations" (Mäki 2013, 338).[8]

In this chapter, we investigate and modify Mäki's account. We argue that scientific imperialism occurs when some novel applications of methods, theories, or research programs are favored at the expense of other methods, or theories, or research programs. For the *identification* of scientific imperialism, we argue that an account of *imperialism of standing is needed.* Yet, we claim that any account of scientific imperialism relies upon a corresponding understanding of scientific progress that cannot be disentangled from it. Imperialistic practices are justified by imperialists, or rebuked by the imperialized, in light of what they each believe to be scientific progress.

The structure of this chapter is the following. First, we introduce the behavioral approach as a relatively recent, influential intellectual development in law and policy that may be regarded as imperialistic. Second, we note that the behavioral approach can be seen a manifestation of a longer trend within the legal sciences and we show why this observation matters for the discussion of scientific imperialism. Third, we bring the current debate on scientific imperialism to bear on these developments. In this process, we both learn more about the behavioral approach but also discover some shortcomings in the philosophical debate on scientific imperialism. Fourth, we propose an amended account of scientific imperialism. The account builds upon Mäki's notion of imperialism of standing by considering the concrete consequences of scientific trespassing. This then calls for the empirical analysis of the position and power that certain theories or research programs can gain. The account, however, also considers the importance of the notion of scientific progress for defining scientific imperialism. Fifth, we analyze the behavioral approach in light of our account of scientific imperialism.

1 What is the behavioral approach?

By behavioral approach we mean the contemporary approach to law and policy based on the latest findings in the behavioral sciences. This approach, which surfaced in the 1990s with increased attention to research in behavioral economics and cognitive psychology, has since become a heterogeneous field of study, encompassing behavioral law and economics, heuristics and law, experimental and empirical studies on law, and a normative approach built upon behavioral insights in law and policy (also called behavioral public policy). Here, "nudges" are clearly the most prominent, visible, and contested example in public policy (Thaler and Sunstein 2008).

The behavioral approach developed to a great extent as the result of criticism of one of the most consequential and contentious developments in the legal sciences in recent times – the economic analysis of law, which is based on the premise that the legal sciences should analyze law by applying neoclassical economic theories and their style of inquiry (Posner 1977; Cooter and Ulen 2004; Shavell 2009). Economic analysis of law has been criticized for being imperialistic from a variety of perspectives (see White 2015; Medema 1998; Epstein 1997; Zelizer 1988; Kelman 1988), and has even declared itself as imperialistic (see Cooter 1981; Becker 1968; and compare Landes and Posner 1993).

The behavioral approach, being a reaction to the perceived deficiencies of the economic analysis of law, promised "new and better understandings of decision and choice" (Sunstein 1997, 1). Behavioralists claimed that a more "realistic" theory was needed, grounded in empirical and experimental works. For them, this also meant that a theory should be able to explain and predict the impact of law and serve as a basis of policy recommendations. In the behavioral approach to law, a particular emphasis is put on experiments. Sunstein, for example, writes that it is important to "instil a culture of experimentation and evaluation" in law- and policymaking (Sunstein 2011, 1362). Reliance on experimental methods – and often more generally, empirical methods – is treated as a proper way of both testing hypotheses and collecting observations of behavior that can be a basis for the formulation of a more "realistic" model of human behavior, as well as a proper and promising way of testing policy solutions.[9] The proximity of the behavioral approach to policy is made clear by a recent textbook on the topic, where it is claimed that "[s]uccessful policy [...] must depend on a thorough understanding of human behavior" (Shafir 2013, 1). Yet the justification for the adoption of experimental methods is made without reference to a specific *discipline* from which they stem. Reliance on experiments and on observational data is understood as more accurate and reliable in providing accounts and explanations of behaviors than the formulation of abstract and general theories or models.[10] The optimism about the capabilities and reach of the behavioral approach is apparent here. Ulen argues that behavioral scholars "edge incrementally closer to accurate descriptions of the forces shaping human behavior and to understanding how those forces can be deliberately harnessed" (Ulen 2014, 2).

For us, the following development is important: behavioralists have the ambition to improve or complement neoclassical economic models as they have been applied to law. The idea is to propose an alternative account of behavior and decision-making in legal settings.[11] Yet in contrast with earlier episodes in the legal sciences, especially in contrast to the rhetoric of the economic analysis of law (at least of some of its proponents, for instance Gary Becker), the behavioral approach does not formulate "imperialistic" postulates that show the intention of completely replacing other research programs. Behavioralists often strike a conciliatory tone (Thaler 2015), and stress the complementarity, rather than the substitutability, of their approach with prior approaches. This does not mean, as will be shown below, that the behavioral approach cannot still be understood as an instance of scientific imperialism.

2 What is special about the legal sciences?

The debate whether to apply behavioral sciences to law can be seen as a novel manifestation of a discussion, lasting for more than 100 years, on the question of the method and scientific character of legal scholarship. This discussion was accompanied by a debate on the possibility of bringing non-legal disciplines to study law. We make this brief historical note because we would like to show why the legal sciences and their relationship to other scientific fields pose a challenge for defining scientific imperialism as a relation between disciplines. We will discuss this challenge in more detail in section three, below.

We identify two different rationales for why references to "outside" sciences within legal scholarship have historically been advocated. First, they were made in order to criticize legal scholarship for its common-sense, non-scientific character. The criticism was often accompanied by attempts to establish a new science of legal phenomena. Second, references to other sciences were made to analyze the impact of law in order to inform law- and policymakers so that they could propose more effective laws and regulations. To understand the role of "outside" sciences in law and their potentially imperialistic character, it is important to take into account these two rationales and their interplay. The behavioral approach is, we argue, mainly a manifestation of the latter rationale.

First, let us focus on the critique of legal scholarship and attempts to treat it as a proper science that started in the 19th century. In the realm of the continental legal tradition, Julius Hermann von Kirchman (1848) famously accused legal dogmatics[12] of having quasi-intellectual character: for him, legal dogmatics was pseudo-science. On the other hand, in the common law tradition, American legal realists (e.g. Oliver Wendell Holmes, John Chipman Gray, Herman Oliphant, Jerome Frank) criticized the "formalistic jurisprudence" and argued that methods of interpretation of legal norms, methods of legal reasoning, as well as methods of adjudication, were based on common sense or intuition, and as such were arbitrary, unjustified, and unscientific. Partly as a result of the critique directed against legal dogmatics, scholars tried to launch the project of establishing a scientific legal theory, understood as a discipline explaining legal phenomena. Some, like Hans Kelsen, argued for the autonomous character of legal theory. Kelsen proposed his pure theory of law, with its own subject matter (system of legal norms) and method (transcendental analysis), with the intent to save the legal sciences from influences of any "alien elements," as he called them (he had in mind mainly the theoretical influence of sociology) (Kelsen 1934/1967). On the other side were legal realists who argued for the application of methods and theories of other (mainly social) sciences in order to analyze law that was – for them – reducible to facts. Legal realists abandoned the idea that legal theory had its specific, autonomous, subject matter and method – and instead called for the embracing of empirical methods, a call that finds its revival today in New Legal Realism (Lang 2015; and Suchman and Mertz 2010).

Second, in the 20th century, the question about the method and scientific character of legal scholarship was accompanied by an interest in the analysis of the *impact* that law had on the behavior and decisions of different actors. An analysis of this kind had been present in the realm of American and Scandinavian legal realism, sociological jurisprudence, but also in the disciplines that emerged in the meantime – the sociology of law and psychology of law and, finally, in the economic analysis of law. What these approaches had in common is the belief in offering a description of human behavior that is general enough both to explain and predict people's reactions and responses to legal norms. One of the discussions that preoccupied scholars was the "gap problem" between law and behavior – that is, the problem of the discrepancy between behaviors required by official norms and behaviors observed in social reality. The challenge for behaviorally oriented legal scholars was to get rid of this discrepancy by making law better suited to the behavioral patterns identified. In the 1950s and 1960s, hope and emphasis were put on empirical sociology as a science that could provide insights into how law could influence human behavior (Macaulay and Friedman 1969). Since the 1970s, it was economics that was supposed to serve this role. Today, we can witness a turn to the behavioral sciences in law- and policymaking.

In the debate on scientific imperialism, some draw a distinction between internally driven imperialism and externally driven imperialism (see Mäki 2013,

335). While we believe that the behavioral approach probably represents a mix of both, what we do want to stress here is the extraordinary receptiveness of the legal sciences to outside influences. The brief overview above should demonstrate that there are diverse rationales for referring to outside theories or approaches (which are often connected to questions about the scientific status of the discipline). Legal scholarship has been methodologically unsteady and at the same time open to external influences, as well as prone to internally driven imperialistic influences. The overview shows further that in legal scholarship we have various legal disciplines that all struggle over scientific status (such as, for instance, legal theory and legal dogmatics), and that it is nearly impossible to point to a clearly defined discipline that could represent legal scholarship.

3 Using scientific imperialism to understand the behavioral approach – and vice versa

The debate on scientific imperialism in the philosophy of science has revolved around the question of the permissibility of the application of scientific theories and methods outside the discipline in which they were initially introduced ("scientific trespassing"). Philosophers of science have attempted to clarify what it means for a theory or a discipline to be applied outside its own field or domain, and whether such an application can be understood as imperialistic. The debate was initiated by John Dupré in his article entitled "Against Scientific Imperialism". For him, scientific imperialism takes place when we have horizontal relations between theories, that is when theories of different kinds deal with the objects at the same structural level. Dupré formulates his general remarks on scientific imperialism on the basis of his analysis of two examples of imperialistic approaches to studying human behavior: economics (Gary Becker's economic analysis of society), and evolutionary biology (Dupré 1995). Dupré characterizes scientific imperialism as an application of a "successful scientific idea" "far beyond its original domain" (Dupré 2001, 74), so that this application cannot "provide much illumination." Clarke and Walsh, inspired by Dupré, propose their definition of scientific imperialism as "illegitimate occupation by one discipline of another discipline's territory". As Mäki (2013) concisely points out, they identify three possible reasons for thinking of an occupation as illegitimate: (a) it violates local autonomy; (b) it exploits local colonized populations by "an unjust extraction of resources"; and (c) this exploitation "unfairly prevents the exploited from realising their potential to develop" (Clarke and Walsh 2009, 200, 201). Clarke and Walsh argue that scientific imperialism can cause science to fail to progress in a way that it otherwise would progress. This claim provokes a response from Kidd, who criticizes them for endorsing an idea of "counterfactual progress" (Kidd 2013).

As mentioned at the beginning of the text, the point of departure for our analysis of scientific imperialism is Uskali Mäki's (2013) account. In his attempt to explicate the notion of scientific imperialism, we share many of Mäki's intuitions, for instance the introduction of the concept of imperialism of *standing* and his remarks on Dupré (1995), Clarke and Walsh (2009) and Kidd (2013). Mäki points to the ambiguities in John Dupré's text, in which Dupré characterizes scientific imperialism as an application of a "successful scientific idea" "far beyond its original domain" (Dupré 2001, 74), so that this application cannot "provide much illumination." Mäki argues instead that "there is nothing imperialistic per se in applying a scientific idea beyond its original domain of application" (Mäki 2013, 327), and that "the idea of frequent failure – should [not] be part of a general characterization of scientific imperialism" (Mäki 2013, 327) because these views will result in excessive scientific caution and risk-aversion. Also, in his discussion with Steve Clarke and Adrian Walsh he emphasizes that "scientific imperialism is neither sufficient nor necessary for [...] poor explanations" (Clarke and Walsh 2009, 333). Likewise, our account is indebted to his proposal to incorporate an institutional perspective into the analysis of scientific imperialism through the introduction of the notion *standing*. However, even though we perceive Mäki's proposal as an important advancement in the debate, there are claims in Dupré (1995), as well as in Clarke and Walsh (2009), which we think must be drawn upon for a convincing account of scientific imperialism. This concerns their initial intuitions about the relationship between imperialistic tendencies in science and notions of scientific progress presumed by the "imperialist".

Mäki introduces a distinction between imperialism of scope, style, and standing. For Mäki, imperialism of scope obtains when "an expansionist discipline seeks to explain phenomena that belong to the perceived domain of another discipline. This is the pursuit of explanatory unification that is disrespectful of disciplinary boundaries" (Mäki 2013, 334). Imperialism of style appears when "the styles and strategies of research, such as the techniques and standards of inquiry and communication, characteristic of one discipline, are transferred to, or imposed on, other disciplines" (Mäki 2013, 334). Imperialism of standing is characterized in the following way: "The academic and non-academic prestige, power, and resources as well as the acknowledged technological and political relevance of one discipline increase at the expense of those of another" (Mäki 2013, 334).

Mäki defines scientific imperialism in disciplinary terms – for him, scientific imperialism obtains through the relation between disciplines. Yet this understanding cannot capture misgivings we may have in the context of references to "outside" scientific findings in law described earlier. Looking at the history of legal scholarship, it is not clear at all which legal discipline could be imperialized – legal dogmatics, legal theory, jurisprudence? The status and boundaries of each of the fields are particularly contested within legal scholarship, their character as scientific is still hotly debated, and any attempt to explain a notion of scientific imperialism in this context will require clarification of these debates (and thus to be able to demarcate disciplines clearly). Later we show why a disciplinary framework is too restrictive when attempting to analyze the potential imperialism of the behavioral approach.

As already mentioned, for Mäki the imperialism of scope takes place when "an expansionist discipline seeks to explain phenomena that belong to the perceived domain of another discipline". Yet in the behavioral approach, findings of the behavioral sciences are incorporated in order to explain and predict how people react to legal norms, and not in order to explain *legal* phenomena of any kind. Law (legal norms) is treated as given, as a part of the *explanans*, and not of the *explanandum*. Therefore, it would be difficult to understand the behavioral approach as replacing a traditional legal analysis and explaining phenomena that previously "belonged" to a specifically "legal" domain (that contains "legal phenomena"). Furthermore, it should be noted that like other historical trends in legal scholarship we mentioned above, the behavioral approach we find today embodies a scientistic attitude. Even though proponents of the behavioral approach do not want to offer explanations of *legal* phenomena or legal norms, they *do* claim that most of the legal scholarship and, more importantly, law-making before the "behavioral turn", was based on mere intuitions or common-sense generalizations, and that it is the behavioral approach which provides scientific advancement and insights to legal scholarship.

Disciplinary framing cannot help us to understand the imperialism of style that takes place not due to the fact that a certain way of inquiry proved successful in any particular discipline. In contemporary references to the behavioral sciences in law, experimental methods are relied upon as the proper way of testing hypotheses and collecting observations of behavior that can be a basis for the formulation of a more "realistic" model of human behavior. Experimental work and the resulting empirical evidence is understood as more accurate and superior compared to the formulation of abstract and general models – like those in neoclassical law and economics. Instead, the experimental method is advocated not because of its relation to any particular *discipline*. Indeed, the superior insight that is supposedly gained is not justified by arguing that it proved successful in, for example, a more advanced discipline, but that the newly introduced approach is "more accurate" or "scientific", or more policy relevant. We think that the adoption of experimental methods in law and policy can reasonably be called imperialistic – but that this would elide Mäki's definition due to its focus on relations between disciplines.

Finally, understanding scientific imperialism in disciplinary terms excludes certain other interesting cases of scientific imperialism. We mention here cases that come into existence because of factors not directly related to disciplines proper – for example, the imperialism of a certain theory that is owed to its success in policymaking and politics. Arguably, one factor contributing to the popularity and influence of the behavioral approach stems from the eager reception of both private and public actors outside academia. Here, these private and public actors developed and sustained an interest in the behavioral approach not because of the scientific advancement it provides (for the explanation of legal phenomena, or for the explanation of mechanisms underlying human behavior and decision-making, for instance), but because of its *usefulness* in direct application – which behavioralists in turn are quick and eager to provide.[13] The behavioral approach is attractive because of its promise to deliver results in the form of allowing policymakers to shape the behavior of their policy-takers with little effort, cost, and hindrance. These developments facilitate scientific imperialism.

The rise of "nudges" in public policy illustrates our point. The so-called "nudge theory" does not provide explanations for why law or policy "work" in a certain context (the mechanisms responsible for the behavioral impact of many nudges are still not properly understood; see Grüne-Yanoff 2015). However, the increased *standing* of nudging (e.g. through support and acknowledgment of policymakers) leads to a growing popularity of a certain brand of behavioral research within legal scholarship. In addition, nudges are often presented as policy innovations, drawing on the authority of science. It is the scientific character of these policy instruments, together with their prized effectiveness, that makes them influential for legal scholars as well as policymakers.[14] Hence, it can be seen that imperialism can be driven not only by epistemic advancements that are external or internal to a discipline, but also by concerns of usefulness in policymaking backed by the invocation of the authority and progress of science. An account of scientific imperialism should be able to accommodate this.

Discarding the disciplinary frame in the debate about scientific imperialism does not, of course, imply downplaying the importance of disciplines in scientific activity or downplaying the usefulness of it for understanding scientific imperialism. It may well be that disciplinary reactions to applications of certain theories or methods to new problems, or fields, can be informative for a discussion on scientific imperialism. In fact, science continues to be materially, socially, and institutionally organized within disciplines. One of the reasons why there is sometimes opposition to new approaches is precisely the danger of undermining a given institutional or organizational structure of disciplines (when, for instance, as Mäki notices, scientific trespassing leads to undermining "local disciplinary monopolies"; Mäki 2013, 331). On the other hand, one of the reasons why a certain theory, research program, or method is able to expand, can similarly lie in the way in which a particular discipline is institutionalized, or organized. However, it is not the case that theories and methods of an entire discipline are transferred to another discipline; such transfers are rather partial. Also, it is not always the case that theories and methods are transferred because they proved "successful" in one particular discipline (see the case of the experimental method in the behavioral approach, discussed above).

4 Scientific imperialism: proposal of an account

4.1 Definition

For the reasons discussed above, we do not propose to analyze scientific imperialism in purely disciplinary terms. Thus, we would like to argue that some novel application X of methods, theories, research programs[15] becomes imperialistic when:

1 X is *favored* (by members of the scientific community) at the expense of other methods, or theories, or research programs in terms of academic and non-academic prestige, power, or resources, and when
2 the attempt of justifying this favoring of X is made by claiming that X is:

   a more "progressive" than applications of other methods, or theories, or research programs (justified by the progress *in* science);[16] and/or
   b more "scientific" than applications of other methods, or theories, or research programs (justified by the progress *of* science),[17]

3 claim (2) is assumed to hold without providing an argument for it.

By novel application we understand the application of methods, or theories, or research programs to problems that have not been analyzed on the basis of them so far (sometimes in the text we call a novel application also an instance of scientific trespassing). Our account emphasizes that instances of scientific imperialism require a reason given by the imperialist of why a novel application is undertaken. However, even if the reason is given, the definiton requires that it is not argued for, nor demonstrated by the imperialist, why the novel application is more progressive. In order for an onlooker to judge an instance of scientific trespassing as unjustified, she must find the justification questionable by disputing that X is in fact more scientific or more progressive. Note that our account builds on the notion of *standing* as introduced by Mäki: it means the favoring of certain scientific theories, or methods, or research programs[18] in terms of academic and non-academic prestige, power, and resources, acknowledged technological and political relevance at the expense of other theories, or methods, or research programs.

Furthermore, we argue that scientific imperialism is an activity that is related both to a certain view on improvement and progress, as well as to a power to realize it (it is in fact favored). Otherwise, without pursuing this vision of improvement, scientific trespassing is only aggressive, or invasive, but not imperialistic. Without the power actually to have an effect – that is, actually to affect the standing between approaches – such an attempt is mere scientific quixoticism.

We propose that the "expense" at which one theory is favored over another should be analyzed both empirically and normatively. In terms of empirical analysis, *improved* or *worsened* standing could be tracked by looking at indicators such as public money distributed to X rather than other approaches, time spent in teaching and research on X, the prominence of X in curricula, etc. Much more sophisticated ways could ascertain how an increase in standing could be identified and measured. Here, the debate about scientific imperialism could be considerably enhanced and expanded by drawing on the sociology of science and science studies more broadly.

The normative analysis could be advanced by comparing the scope of the current standing of theories, or methods, or research programs with a standard of appropriate *distribution of standing* across disciplines. This standard could be defined in light of someone's views on the organization of science, as well as in light of the view on the epistemic loss related to the institutional favoring of one approach. We argue that the term "at the expense" in our definition denotes an asymmetrical relation: being in this relation is beneficial for one scientific approach while it is harmful to another approach. To put it simply, when we deal with the phenomenon of scientific imperialism, we always identify loss (loss of opportunities, loss of funding, loss of societal relevance) experienced by a certain theory, research program, or approach. This loss has an epistemic dimension – it leads to forgoing a certain type of research and knowledge. If someone does not agree with the distribution of standing resulting from the imperialistic scientific trespassing, then one can oppose it as unjustified in the light of one's view on the appropriate distribution of standing.

In order to evaluate the justifications given for instances of scientific trespassing, one needs to engage in a philosophical debate on scientific progress. We believe that the role of the philosophy of science (in understanding scientific imperialism) is to analyze which notions of *scientific progress* are being relied upon when someone makes the charge that a novel application is more *progressive* than other theories, or research programs. In the light of our proposal, philosophy of science helps to reconstruct the notions of scientific progress that are presumed in a given account of scientific imperialism. It should clarify criteria that are used by those making the charge of scientific imperialism, as well as by participants in the debate to state when a theory, or a method, or a research program is more *progressive* than other theories, or methods, or research programs, and whether it is more *scientific* than other approaches. Our account then allows for some charges of scientific imperialism having less plausible justifications than others.

Scientific progress (*progress of science*) can be examined along diverse dimensions and aspects of science: non-epistemic (economical, professional, educational, methodical) and epistemic (increase and advancement of knowledge), following Niiniluoto (1980). As Niiniluoto notes:

> these types of progress have to be conceptually distinguished from advances in other human activities, even though it may turn out that scientific progress has at least some factual connections with technological progress (increased effectiveness of tools and techniques) and social progress (economic prosperity, quality of life, justice in society.
>
> (Niiniluoto 2015)

Progress within science (that is, the scientific progress of a research program, or a theory) can be judged by employing an account of scientific progress understood as the increase and advancement of scientific knowledge. Following Kitcher (1993), we argue that there is a variety of dimensions of progress within science (cognitive, conceptual, explanatory).[19] The question of whether an instance of scientific trespassing is correctly characterized as justified and unjustified requires making explicit an account and dimension of *scientific progress* in the light of which the expanding theory, or method, or research program is supported (at the expense of another).

Let us summarize the ways in which scientific imperialism can be criticized and opposed in the light of our account. First, when analyzing the expense at which one application of method, or theory, or research program is favored over another, one has to adopt a standard of appropriate distribution of standing. If someone opposes this standard, then this person can judge the novel application as *unjustified* in the light of her view on the appropriate distribution of standing in the organization of science, as well as due to the epistemic loss related to the institutional favoring of one approach. Second, one can analyze whether a novel application is *unjustified* according to her views on scientific progress (*progress of science*, or *progress in science*). A critic can question the notion of scientific progress an imperialist is putting forward upon which a novel application relies. This can be done by either arguing to replace it by another notion, by demonstrating that the imperialistic approach is not progressive in the light of the notion of scientific progress endorsed by its proponents, or by rejecting entirely the very notion of scientific progress.

Our account does not exclude the possibility of having justified instances of scientific imperialism. It is conceivable that the scientific community manages to provide reasons for accepting an epistemic loss resulting from novel applications, or for accepting the notions of progress endorsed by an imperialist. However, our definition says that we face an instance of scientific imperialism if the reason for trespassing is not argued for or it is not demonstrated by an imperialist, why the novel application is more progressive. This means that even in cases which justifications are provided, scientific imperialism will be always unfair. Our understanding of unfairness in this context is close to that of Clarke and Walsh. They hold that the favoring of a particular theoretical perspective is grounds for criticizing scientific imperialism because "other theoretical perspectives are not given a fair hearing" (Clarke and Walsh 2013, 345).

We would like to emphasize that the above-mentioned types of critique can overlap. Scientific imperialism in our view presupposes an understanding of progress of science or progress in science. Such convictions (concerning the importance and role of science) can have implications on the view of what constitutes an appropriate distribution of standing, and therefore, how scientific approaches should "treat" each other.[20] For example, a behavioralist who thinks of law and economics as unprogressive might wish for law and economics to feature significantly less prominently in the education of the next generation of legal students. However, apart from implying a stance on the distribution of standing, such convictions also imply a stance on what counts as fair interaction between scientific approaches. To stick with the example, the behavioralist here might also hold that it is advantageous that discarded approaches are to be replaced by the more "progressive" or more "scientific" approach. Yet others might have different convictions and think that scientific approaches generally deserve a fair hearing.

In the debate on the notion of scientific imperialism the link to the notion of scientific progress is sometimes made. Dupré notices that imperialistic tendencies that manifest themselves in claims that a particular theory provides the key in the understanding of a given phenomenon, are often related to the attempts to achieve explanatory unification that is presented as "unqualified scientific good". He criticizes such tendencies for introducing inappropriate methodology for studying a given phenomenon. Furthermore, as Clarke and Walsh point out, Dupré attacks evolutionary psychology and neoclassical economics for assuming that progress in science will be made as a result of explanatory unification. They mention that on Dupré's view:

> we should make no overarching assumptions about the trajectories along which particular scientific disciplines will or should progress, but rather should focus closely on empirical detail, and decide between competing explanations, on their own explanatory merits, regardless of whether or not these explanations happen to be part of attempts to unify science.
>
> (Clarke and Walsh 2013, 343)

This is also the view that they defend. Clarke and Walsh believe that there may be a progress in science, even though "there is no one definitive account of progress [...] What we decide to count as progress in science will be a matter of how we decide to weigh the various backward-looking factors that contribute to progress" (Clarke and Walsh 2013, 345). Mäki uses the notion of epistemic scientific progress, of advancement in knowledge, "including explanatory knowledge about the world: growth of explanatory unification" (Mäki 2013, 336), in order to assess scientific imperialism. He proposes four constraints on scientific imperialism (ontological, epistemological, axiological, institutional) that set conditions for acceptable scientific imperialism that contributes to scientific progress in the sense advocated by him. All of the preceding accounts (Dupré, Mäki, Clarke and Walsh) are examples of how participants in the debate of scientific imperialism bring their convictions about scientific progress to the discussion.

In our account the notion of scientific progress already forms part of any charge of scientific imperialism, as we argue that those novel applications of theories, methods, or research programs are imperialistic that are supported as being more progressive. The critique of novel applications can involve questioning their progressive character, or the very notion of scientific progress being endorsed by the "imperialists".

In the next section, we consider whether the behavioral approach can be seen as a possible instance of scientific imperialism.

4.2 The behavioral approach and scientific imperialism: behavioral imperialism?

We can identify two types of reasons why references to the behavioral sciences are made within legal and policy contexts: *progress of science*, and *progress in science*. Behavioral sciences are brought to law- and policymaking because the authority of science is supposed to provide law and policy with knowledge of the regularities and patterns of human behavior which can then be used in order to influence behavior in an effective manner, hence denoting the *progress of science*. At the same time, the behavioral sciences are also advocated as more progressive – seen as an advancement in comparison with alternative (mainstream economic) theories applied to law and policy, hence denoting the *progress within science*.

In order to state whether the behavioral approach is imperialistic in view of our account, an empirical analysis would have to be conducted to grasp the expense at which it is favored, as well as to endorse a notion of scientific progress in order to justify applications of behavioral sciences in legal settings. It is beyond the scope of this chapter to perform empirical analysis, or to fully reconstruct the view of scientific progress that is presumed within the behavioral approach and that is relied upon to justify the expanding reliance on behavioral theories and methods. Similarly, we would have to commit ourselves to a view of progress of science from which the position of the behavioralists could then be critically assessed.

However, at this stage of our investigation we can formulate the following working hypothesis: proponents of the behavioral approach understand scientific progress in both epistemic (advancement in explanatory and predictive power) and non-epistemic (increase in policy relevance and policy impact) ways. If the behavioral approach is not progressive in the light of a notion of scientific progress that is being endorsed – for instance, if proponents of the behavioral approach support a view of progress in science as advancing explanatory progress, but it turns out that they cannot actually offer greater explanatory and predictive power than other theories or research programs – and if the behavioral approach is nevertheless favored at the expense of other approaches, then the behavioral approach can be criticized as *unjustified in the light of the view on scientific progress*. Our account of scientific imperialism allows the treatment of the behavioral approach as *unjustified* also in case one questions the underlying idea of scientific progress as advancement in explanatory and predictive power, or as advancement in practical (policy) impact of research.

Nevertheless, we venture to make two hypotheses here: in the case of actual effects, we do observe a strong intellectual movement that gradually supplants (rather than complements) traditional ways to engage in legal and policy analysis. Likewise, we have provided some textual evidence that shows the underlying view of progress of science that behavioralists hold and with which they justify scientific trespassing. Second, and taking our account seriously, we would have to commit to a view of progress of science and progress in science in order to say whether the justification of the behavioralists has merit. One way in which such an argument (which stems from an underlying stance on what constitutes progress in science) could be made is the following: for us, it is not obvious that the social sciences should strive similarly towards explanatory unification as is often called for in the natural sciences. We believe that in the case of the social sciences, we should not strive to achieve "explanatory unification," but rather that it is more important to retain a "cautionary pluralism" of social scientific approaches (similar to what Dupré calls "horizontal pluralism").

5 Conclusion

Our analysis of the application of the behavioral sciences to law and policy has helped us to propose a refined account of scientific imperialism.

Past and recent developments in legal scholarship lead us to question the approach to scientific imperialism as a purely interdisciplinary phenomenon. We instead stress the actual impact of changes in the distribution and character of *standing* as key to identifying instances of scientific imperialism. Further, we emphasize that only these instances of scientific trespassing can be called imperialistic that endorse and pursue a certain understanding of scientific progress and are being justified by imperializers as more progressive.

To say conclusively whether the behavioral approach is imperialistic or not requires further (empirical and philosophical) work, yet the groundwork for such an investigation has been laid here. For now, we have contributed conceptual clarifications and suggestions for how advancement can be made on the conceptual analysis of the notion of scientific imperialism.

The role for philosophy of science here is clear: what is needed is a debate about the conditions of scientific progress upon which any conception of scientific imperialism is built.

The role played by sociology of science should also be clear: what is needed for the identification of instances of scientific imperialism is an empirical matter as well – and approaches from the broad field of science studies can help to suggest indicators of how to approximate the "imperial standing" of scientific approaches.

We hope that our account can be helpful in shedding new light on the debate on scientific imperialism, and also explain existing positions in it, as well as clear up some misunderstandings. Clarke and Walsh are worried that scientific imperialism can be problematic for epistemic reasons: "the adoption of one theory, due to its successes in another explanatory domain, leaves us liable to dismiss rival theories, without properly testing their ability to account for reality" (Clarke and Walsh 2013, 345). In the light of our account, we could explicate their worry as a critique of the expense at which one theory or research program is favored over other scientific approaches. We stress that in order to fully spell out this type of critique, it should be clarified what notion of the fair distribution of *standing* one endorses. Furthermore, when Dupré opposes an idea of scientific unification while analyzing examples of applications of economics to new fields, our account can help clarify and understand the source of his anxiety. Dupré opposes these applications because he does not believe that progress in science stems from the realization of the pursuit of unification, and he is worried by the extent to which applications of economics and evolutionary biology, advocated in the spirit of unificatory crusade, are favored at the expense of other approaches to analysis of human behavior. Mäki, on the other hand, believes that science progresses through explanatory unification. His normative evaluation of scientific imperialism is proposed as "constraints on the pursuit of explanatory unification across disciplinary boundaries" (Mäki 2013, 336). A legitimate instance of scientific trespassing for him is one that is argued for in the light of these conditions. We do not oppose this view – our account permits that this is a reasonable approach to justifying instances of scientific trespassing. What we would like to stress, however, is this: it is possible to question this stance on the grounds of a different understanding of scientific progress.

Notes

1 "[T]he behavioral perspective occupies a respectable and increasingly popular niche in many graduate programs in economics, business, law, policy, and the social sciences more generally. And thus we have arrived at a point where it is only natural to explore how best to incorporate elements of the behavioral perspective into policy thinking" (Shafir 2013, 1). "My belief is that behavioral law is one of the most important developments – and probably *the* most important – in legal scholarship of the modern era" (Ulen 2014, 14).

2 See the Social and Behavioral Science Team (SBST) of the US White House; the Behavioural Insights Team in the UK. Increasingly, international organizations commission behavioral studies, for example, the EU's Directorate-General for Health and Consumers (DG SANCO) or the recently formed Global Insights Initiative team at the World Bank (now called eMBeD). At the same time, see several associations bring together like-minded behavioralists, such as The European Nudging Network (TEN) and other national associations (such as the Behavioral Policy & Science Association).

3 Other seminal texts for the behavioral approach are Camerer et al. (2003); Sunstein (2000); Korobkin and Ulen (2000).

4 Shafir (2013) argues that there are "many ways in which our new understanding, this new view of the human agent, might help design and implement better public policy" (Shafir 2013, 9). Strassheim et al. (2015) analyze behavioral expertise comparatively; they show that the behavioral approach was partly a response to the criticism of "new public management" and "evidence-based policy" approaches arguing that the behavioral approach "promised a renewed state-science-citizen relationship in which scientific experts and political actors would become collaborators".

5 See the contributions of Cserne, Feldman, and Lobel in Alemanno and Sibony (2015); Nourse and Shaffer (2009); Mitchell (2014, 2002); Posner (1998); Arlen (1998).

6 Thus far, economics imperialism (e.g. Kuorikoski and Lehtinen 2010; Mäki 2009) is the most-debated example. See also Clarke and Walsh 2009; Davis 2012; Dupré 1995. Other accounts include, e.g. geographical economics (Mäki and Marchionni 2011; Marchionni 2012), and the case of evolutionary psychology in international relations (Bell 2006).

7 See the debate between Clarke and Walsh (2009, 2013), Kidd (2013) and Mäki (2013), inspired by the text of Dupré (1995).

8 We regard scientific imperialism, like most commentators in the debate, as an inherently pejorative term that defies a purely neutral or descriptive definition. However, we will not explicitly draw upon political metaphors, which is an approach that other participants in the debate have taken (Stigler 1984; Dupré 1995).

9 See e.g.: "regulations [should be] written and implemented in ways that lend themselves to experimental evaluation and creation of independent review to assess the *effectiveness* of regulations" (Sunstein 2011, 1391, our emphasis).

10 Ulen claims that behavioral findings used in the legal academy and well known within the social science community "are not theoretical; they are empirical" (Ulen 2014, 2). Korobkin and Ulen argue that behavioral law and economics can develop "without a *grand, overarching theory of behavior* so long as one has a due regard for the relevant decision-making capacities of the actors in that specific setting. By borrowing from psychological and sociocultural theories in addition to economics, the law-and-behavioral-science approach consciously chooses to emphasize its *external usefulness* in analyzing legal problems rather than either its internal elegance or universal applicability" (Korobkin and Ulen 2000, 1058, our emphasis). See also: van Gestel and van Dijck 2011; and Ranchordás 2013.

11 Their insistence is to continue using "the tools of traditional economic analysis" (Jolls et al. 1998, 1474), being "deeply constructive" (ibid., 1475), and "enriching the traditional economic framework" and not to "undermine it" (ibid., 1475). Opinions diverge, however, on whether and to what extent economic analysis of law is compatible with behavioral approaches. See e.g. Nourse and Shaffer (2009), who also regard the behavioral approach as correcting law and economics. It is also claimed that expected utility theory underlies neoclassical law and economics, whereas behavioral law and economics is based on the alternative theory of decision-making – prospect theory. See Arlen and Talley 2008.

12 Legal dogmatics, mainly in continental tradition, is a branch of legal sciences analyzing a law in force enacted by a lawmaker.

13 See e.g. "Low-cost regulatory policies, such as disclosure and simplification, may be justified even if we do not have a clear understanding, in the abstract, of whether the relevant behavior is mostly a product of loss aversion or social influences" (Sunstein 2011, 1362).

14 We would like to argue that this scientistic attitude should be included in the account of scientific imperialism. We include it in our account and argue that scientific imperialism is often being justified by its proponents as more "scientific". This claim presupposes a certain notion of progress of science. See more in sections 4.1. and 4.2.

15 By novel, we mean that specific applications are being brought to problems that have not been analyzed by these applications before. We used the short-hand term "scientific trespassing" to describe these instances (see also Hirschman 1981). Another word used in the literature is the term "horizontal relations" by Dupré.

16 The analysis of scientific progress also has a normative component. Progress is a normative concept that should be distinguished from a notion of "change" and of "development" (Niiniluoto 1995). Progress within science means improvement judged through the criteria for "good science".

17 For example, when a scientific approach is brought to practically oriented fields, it is sometimes claimed that it will make them more scientific. Here we have a move from non-science to science (justification through the *progress of science*). However, if, for instance, prospect theory is claimed to replace or complement expected utility theory, it is believed that it will lead to scientific progress understood as *progress within science*. Here we have a move from one scientific theory to a more advanced one. These reasons can also overlap.

18 Mäki speaks about disciplines, but for the reasons mentioned above we resign from a disciplinary frame in our analysis.

19 See also Mäki (2002) and his list of perspectives from which scientific progress can be analyzed.

20 This account of standing presupposes that the *distribution* of standing is in some respect a zero-sum game: scientific approaches compete for scarce resources and attention. However, the notion of standing is not confined to such questions of distributional fairness. Instead, it can also encompass questions of relational fairness: not giving a fair hearing to a specific scientific approach could be unfair regardless of

how resources and attention are distributed but instead be unfair because it unfairly treats a competing approach as having a less-than-equal status.

References

Alemanno, Alberto and Anne-Lise Sibony (eds). 2015. *Nudge and the Law: A European Perspective*. Oxford: Hart Publishing.

Alexander, Gregory S. 2002. "Comparing the Two Legal Realisms: American and Scandinavian". *Journal of American Comparative Law* 50: 131–174.

Arlen, Jennifer. 1998. "Comment: The Future of Behavioral Economic Analysis of Law". *Vanderbilt Law Review* 51: 1765–1770.

Arlen, Jennifer H. and Eric L. Talley (eds). 2008. *Experimental Law and Economics*. Cheltenha: Edward Elgar Publishing.

Becker, Gary S. 1968. "Crime and Punishment: An Economic Approach". *The Journal of Political Economics* 76(2): 169–217.

Bell, Duncan. 2006. "Beware of False Prophets: Biology, Human Nature and the Future of International Relations Theory". *International Affairs* 82: 493–510.

Camerer, Colin, Samuel Issacharoff, George Loewenstein, Ted O'Donoghue and Matthew Rabin. 2003. "Regulation for Conservatives: Behavioral Economics and the Case for 'Asymmetric Paternalism'". *University of Pennsylvania Law Review* 151: 1211–1254.

Clarke, Steve and Adrian Walsh. 2009. "Scientific Imperialism and the Proper Relations Between the Sciences". *International Studies in the Philosophy of Science* 23: 195–207.

Clarke, Steve and Adrian Walsh. 2013. "Imperialism, Progress, Developmental Teleology, and Interdisciplinary Unification". *International Studies in the Philosophy of Science* 27(3): 341–351.

Cooter, Robert. 1981. "Law and the Imperialism of Economics: An Introduction to Economic Analysis of Law and a Review of the Major Books". *UCLA Law Review* 29: 1260–1269.

Cooter, Robert and Thomas Ulen. 2004. *Law and Economics*. Fourth edition. Boston, MA: Addison Wesley.

Davis, John B. 2012. "Mäki on Economics Imperialism". In Aki Lehtinen, Jaakko Kuorikoski and Petri Ylikoski (eds). *Economics for Real*. London: Routledge, pp. 203–219.

Dupré, John. 1995. "Against Scientific Imperialism". In *PSA 1994: Proceedings of the 1994 Biennial Meeting of the Philosophy of Science Association*, vol. 2. East Lansing, MI: Philos Sci Association, pp. 374–381.

Dupré, John. 2001. *Human Nature and the Limits of Science*. Oxford: Oxford University Press.

Epstein, Richard A. 1997. "Law and Economics: Its Glorious Past and Cloudy Future". *The University of Chicago Law Review* 64: 1167–1174.

Fricker, Miranda. 2007. *Epistemic Injustice: Power and the Ethics of Knowing*. Oxford: Oxford University Press.

Grüne-Yanoff, Till. 2015. "Why Behavioural Policy Needs Mechanistic Evidence". *Economics and Philosophy*: 1–21.

Haack, Susan. 2011. *Defending Science – Within Reason: Between Scientism and Cynicism*. New York: Prometheus Books.

Haack, Susan. 2012. "Six Signs of Scientism". *Logos and Episteme* 3(1): 75–95.

Hacker, Philipp. 2015. "The Behavioral Divide. A Critique of the Differential Implementation of Behavioral Law and Economics in the US and the EU". *European Review of Contract. Law* 11(4): 299–345.

Hirschman, Albert O. 1981. *Essays in Trespassing: Economics to Politics and Beyond.* Cambridge: Cambridge University Press.
Hookway, Christopher. 2010. "Some Varieties of Epistemic Injustice: Reflections on Fricker". *Episteme* 7(2): 151–163.
Jolls, Christine, Cass R. Sunstein and Richard Thaler. 1998. "A Behavioral Approach to Law and Economics". *Stanford Law Review* 50: 1471–1550.
Kelman, Mark. 1988. "On Democracy-Bashing: A Skeptical Look at the Theoretical and 'Empirical' Practice of the Public Choice Movement". *Vanderbilt Law Review* 74: 199–273.
Kelsen, Hans. 1967. *Pure Theory of Law.* Los Angeles: University of California Press.
Kidd, James I. 2013. "Historical Contingency and the Impact of Scientific Imperialism". *International Studies in the Philosophy of Science* 27(3): 315–324.
Kirchman, Julius Hermann. 2000 [1848]. *Die Wertlosigkeit der Jurisprudenz als Wissenschaft.* Heidelberg: Manutius Verlag, 2. Auflage.
Kitcher, Philip. 1993. *The Advancement of Science.* Oxford: Oxford University Press.
Korobkin, Russell and Thomas Ulen. 2000. "Law and Behavioral Science: Removing Rationality Assumption from Law and Economics". *California Law Review* 88: 1051–1144.
Kuorikoski, Jaakko and Aki Lehtinen. 2010. "Economics Imperialism and Solution Concepts in Political Science". *Philosophy of the Social Sciences* 40: 347–374.
Landes, William and Richard Posner. 1993. "The Influence of Economics on Law: A Quantitative Study". *Journal of Law and Economics* 36(1), Part 2: 385–424.
Lang, Andrew. 2015. "New Legal Realism, Empiricism, and Scientism: The Relative Objectivity of Law and Social Science". *Leiden Journal of International Law* 28(2): 231–254.
Lepenies, Robert and Magdalena Małecka. 2015. "The Institutional Consequences of Nudging–Nudges, Politics, and the Law". *Review of Philosophy and Psychology* 6: 427–437.
Macaulay, Stewart and Lawrence M. Friedman. 1969. *Law and the Behavioral Sciences.* Indianapolis, IN: The Bobbs-Merrill Publishing Company.
Mäki, Uskali. 2002. "Scientific Progress: Complexities of a Contestable Concept". In Stephan Böhm, Christian Gehrke, Heinz D. Kurz and Richard Sturn (eds). *Is There Progress in Economics?* Cheltenham: Edward Elgar, pp. 123–130.
Mäki, Uskali. 2009. "Economics Imperialism: Concept and Constraints". *Philosophy of the Social Sciences* 9: 351–380.
Mäki, Uskali. 2013. "Scientific Imperialism: Difficulties in Definition, Identification, and Assessment". *International Studies in the Philosophy of Science* 27(3): 325–339.
Mäki, Uskali and Caterina Marchionni. 2011. "Is Geographical Economics Imperializing Economic Geography?" *Journal of Economic Geography* 11(4): 645–665.
Marchionni, Caterina. 2012. "Geographical Economics and its Neighbors – Forces Towards and Against Unification". *Handbook of the Philosophy of Science: Philosophy of Economics* 13: 425–458.
Martin, Michael. 1997. *Legal Realism American and Scandinavian.* New York: P. Lang.
Medema, Steven G. 1998. "The Trial of Homo Economicus: What Law and Economics Tells Us about the Development of Economic Imperialism". In John B. Davis (ed.). *New Economics and its History.* Durham, NC: Duke University Press, pp. 122–142.
Mitchell, Gregory. 2002. "Why Law and Economics' Perfect Rationality Should Not be Traded for Behavioral Law and Economics' Equal Incompetence". *Georgetown Law Journal* 91: 83–119.
Mitchell, Gregory. 2014. "Alternative Behavioral Law and Economics". In *Oxford Handbook of Behavioral Economics and Law.* Oxford: Oxford University Press, pp. 167–191.
Niiniluoto, Ilkka. 1980. "Scientific Progress". *Synthese* 45: 427–464.

Niiniluoto, Ilkka. 1995. "Is There Progress in Science?" In H. Stachowiak (ed.). *Pragmatik, Handbuch pragmatischen Denkens*. Band V, Hamburg: Felix Meiner Verlag, pp. 30–58.

Niiniluoto, Ilkka. 2015. "Scientific Progress". In Edward N. Zalta (ed.). *Stanford Encyclopedia of Philosophy*. http://plato.stanford.edu/entries/scientific-p-rogress/.

Nourse, Victoria and Gregory Shaffer. 2009. "Varieties of New Legal Realism: Can a New World Order Prompt a New Legal Theory?" *Cornell Law Review* 95: 61–138.

Posner, Richard. 1977. *Economic Analysis of Law*. Second edition. Boston, MA: Little, Brown.

Posner, Richard. 1998. "Rational Choice, Behavioral Economics, and the Law". *Stanford Law Review* 50: 1551–1575.

Pound, Roscoe. 1908. "Mechanical Jurisprudence". *Columbia Law Review* 8: 605–623.

Ranchordás, Sofia. 2013. "The Whys and Woes of Experimental Legislation". *Theory and Practice of Legislation* 1(3): 415–440.

Shafir, Eldar. 2013. *The Behavioral Foundations of Public Policy*. Princeton, NJ: Princeton University Press.

Shavell, Steven. 2009. *Foundations of Economic Analysis of Law*. Cambridge, MA: Harvard University Press.

Stigler, George J. 1984. "Economics – The Imperial Science?" *Scandinavian Journal of Economics* 86: 301–313.

Strassheim, Holger, Arlena Jung and Rebecca-Lea Korinek. 2015. "Reframing Expertise: The Rise of Behavioural Insights and Interventions in Public Policy". In A. Berthoin Antal, M. Hutter and D. Stark (eds). *Moments of Valuation. Exploring Sites of Dissonance*. Oxford: Oxford University Press, pp. 249–268.

Strassheim, Holger and Pekka Kettunen. 2014. "When Does Evidence-based Policy Turn into Policy-based Evidence? Configurations, Contexts and Mechanisms". *Evidence and Policy* 10(2): 259–277.

Suchman, Mark C. and Elizabeth Mertz. 2010. "Toward a New Legal Empiricism: Empirical Legal Studies and New Legal Realism". *Annual Review of Law and Social Science* 6: 555–579.

Sunstein, Cass R. 1997. "Behavioral Analysis of Law". *University of Chicago Law Review* 64(4): 1175–1195.

Sunstein, Cass R. 2000. "Introduction". In Cass Sunstein (ed.). *Behavioral Law and Economics*. Cambridge: Cambridge University Press, pp. 1–10.

Sunstein, Cass R. 2011. "Empirically Informed Regulation". *University of Chicago Law Review* 78(4): 1349–1429.

Thaler, Richard H. 2015. *Misbehaving: The Making of Behavioral Economics*. New York: W.W. Norton & Company.

Thaler, Richard H. and Cass R. Sunstein. 2008. *Nudge: Improving Decisions about Health, Wealth, and Happiness*. New Haven, CT: Yale University Press.

Ulen, Thomas. 2014. "The Importance of Behavioral Law". In Elyan Zamir and Doron Teichman (eds). *The Oxford Handbook of Behavioral Economics and the Law*. Oxford: Oxford University Press.

Van Gestel, Rob and Gijs van Dijck. 2011. "Better Regulation through Experimental Legislation". *European Public Law* 17(3): 539–553.

White, Mark D. 2015. "Introduction". In Mark D. White (ed.). *Law and Social Economics. Essays in Ethical Values for Theory, Practice, and Policy*. Basingstoke: Springer.

Zelizer, Viviana A. 1988. "Beyond the Polemics on the Market: Establishing a Theoretical and Empirical Agenda". *Sociological Forum* 3(4): 614–634.

14 Imperializing epistemology

Shortcomings of the naturalistic turn

Manuela Fernández Pinto

Introduction

The main purpose of this chapter is to highlight some of the shortcomings of the naturalistic turn in epistemology. I focus on a particular way of naturalizing epistemology, i.e., *active cooperative naturalism*, in which philosophers actively engage in scientific inquiry looking for answers to their epistemological questions. This way of naturalizing epistemology is taken as a case of scientific imperialism, where the explanatory scope of a scientific discipline is pursued beyond its original domain. I take the scientific imperialism framework as a useful tool for understanding and evaluating the appropriateness of such attempts at naturalizing epistemology. Following Mäki (2009, 2013), I claim that appropriate cases of scientific imperialism should follow an epistemological constraint, advising against the dogmatic commitment of background assumptions.

I examine social epistemology as a case study, in which philosophers have developed rational choice models and computer simulation models (Fernández Pinto 2016). I argue that naturalizing epistemology through active cooperation in these cases has an important shortcoming. The uncritical adoption of background assumptions from other scientific disciplines questions the epistemological appropriateness of such an imperialistic move, especially given that the background assumptions imported from foreign disciplines include epistemological assumptions that limit in an imperceptible way the development of a theory of knowledge. This is due at least in part to the fact that active cooperation requires appropriate spaces for criticism and relevant expertise, which has been missing in philosophy. To conclude, I present some of the possible ways in which the challenge to active cooperation might unfold.

1 The naturalization of epistemology

At the core of the naturalistic approach to epistemology is first a challenge to the foundationalist project of traditional epistemology, which aimed at establishing *a priori* conditions for justified true belief, and second the acknowledgment of the importance of empirical input for addressing epistemological questions. The naturalistic project in epistemology is, however, pluralist in character. As Pacherie acknowledges:

> one should not take the label 'naturalistic epistemology' to be referring to a single, well-defined, doctrine. Rather, this label functions as an umbrella term covering a set of approaches that question in more or less radical ways the tenets of classical epistemology and insist on the relevance of empirical research to epistemological investigations.
>
> (Pacherie 2002, 299)

Historically, the naturalistic turn in epistemology can be traced back to Quine's famous 1969 article, "Epistemology Naturalized". Here Quine examines the failure of the foundationalist project, which he divides in two parts: (i) the conceptual reduction, i.e., the reduction of the meaning of theoretical terms to the meaning of terms of sense experience; and (ii) the doctrinal reduction, i.e., the reduction of theoretical truths to sense experience. Following Quine, we know of the impossibility of the doctrinal reduction since Hume's problem of induction: given that deductive derivation is not available in matters of fact and that induction does not preserve certainty, we cannot derive theory from observation with certainty. Now, Quine argues that conceptual derivation is also impossible. Given semantic holism, i.e., if our statements about the world face the tribunal of experience, not in isolation, but as a coherent whole (Quine 1951, 107), an indeterminacy of translation follows: there are mutually incompatible ways of translating theoretical statements into empirical or observational statements. Accordingly, Quine argues, both the conceptual and the doctrinal reduction fail and with them the foundationalist project.[1]

For Quine, the collapse of the foundationalist project in epistemology opened the door for a more naturalized approach to the study of knowledge, one in which epistemology becomes a part of science, or in other words, where *a priori* inquiry turns empirical: "We are after an understanding of science as an institution or process in the world, and we do not intend that understanding to be any better than the science which is its object" (Quine 1969, 84). After the naturalistic turn, the metaphor of epistemology as the grounding foundation upon which the edifice of science ought to be constructed was replaced by Neurath's metaphor of the boat at sea: "We are like sailors who on the open sea must reconstruct their ship but are never able to start afresh from the bottom. Where a beam is taken away a new one must at once be out there, and for this the rest of the ship is used as support" (Neurath 1973 [1921], 199).

The project of naturalizing epistemology thus leaves aside the search for *a priori* conditions for knowledge. Instead naturalists assume that knowledge is already attainable and proceed to analyze the conditions under which knowledge is actually produced. In Lorraine Code's words, "[naturalists] abandon any quest for a priori, necessary and sufficient conditions for knowledge in general, to examine how epistemic agents actually produce knowledge, variously, within the scope and limits of human cognitive powers as these powers are revealed in the same projects of inquiry" (1996, 1).

Once the foundationalist aims of epistemology are set aside, the door opens for epistemology to draw from scientific accounts of knowledge formation. In particular, if one considers epistemology to be concerned with the human individual's process of belief formation, the naturalistic turn seems to open the door for an epistemology tied to psychology. Accordingly, the original Quinean project focused on psychology as the adequate place for epistemological inquiry: "Why not settle for psychology? … If we are out to understand the link between observation and science, we are well advised to use any available information, including that provided by the very science whose link with observation we are seeking to understand" (Quine 1969, 76).

Quine went even further. His project of naturalized epistemology was not only about an epistemology that was informed by psychology, but about an epistemology that was *part* of psychology. Hilary Kornblith (1994) captures this radical aspect of Quine's view in the *replacement thesis* or the idea that all epistemological questions are ultimately psychological questions. As Quine famously stated:

> Epistemology, or something like it, simply falls into place as a chapter of psychology and hence of natural science. It studies a natural phenomenon, viz., a physical human subject. This human subject is accorded a certain experimentally controlled input … and in the fullness of time the subject delivers as output a description of the three-dimensional external world and its history.
>
> (Quine 1969, 82–83)

Even though the original Quinean project of naturalized epistemology focused on psychology as the scientific discipline of which epistemology should be part, contemporary naturalist philosophers have questioned the preeminence of psychology as the naturalizing base for epistemological inquiry. Lorraine Code, for example, claims that "[e]ven psychology's impressive successes in showing how people go about knowing do not show that epistemology and psychology converge to the extent that epistemology reduces to psychology. There are too many left-overs" (1996, 4). Other scientific disciplines might also contribute, or even be better for, informing epistemological inquiry, especially if one considers psychology's account of knowledge formation to be limited, for example, for being too individualistic (Kim 1988; Code 1996; Hands 2001).

2 Imperializing epistemology through active cooperation

Contemporary philosophers have endorsed the naturalistic turn to different extents. If traditional epistemologists are on one side of the spectrum, where they remain detached from any scientific input in their epistemological inquiries, maintaining the *a priori* or "armchair" methodology criticized by Quine (e.g., Fumerton 1994), on the other side of the spectrum we would have the sort of *replacement naturalism* that many have attributed to Quine (Kornblith 1994; Feldman 2012), where epistemology completely abandons its normative aims and is thus replaced by purely descriptive scientific inquiry (e.g., Kornblith 1999). Most epistemologists and philosophers of science are, however, in a position between these two extremes. They have rejected Quine's *replacement naturalism*, and instead opted for a more *cooperative naturalism*, according to which epistemology still pursues the normative questions proper of philosophy, but taking into account the empirical results from scientific research (Feldman 2012). In particular, while rejecting the foundationalist project of establishing *a priori* conditions for justified true belief, these moderate philosophers still endorse the normative aims of epistemology as the evaluation of different processes of belief formation (Pacherie 2002, 305), while informing their views with the input of current scientific findings. Following Nanay (2015):

> What it means for a philosophical research program to be naturalist is that it integrates the philosophical and the scientific in such a way that the philosophical does not automatically trump the scientific. A naturalist approach is always open to re-evaluating our concepts and intuitions in the light of stronger scientific evidence. And naturalistic accounts are not only informed and constrained by scientific findings, but they are also, at least in principle, in the position to be relevant to further scientific research.
>
> (Nanay 2015, 226)

Most contemporary philosophers of science would consider themselves "naturalists" in this weaker sense of the term.

Given the preference for cooperative naturalism among contemporary epistemologists and philosophers of science, I will set the more radical replacement thesis aside, and focus on cooperative naturalism and its implications for epistemology. To begin with, cooperative naturalism might encompass a variety of positions, all coherent with the idea of a *scientifically informed* epistemology. In particular, one can locate different versions of cooperative naturalism in a spectrum between the more *passive* and the more *active* forms of cooperation, understanding these as follows:

1. *Passive cooperation*: philosophers remain completely detached from the scientific process, but they use the results of scientific inquiry available to inform their epistemic views.
2. *Active cooperation*: philosophers are completely engaged in scientific inquiry. They conduct scientific research themselves, using the conceptual and methodological tools proper of scientific research.

In contrast to passive cooperation, in which philosophers do not interfere in the scientific process and just take advantage of scientific results as they become available, active cooperation is a hands-on approach, in which philosophers directly engage in obtaining answers to their epistemological questions through scientific inquiry.

Naturalizing epistemology through active cooperation immediately raises important concerns regarding the appropriateness of importing scientific theory and methodology into epistemology and actively engaging in scientific inquiry for addressing epistemological questions. Philosophers of science have already raised concerns about extending the explanatory scope of a scientific theory beyond its disciplinary boundaries – the phenomenon of *scientific imperialism* (Dupré 1995; Mäki 2009). If one understands active cooperation as a case of scientific imperialism, in which scientific theories and methodologies are imported to epistemology beyond their traditional disciplinary boundaries, the recent literature on scientific imperialism becomes a valuable tool for understanding some of the limitations of naturalized epistemology.

Broadly speaking, cases of scientific imperialism emerge when a scientific discipline extends its explanatory scope beyond what are considered its target phenomena. Although some consider scientific imperialism to be necessarily detrimental to scientific inquiry (e.g., Dupré 1995), I tend to agree with Mäki (2009) that, as an epistemic expansion, there is nothing wrong in principle with scientific imperialism. The history of science has plenty of examples in which the expansion of scientific disciplines into new realms has certainly rendered epistemic gain. To give a famous example, remember Galileo pointing his telescope to the sky and attempting to explain celestial phenomena in the same way in which terrestrial phenomena were explained. Extending the explanatory scope in this case provided empirical evidence for a more accurate astronomical theory than the geocentric theory defended at the time, leading to clear epistemic gain. Scientific imperialism, however, has also been questioned in cases where the epistemic gain of the explanatory expansion is not clear. For instance, John Dupré has questioned economics imperialism precisely on these terms: "as is very clearly illustrated by incursions of economics into various domains, alien intellectual strategies may import inappropriate and even dangerous assumptions into the colonized domains" (1995, 380).

In part, whether a case of scientific imperialism turns out to be epistemically fruitful, as in the case of Galileo, or not, is an empirical question, which depends on the disciplines involved, the phenomena to be explained, and the reliability of the methodology. However, in general, we expect fruitful cases of scientific imperialism to follow certain normative guidelines. Mäki (2009, 2013) suggests a list of four such normative constraints regarding the ontological, epistemological, axiological, and institutional aspects of the imperialistic move. Particularly relevant for the purpose of this chapter is the *epistemological constraint*, which looks to reject cases of scientific imperialism where the imperializing discipline imposes dogmatically its explanatory framework.

> The epistemological constraint I am proposing […] advises against dogmatic commitment and recommends a strong sense of fallibility and openness to critical conversation across disciplinary boundaries. Personal and "strategic" commitment to a theory may do no harm, but only provided it is accompanied by tolerance and pluralism that derive from a deeper commitment to the uncompromised principle of fallibilism.
>
> (Mäki 2009, 373)

Attempts at naturalizing epistemology, especially through active cooperation, raise concerns about the appropriateness of the scientific theories and methodologies that philosophers adopt for addressing epistemological questions, primarily because there is no reason to assume that naturalizing per se would lead to an adequate epistemology. As in any other scientific discipline, research results need not only follow scientific standards, but also be subjected to a process of criticism among experts (Longino 2002). More importantly, a concern emerges regarding the use of concepts, tools, and methodologies developed in foreign scientific domains for the study of different phenomena.

Accordingly, we can use the scientific imperialism framework to examine the naturalization of epistemology through active cooperation. When is the adoption of scientific concepts, methodologies, theories, etc., appropriately used to answer epistemological questions? In the next section, I take social epistemology as a case study, to show that the naturalization of epistemology through active cooperation has some limitations when trying to provide accounts of scientific knowledge formation.

3 Social epistemology and the shortcomings of the naturalistic turn

In the late 1980s and early 1990s, social epistemology emerged as a subfield of philosophy dedicated to the study of the social dimensions of knowledge, with a particular emphasis on scientific knowledge. Motivated in part by the challenges that historians and sociologists of science had posed to traditional rationalist accounts of scientific knowledge, philosophers of science opened their views to an understanding of knowledge in terms of its social constituents (e.g., Fuller 1988; Rouse 1987; Longino 1990; Kitcher 1993; Solomon 1994).

Broadly speaking, social epistemology is characterized as the study of the social dimensions of knowledge, including other cognitive processes and products such as true belief, justified belief, understanding, scientific knowledge, and the like (Cox and Goldman 1994; Goldman 2006; Kusch 2011; Miller, forthcoming). Social epistemologists posed a serious challenge to naturalized epistemology. Central to this challenge is the question of what is considered *natural*. For Quine, naturalized epistemology was better understood as a chapter of psychology, not only because both were interested in the same *natural* phenomenon, i.e., the human subject, but also because the human subject exhausted the realm of epistemological inquiry. Following a long philosophical tradition going back at least to Descartes, Quine takes the human individual as the locus of knowledge formation. However, the individualism behind Quine's naturalized epistemology has been challenged by philosophers who understand our epistemic processes to be fundamentally social and who furthermore consider that sociality, and the epistemic practices that derive from it, are also *natural* phenomena worthy of scientific inquiry as much as the individual human subject (e.g., Ross 2011). In this sense, social epistemologists have expanded the Quinean project of naturalized epistemology so that epistemology can benefit from other scientific disciplines different from psychology.

In addition, prominent approaches to social epistemology engage in a naturalization of epistemology *through active cooperation.* [2] This means that instead of drawing from already available scientific results, social epistemologists have been actively involved in the application of scientific tools and methodologies for addressing epistemological questions, and thus in the process of obtaining the relevant scientific results. Rather than waiting or conforming to the scientific research available, these social epistemologists have directly engaged in scientific inquiry. Perhaps one of the most prominent lines of research in this respect consists in adopting and adapting models and modeling techniques from scientific disciplines to explore epistemological questions – for instance, questions regarding the optimal distribution of cognitive labor in science, or the optimal organization of information transmission.

Some of the first approaches to social epistemology involved the use of rational choice theory and analytic models to show that a scientist's non-epistemic interests, such as personal credit, can play a role in diversifying research efforts within the scientific community and thus can contribute to the success of science (Kitcher 1990; Goldman and Shaked 1991; Brock and Durlauf 1999; Strevens 2003). More recently, computer simulations have become a central tool for studying the organization of complex systems in a variety of scientific fields (e.g., cosmology, climate science, political science, and economics). Social epistemologists have begun to follow this trend, using computer simulations to model the social organization of scientific research (see e.g., Zollman 2007; De Langhe and Greiff 2009; Hegselmann and Krause 2009; Weisberg and Muldoon 2009). Here we can see two important attempts at naturalizing epistemology through active cooperation, importing methodological and conceptual tools from other scientific disciplines to address philosophical questions: first, through rational choice models, and second, through computer simulation models.

3.1 Rational choice epistemology

In the late 1980s, social epistemologists began using the conceptual and methodological tools offered by economics to study the organization of epistemic systems, in particular the problem of the division of cognitive labor.[3] Originally stated by Kuhn (1977), the problem of the division of cognitive labor emerged as a tension between individual and collective rationality: if each scientist made a rational decision about her research options at the individual level, all scientists would opt for the same research path, failing to attain an adequate division of cognitive labor at the collective level. Kitcher (1990) and Goldman and Shaked (1991) used analytic models in rational choice theory to explain how scientists can make epistemic achievements through an efficient division of cognitive labor, despite following non-epistemic interests, such as the aim for personal credit.

Kitcher's game theory model showed that a self-interested scientist was more likely to choose a research project that might not seem the best one for achieving certain epistemic goals, but which is nonetheless better, given the choices made by other scientists. In this case, the scientist's non-epistemic interests, for example the search for personal credit, motivates the decision to pursue the less promising research project, which leads to a better division of cognitive labor at the collective level, i.e., instead of all scientists pursuing the same project, they divide their efforts. In a similar vein, Goldman and Shaked (1991) developed a decision theory model to show that a credit-seeking agent, whose expected utility only considers her professional success, would usually choose research projects that increase the possession of truth. Thus, they conclude that professional success and truth acquisition are not necessarily incompatible.

Social epistemologists, such as Kitcher and Goldman, appeal to economic models to address questions traditionally pertaining to the philosophy of science. Goldman in particular is explicit about adopting scientific models for such purpose: "We wish to extend the economic paradigm to certain problems in epistemology and the philosophy of science" (Goldman and Shaked 1991, 31). Moving away from the privileged status that psychology had in Quinean naturalized epistemology, some social epistemologists chose rational choice models from economics as the naturalizing base of their epistemologies. As Wade Hands acknowledges, "[e]conomics is no longer just a *subject for* science theory, it has now become an important resource *to be used in* science theory" (2001, 354).

The use of rational choice models in epistemology does not come without concerns, for this imperialistic move from economics to epistemology raises a number of questions: What has social epistemology gained and lost with the use of economics and rational choice models? Have we attained better explanations regarding the social dimensions of scientific knowledge? In particular, have we achieved better normative insights about how to organize scientific research for epistemic gain? More importantly, what is economics importing to epistemology? Are there fundamental background assumptions that epistemologists are uncritically relying on when using these models? All of these questions relate to the epistemological appropriateness of this case of economics imperialism.

Inevitably, extending a discipline's explanatory scope into new domains, and in this case extending a modeling technique, necessarily entails importing the background assumptions embedded in the particular type of models that are being used. In the case of rational choice models these assumptions include, for instance, a view of rationality as utility maximization, an understanding of sociality as the aggregate of homogenous individual agents, agents having perfect information from the get-go, among others. More broadly, economic models that address questions pertaining to the organization of scientific research seem to assume that science can be understood as a market, where scientific knowledge is traded as a commodity. Acknowledging that the analogy with the marketplace is not perfect, Goldman makes this point explicit: "Scientific agents, scholarly inquirers generally, act in some ways like vendors, trying to 'sell' their findings, theories, analyses, or arguments to an audience of prospective 'buyers'" (Goldman and Shaked 1991, 31). Accordingly, scientists are taken to maximize their utility in terms of seeking to achieve an epistemic gain that is treated homogenously as anything that contributes to truth acquisition.

Of course, many of these assumptions are problematic, as many philosophers have already pointed out (Mirowski 1995; Ylikoski 1995; Hands 1997; Roorda 1997; Sent 1997; Solomon 2001). Some contend, for example, a view of rationality as utility maximization (Anderson 1993; Longino 2002), and others strongly warn us against the idea of science as a market (Mirowski 2011). In the context of a naturalized epistemology, where the imported models are introducing at the same time solutions for epistemological questions and background assumptions that set our theory of knowledge, having problematic assumptions leads to a dilemma. If there are good epistemological reasons to disagree with the underlying assumptions of rational choice models, and the models are at the same time central to solving certain epistemological questions, one is left with two possibilities: either one attempts to answer the objections to the assumptions and remains committed to the models, or one accepts that the assumptions are problematic and looks for different ways of naturalizing epistemology. The former case requires a critical engagement with the objections, while the latter requires new modeling techniques that presumably have fewer problematic assumptions. Notice that in both cases, the apparent vicious circle stemming from the naturalization of epistemology turns out to be a virtuous one, where both epistemology and science are self-critical and reconstructive, which also entails the fulfillment of Mäki's epistemological constraint against dogmatically imposing the acceptance of background assumptions.

Rational choice epistemologists, however, have done little about critically engaging with their opponents or attempting to vindicate the background assumptions of their models, leaving the dilemma untouched (Fernández Pinto 2016). In fact, some have even explicitly rejected some of these assumptions (e.g., Kitcher 2011, 208), despite clearly implementing them in their epistemological models. In this sense, much of rational choice epistemology has not fulfilled Mäki's epistemological constraint, which casts doubt on the epistemological appropriateness of this imperialistic move. This is not to say that work on rational choice epistemology has to slow down, but rather to highlight that new contributions to the topic seem to improve details of the models, leaving the background assumptions untouched (e.g., Brock and Durlauf 1999; Strevens 2003). Other social epistemologists, however, have opted for the second strategy and have tried to look for other ways of naturalizing epistemology to address similar questions as rational choice epistemologists.

3.2 Computer simulation epistemology

Following the current trend for studying the social organization of complex systems (Mainzer 2007; Miller and Page 2007), social epistemologists have started using computer simulation models, such as social network models and agent-based models, to naturalize epistemology. The change to computer simulations has been driven in part by the problems found in the original rational choice models, particularly with respect to their underlying assumptions. Muldoon and Weisberg (2011), for example, have shown that if the assumptions about perfect information in Kitcher's model are relaxed, the results no longer hold. Given that real scientists are not expected to have perfect information and that information is crucial for deciding which projects to pursue, Weisberg and Muldoon take this to be an important limitation of Kitcher's rational choice approach, and instead advocate the use of computer simulation models which allow "for greater representational flexibility for modeling epistemic situations that are common to modern scientific communities" (Weisberg and Muldoon 2009, 227). Thus, the move to computer simulations is seen as an opportunity to revise some of the problematic assumptions of rational choice models – e.g., agents do not need to have perfect information, different types of agents and different types of epistemic gains can be programmed into the model, and so on. In this way, social epistemologists using computer simulations acknowledge the limitations of rational choice models and opt for a different type of model for epistemological inquiry.

For instance, Kevin Zollman (2007) adopts a social network model developed by two economists (Bala and Goyal 1998) in order to examine optimal ways of transmitting information within the scientific community and their relation to epistemic success or reliability. Unexpectedly, the model shows that "in many cases a community made up of less informed individuals is more reliable at learning the correct answers" (Zollman 2007, 575). Adapting a fitness landscape model from evolutionary biology, Weisberg and Muldoon (2009) address the problem of the division of cognitive labor using an epistemic landscape model where two types of scientists (followers and mavericks) explore the landscape differently and attain different types of epistemic gain (epistemic progress or epistemic significance). They conclude that "optimum research communities are going to be composed of a healthy number of followers with a small number of mavericks" (ibid., 251).

As with the case of rational choice models, importing the use of computer simulation models into social epistemology goes hand in hand with importing background assumptions from foreign scientific disciplines that might pass unnoticed. Zollman's social network model, for example, is based on a previously developed economic model of learning in social networks (Bala and Goyal 1998), which still preserves some of the assumptions of rational choice models, e.g., rationality is understood as utility maximization, agents still have perfect information regarding certain expected payoffs, etc. Similarly, Weisberg and Muldoon's (2009) model is based on hill-climbing models developed for artificial intelligence research. Hill climbing is a search algorithm for finding the maximum of an evaluation function. Implemented in a computer simulation model, the hill-climbing algorithm allows the agents to evaluate their neighborhood and move in the direction of increasing value until the agent finds a peak (Russell and Norvig 1995). Hill climbing is taken to be a good algorithm for finding local maxima, but agents can easily get stuck when trying to find a global maximum, when encountering ridges – i.e., a sequence of local maxima – or when they are on a plateau – i.e., a flat area on the landscape (1995, 122–123). Accordingly, one of the main problems that Weisberg and Muldoon encounter in their model is that a certain type of scientist easily gets stuck in flat areas of the landscape, a problem that stems from the type of model they have chosen.

These cases illustrate two possible scenarios in the naturalization process. On the one hand, Zollman's social network model attempts some methodological improvements (moving from game theory models to computer simulations for more flexibility), while maintaining some of the same background assumptions of rational choice models. So even if the model seems to provide new insights regarding the social organization of science, thus leading to epistemic gain, the model still introduces background assumptions that are considered problematic, failing to address the epistemological constraint. Weisberg and Muldoon's model, on the other hand, seems to import new background assumptions, such as the hill-climbing algorithm, which ought to be assessed. The initial dilemma reappears: either engage with the background assumptions or move on to more promising modeling techniques. For now the door is open for the dilemma to be solved either way. Of course, there is nothing wrong with the dilemma per se, for it seems to follow naturally from the process of appropriate criticism within the scientific community, as well as the epistemological constraint in cases of scientific imperialism. However, in the case of a naturalization of epistemology through active cooperation, the dilemma seems to present an important challenge.

I claim that the epistemological constraint is an especially significant normative guideline for naturalizing epistemology and that solving the dilemma thus becomes especially relevant in social epistemology. According to the epistemological constraint, background assumptions must be open to criticism and revision. However, notice that in any scientific discipline, background assumptions provide, among other things, the theoretical framework to address scientific questions, or in other words, they provide the theory of knowledge from which disciplines engage in scientific inquiry. Accordingly, background assumptions include a common understanding regarding what counts as evidence for a claim, what counts as an adequate explanation of the phenomena, when the scientist is justified in making a claim, etc. Broadly speaking, background assumptions include the epistemological assumptions that any scientist needs before engaging in active scientific inquiry.

Now, this presents a particularly interesting problem in any project of naturalized epistemology, for some of the background assumptions of the scientific discipline that philosophers are using for the naturalization of epistemology are epistemological assumptions that will limit the scope of the epistemology they are aiming to develop. Moreover, they will limit the scope of such epistemology *imperceptibly*, unless we make these assumptions explicit and submit them to the appropriate criticism. For instance, rational choice epistemology imports an assumption of rationality as utility maximization, which limits its understanding of the decision-making process in scientific research to a process of utility maximization. In this way, a theory of the process of knowledge production has importantly been reduced to a very particular way of understanding decision-making. If we do not follow the epistemological constraint here, we are in danger of uncritically accepting an epistemology, i.e., a theoretical framework about knowledge production, in using rational choice models to answer our epistemological questions. Thus, the epistemological constraint becomes especially important for naturalized epistemology.

One can take this to be a consequence of the fact that the project of naturalized epistemology also includes a *reciprocal containment* of "epistemology in natural science and natural science in epistemology" (Quine 1969, 83). In this sense, epistemology follows what our best scientific theories tell us about the process of knowledge production, but at the same time studies those same theories and their assumptions. Hence, in the process of reconstructing the boat at sea, philosophers must be overtly aware of the background assumptions that they are importing to epistemology, for these assumptions are constraining our theory of knowledge.

4 Moving ahead

Any attempt at naturalizing epistemology through active cooperation comes together with the use of certain background assumptions that are taken for granted. In a sense, this is inevitable. As it has been shown repeatedly, scientific theorizing and researching do not happen in an epistemological vacuum. Every scientific discipline works with background assumptions that inform its scientific process (Longino 1990). The problem, as Mäki warns us, is to hold these assumptions dogmatically. If we believe in scientific knowledge as the result of a process of effective criticism among qualified experts, then background assumptions must be open to criticism and revision as well. This is especially important in the context of naturalized epistemology, given that background assumptions frequently include epistemological assumptions needed for scientific inquiry. Thus, to impede that these epistemological assumptions limit imperceptibly our theory of knowledge, epistemologists must be overtly aware of the assumptions they are importing from other scientific disciplines.

How shall we proceed? A disciplinary challenge emerges with the naturalization of epistemology through active cooperation here. Even if epistemologists were really open-minded about them, background assumptions are not easy to identify, evaluate, or revise, especially if the philosophical community that is supposed to provide the disciplinary venues for critical assessment does not have the scientific expertise needed for such purpose. The imperialistic move faces a disciplinary shortcoming here, given that the imperialism is internally driven.[4]

The danger here is accepting active cooperation strategies without the appropriate scientific expertise, in which case scientific imperialism turns into a sort of *scientism*. This happens if scientific approaches to the study of the social dimensions of scientific knowledge are adopted just for the sake of being scientific, without the appropriate venues for effective criticism needed for the discussion and revision of such approaches. As I already stated, the naturalization of epistemology per se need not lead us to an adequate epistemology. Whether a particular approach to naturalized epistemology is adequate or not is an empirical question that can only be appropriately answered if the venues for effective criticism among experts are in place.

Do we have the required disciplinary framework? Some might argue we do, given that many philosophers of science hold undergraduate and master's degrees in a variety of scientific disciplines, and given that some PhD programs in philosophy of science now require their students to obtain some scientific training. Others might argue we do not, given that, in general, disciplinary training in epistemology or philosophy of science does not include the development of the methodological skills needed for the critical assessment of the models we have mentioned. Although the answer remains unclear, the point here is that we need such a disciplinary framework if we want internal critical engagement of active cooperation strategies in social epistemology. However, other possibilities also emerge.

I would like to highlight four possible answers here, which also show plausible future developments for the naturalization of epistemology.

1 *Disciplinary training*: make active cooperation acceptable by making epistemology more scientific and by training experts accordingly.
2 *Interdisciplinary interaction*: make active cooperation acceptable by encouraging collaboration between philosophers and scientists.
3 *Passive naturalism*: reject active cooperation and encourage the use of scientific results as they develop according to their own scientific process.
4 *Replacement naturalism*: reject active cooperation and leave epistemology to the scientists.

To conclude, let me briefly assess these four possible responses. Disciplinary training, which is an open option that some are already implementing, is expected to face important resistance from the broader philosophical community, which traditionally remains detached from descriptive and empirical inquiry. Some might consider this a reason in favor of, rather than against, disciplinary training, but it is important to acknowledge the broader disciplinary challenges that it might entail, e.g., further marginalizing philosophy of science from mainstream philosophy. In any case, given the previous analysis, if this strategy were to succeed, it would require reaching a critical mass of experts to warrant the disciplinary framework needed.

Although much of the chapter has taken a rather negative tone, I consider that there is much to gain from a fruitful interaction between social epistemologists and scientists, especially with respect to questions regarding the social organization of science that might have an important societal impact (e.g., how to distribute resources for medical research, climate change, etc.). Accordingly, I consider that interdisciplinary interaction seems a most promising venue for active cooperation. Mixing scientific and philosophical expertise could, in principle, help us uncover the epistemological assumptions behind scientific models, and thus make transparent the epistemological commitments for philosophers who engage in active cooperation. At least, this strategy could nicely complement and strengthen disciplinary training, filling the gaps where philosophers' scientific training is lacking.

Passive naturalism should remain an open option in social epistemology; after all, it has been the most successful and popular way in which philosophers of science have naturalized epistemology so far. Finally, it remains an open empirical question whether replacement naturalism can succeed or not, i.e., whether epistemology is to become a purely scientific enterprise or not. If this turns out to be the case, epistemology as proper philosophical inquiry would probably disappear. As that question unfolds, it seems that for the time being philosophers still have significant contributions to make in understanding the process of knowledge production, especially when aided by scientific expertise and framed in an interdisciplinary research environment.

Acknowledgments

I would like to thank Uskali Mäki and Adrian Walsh for their collaboration in organizing the workshop on scientific imperialism held at the University of Helsinki on April 2015, which led to this volume. Thanks also to my colleagues at TINT and the Department of Philosophy at Universidad de los Andes, especially Carlos Manrique and Andrés Páez, who provided insightful comments on earlier drafts of this chapter.

Notes

1 As many have pointed out, Quine's critique of the foundationalist project presupposes the rejection of the synthetic/analytic distinction (Quine 1951), and the recognition that both meaning and evidence ultimately rely on sensory experience. For objections to Quine's argument against the analytic/synthetic distinction, see for example, Strawson and Grice (1956), Almeder (1990), Papineau (2015).

2 Another way in which some philosophers have engaged in active cooperation is through the growing program of experimental philosophy (or x-phi), in which philosophers many times actively participate in the formulation and development of scientific – mainly psychological – experiments to discern the ways in which humans think about certain topics, or in other words, to test human intuitions about a certain topic (Fischer and Collins 2015, 4). Whether the x-phi program as a whole is naturalistic or not is still controversial (Nanay 2015).

3 In this sense, social epistemology can be considered another discipline to fall prey to the prominent imperialistic tendencies of economics as famously promulgated by economists of the Chicago School, such as Becker (1976) and Stigler (1984). For a more detailed account of economics imperialism in social epistemology, see Hands (2001); Mirowski (2004); Mäki (2005); and Fernández Pinto (2016).

4 For a discussion about the distinction between internally and externally driven imperialism, see Mäki (2013, 335).

References

Almeder, Robert. 1990. "On Naturalizing Epistemology". *American Philosophical Quarterly* 27(4): 263–279.

Anderson, Elizabeth. 1993. *Value in Ethics and Economics*. Cambridge, MA: Harvard University Press.

Bala, Venkatesh and Sanjeev Goyal. 1998. "Learning from Neighbours". *Review of Economic Studies* 65: 595–621.

Becker, Gary S. 1976. *The Economic Approach to Human Behavior.* Chicago, IL: University of Chicago Press.

Brock, W.A. and S.N. Durlauf. 1999. "A Formal Model of Theory Choice in Science". *Economic Theory* 14: 113–130.

Code, Lorraine. 1996. "What is Natural about Epistemology Naturalized?" *American Philosophical Quarterly* 33: 1–22.

Cox, J.C. and Alvin I. Goldman. 1994. "Accuracy in Journalism: An Economic Approach". In Frederick F. Schmitt (ed.). *Socializing Epistemology.* Lanham, MD: Rowman & Littlefield Publishers.

De Langhe, Rogier and Matthias Greiff. 2009. "Standards and the Distribution of Cognitive Labor". *The Logic Journal of the IGPL* 18: 278–293.

Dupré, John. 1995. "Against Scientific Imperialism". *PSA: Proceedings of the Biennial Meeting of the Philos Sci Association* 2: 374–381.

Feldman, Richard. 2012. "Naturalized Epistemology". In Edward N. Zalta (ed.). *The Stanford Encyclopedia of Philosophy* (Summer 2012 Edition). http://plato.stanford.edu/archives/sum2012/entries/epistemology-naturalized/.

Fernández Pinto, Manuela. 2016. "Economics Imperialism in Social Epistemology: A Critical Assessment". *Philosophy of the Social Sciences* 45(5): 443–472.

Fischer, Eugen and John Collins. 2015. "Rationalism and Naturalism in the Age of Experimental Philosophy". In Eugen Fischer and John Collins (eds). *Experimental Philosophy, Rationalism, and Naturalism: Rethinking Philosophical Method.* London and New York: Routledge, pp. 3–33.

Fuller, Steve. 1988. *Social Epistemology.* Bloomington, IN: Indiana University Press.

Fumerton, Richard. 1994. "Skepticism and Naturalistic Epistemology". *Midwest Studies in Philosophy* 19: 321–340.

Goldman, Alvin. 2006. "Social Epistemology". In *Stanford Encyclopedia of Philosophy.* http://plato.stanford.edu/entries/epistemology-social/.

Goldman, Alvin and Moshe Shaked. 1991. "An Economic Model of Scientific Activity and Truth Acquisition". *Philosophical Studies* 63: 31–55.

Hands, D.Wade. 1997. "Caveat Emptor: Economics and Contemporary Philosophy of Science". *Philosophy of Science* 64: S107–S116.

Hands, D. Wade. 2001. *Reflection without Rules: Economic Methodology and Contemporary Science Theory.* Cambridge: Cambridge University Press.

Hegselmannn, Rainer and Ulrich Krause. 2009. "Deliberative Exchange, Truth, and Cognitive Division of Labour: A Low-Resolution Modeling Approach". *Episteme* 6: 130–144.

Kim, Jaegwon. 1988. "What is Naturalized Epistemology?" In James E. Tomberlin (ed.). *Philosophical Perspectives* 2. Asascadero, CA: Ridgeview Publishing Co., pp. 381–406.

Kitcher, Philip. 1990. "The Division of Cognitive Labor". *The Journal of Philosophy* 87: 5–22.

Kitcher, Philip. 1993. *The Advancement of Science: Science without Legend, Objectivity without Illusions.* New York: Oxford University Press.

Kitcher, Philip. 2011. *Science in a Democratic Society.* Amherst, NY: Prometheus Books.

Kornblith, Hilary. 1994. *Naturalizing Epistemology.* Second edition. Cambridge, MA: MIT Press.

Kornblith, Hilary. 1999. "In Defense of a Naturalized Epistemology". In John Greco and Ernest Sosa (eds). *The Blackwell Guide to Epistemology*. Malden, MA: Blackwell, pp. 158–169.

Kuhn, Thomas S. 1977. *The Essential Tension: The Selected Studies in Scientific Tradition and Change*. Chicago, IL: University of Chicago Press.

Kusch, Martin. 2011. "Social Epistemology". In S. Bernecker and D. Pritchard (eds). *The Routledge Companion to Epistemology*. New York: Routledge.

Longino, Helen. 1990. *Science as Social Knowledge: Values and Objectivity in Scientific Inquiry*. Princeton, NJ: Princeton University Press.

Longino, Helen. 2002. *The Fate of Knowledge*. Princeton, NJ: Princeton University Press.

Mainzer, Klaus. 2007. *Thinking in Complexity: The Computational Dynamics of Matter, Mind, and Mankind*. Berlin: Springer.

Mäki, Uskali. 2005. "Economic Epistemology: Hopes and Horrors". *Episteme* 1: 211–222.

Mäki, Uskali. 2009. "Economics Imperialism: Concepts and Constraints". *Philosophy of the Social Sciences* 39: 351–380.

Mäki, Uskali. 2013. "Scientific Imperialism: Difficulties in Definition, Identification, and Assessment". *International Studies in the Philosophy of Science* 27: 325–339.

Miller, Boaz. (forthcoming). "Social Epistemology". In *Internet Encyclopedia of Philosophy*.

Miller, John H. and Scott E. Page. 2007. *Complex Adaptive Systems: An Introduction to Computational Models of Social Life*. Princeton, NJ and Oxford: Princeton University Press.

Mirowski, Philip. 1995. "Philip Kitcher's Advancement of Science: A Review Article". *Review of Political Economy* 7(2): 227–241.

Mirowski, Philip. 2004. "The Scientific Dimensions of Social Knowledge and their Distant Echoes in 20th-Century American Philosophy of Science". *Studies in History and Philosophy of Science* 35: 283–326.

Mirowski, Philip. 2011. *Science-Mart: Privatizing American Science*. Cambridge, MA: Harvard University Press.

Muldoon, Ryan and Michael Weisberg. 2011. "Robustness and Idealization in Models of Cognitive Labor". *Synthese* 183: 161–174.

Nanay, Bence. 2015. "Experimental Philosophy and Naturalism". In Eugen Fischer and John Collins (eds). *Experimental Philosophy, Rationalism, and Naturalism: Rethinking Philosophical Method*. London and New York: Routledge, pp. 221–239.

Neurath, Otto. 1973 [1921]. "Anti-Spengler". In Marie Neurath and Robert S. Cohen (eds). *Empiricism and Sociology*. Dordrecht: Reidel, pp. 158–213.

Pacherie, Elisabeth. 2002. "Naturalistic Epistemologies and Normativity". *Croatian Journal of Philosophy* 2(6): 299–317.

Papineau, David. 2015. "The Nature of a priori Intuitions". In Eugen Fischer and John Collins (eds). *Experimental Philosophy, Rationalism, and Naturalism: Rethinking Philosophical Method*. London and New York: Routledge, pp. 51–71.

Quine, W.V.O. 1951. "Two Dogmas of Empiricism". *Philosophical Review* 60: 20–43. Reprinted in: *From a Logical Point of View*. Cambridge, MA: Harvard University Press, 1953, pp. 20–46.

Quine, W.V.O. 1969. "Epistemology Naturalized". In *Ontological Relativity and Other Essays*. New York and London: Columbia University Press.

Roorda, J. 1997. "Kitcher on Theory Choice". *Erkenntnis* 46: 215–239.

Ross, Don. 2011. "Naturalism: The Place of Society in Nature". In Ian Jarvie and Jesús Zamora Bonilla (eds). *The Sage Handbook of the Philosophy of Social Sciences*. London: Sage, pp. 121–136.

Rouse, Joseph. 1987. *Knowledge and Power: Toward a Political Philos Sci.* Ithaca, NY: Cornell University Press.
Russell, Stuart and Peter Norvig. 1995. *Artificial Intelligence: A Modern Approach.* Third edition. Upper Saddle River, NJ: Pearson Education, Inc.
Sent, Esther-Mirjam. 1997. "An Economist's Glance at Goldman's Economics". *Philosophy of Science* 64: S139–148.
Solomon, Miriam. 1994. "Social Empiricism". *Noûs* 28: 325–343.
Solomon, Miriam. 2001. *Social Empiricism.* Cambridge, MA: MIT Press.
Stigler, George J. 1984. "Economics: The Imperial Science?" *Scandinavian Journal of Economics* 86(3): 301–313.
Strawson, Peter F. and Paul Grice. 1956. "In Defense of a Dogma". *The Philosophical Review* 65(2): 141–158.
Strevens, Michael. 2003. "The Role of the Priority Rule in Science". *The Journal of Philosophy* 100: 55–79.
Weisberg, Michael and Ryan Muldoon. 2009. "Epistemic Landscapes and the Division of Cognitive Labor". *Philosophy of Science* 76(2): 225–252.
Ylikoski, Petri. 1995. "The Invisible Hand and Science". *Science Studies* 8: 32–43.
Zollman, Kevin. 2007. "The Communication Structure of Epistemic Communities". *Philosophy of Science* 74(5): 574–587.
Zollman, Kevin. 2010. "The Epistemic Benefit of Transient Diversity". *Erkenntnis* 72: 17–35.

15 Imperialism and its discontents

Competing cultures of evidence in the discovery of dark energy

Genco Guralp

Introduction

The questions of what forms of interdisciplinary relations constitute "scientific imperialism," and what are the mechanisms of this process, have increasingly been debated in contemporary philosophy of science. The scientific context of this debate has largely been the behavioral sciences such as economics, or life sciences such as evolutionary biology or neuroscience, although traditionally physics has also been considered as an "imperialistic" science.[1] Furthermore, recent discussions focus mostly on the *theoretical* aspect of scientific knowledge. The issue of imperialism in the context of *experimental* science has not been fully explored. In this chapter, I aim to discuss the question of imperialism by focusing on two natural sciences – namely, particle physics and astronomy – in the context of an experimental episode that led, in its culmination, to the discovery of *dark energy*.

As there exists neither an agreed upon philosophical definition of imperialistic science, nor an established usage of the term in the sciences themselves (with the possible exception of economics), it might seem premature to present a case study on this matter.[2] However, even though a case study of imperialism cannot proceed in as straightforward a fashion as in topics such as, say, *discovery* or *explanation*, examining several cases that seem to exhibit imperialism is indispensable for the very problem of establishing a definition. Cases of interdisciplinary breach, where one finds members of a scientific community reacting to members of another whom they perceive as invaders, should be examined to understand what forms the invasion and the reaction take, and to see whether the concept of imperialism proves suitable and fruitful to describe the episode.

The case that I examine in this chapter exhibits several features which suggest that it can be understood as constituting a case of scientific imperialism. When a research group of particle physicists began an experimental program for measuring the deceleration of the universe using supernova brightnesses, the astronomical community approached their research with skepticism. As I will show below, this skepticism stemmed mostly from the perception of particle physicists as "outsiders" invading the astronomers' turf. This critical stance of the astronomers eventually led to their decision to form their own scientific team to compete with the "physicists," a term they used in an unmistakably pejorative way. When the episode ended, both groups reached the same conclusion, and researchers from each team were awarded Nobel Prizes for their work.

In several of his publications, Uskali Mäki presents a general framework for addressing the issue of imperialism which provides a convenient setting to discuss the case at hand (see, Mäki 2013; Mäki and Marchionni 2011). In contrast with John Dupré's account,[3] which sees scientific imperialism as a negative phenomenon that "is always to be resisted" (Mäki 2013, 338), Mäki offers a neutral account on the basis of which one can distinguish "between good and bad scientific imperialism in terms of normative constraints" (Mäki 2013, 337). This type of emphasis on a *descriptive* instead of a strictly *normative* account of imperialism is especially useful, for, as I will argue, the imperialistic tendencies of the particle physicists actually led to progress of knowledge in this case, and to this extent, imperialism proved to have a positive effect on science.[4] Still, a key element in Mäki's account, namely *imperialism of scope*, is mainly geared towards *theoretical* science with its emphasis on an *explanatory constraint*, and it will need to be expanded in certain respects to capture the experimental side of scientific knowledge as well, or so I shall argue.

The primary aim of this chapter, then, is to extend the account of scientific imperialism that Mäki offered to experimental practice, using the discovery of dark energy as a test case. In what follows, I begin with a short history of the experimental problem that lies at the origin of the research of both groups. I then move on to the work of the particle physicists, *the Berkeley group*, and examine how they approached the problem. Here, I study their decision process for embarking on an astronomical research program and its consequences. In the third section, I study the astronomical community's reaction to physicists' work and focus on how they formed their competing research program. Next, I examine Mäki's work on scientific imperialism and propose to expand it so that it can also account for imperialistic breaches in the context of experimental knowledge. More specifically, I argue that in the context of experimental inquiry, Mäki's *imperialism of scope*, defined on the basis of explanatory unification, is not applicable. Rather, the expansionist tendency in experimentation stems from the *evidential significance* of the measurement in question, and this notion should be added to considerations of scope. I propose to study *evidential significance* via the notion of *evidential culture* that was introduced by Trevor Pinch and Harry Collins in their research on solar neutrino and gravitational wave experiments, respectively, and contrast the evidential culture of particle physicists with that of astronomers (Pinch 1985; Collins 2004, 2014). I conclude by arguing that in the contemporary research setting of modern natural sciences where singular experimental episodes are less and less capable of establishing a result with confidence by themselves, imperialistic science can have a positive effect on scientific progress through increasing *robustness*. Imperialism in science can lead to robust results, provided that the host discipline has local epistemic resources for it to respond to the imperialistic breach, as was the case with contemporary astronomy during the discovery of dark energy.

1 Genesis of the problem

In 1929, Edwin Hubble published his historic result that a linear relationship exists between the velocities and the distances of extra-galactic nebulae. Although he himself was skeptical, many theoreticians of the era quickly seized on this work as empirical confirmation of the expanding universe hypothesis.

From this point on, Hubble's research program aimed to test whether, or to what extent, this relationship held for larger distances as well, using bigger telescopes capable of collecting light from fainter sources. This program constituted the main problem horizon of the experimental work in cosmology for years to come. By extending the velocity-distance relation to larger scales with more precise instruments, researchers hoped to be able to distinguish empirically between various cosmological models. The standard description of these measurements has been given in terms of a quantity known as the *Hubble constant*, which is the constant of proportionality between the velocity and the distance of objects at cosmological distances. At any given time, measuring the Hubble constant required the most sophisticated technology available.[5] As a result, up until very recently, this type of measurement was not open to different research groups working simultaneously, but to a single scientist, or a team of scientists, who had access to the best equipment available.[6] Notably, after Hubble's death, the research program of improving the distance scale measurements by investigating more distant astronomical objects passed on to Allan Sandage, Hubble's assistant at Mt Palomar. Until the late 1980s, Sandage was the leading authority in this research program.

Even though Sandage's program was the continuation of Hubble's, there was one significant difference between them. During Hubble's era, theoretical cosmologists favored an expanding universe model, based on Einstein's general relativity. A significant aspect of this model (which then came to be known as the Big Bang) was that it presupposed the early universe to be hotter and denser than present. In other words, it pictured the universe as a dynamically evolving system. However, in 1948, three Cambridge astronomers proposed a steady state model, which assumed the universe to be static.[7] This theoretical context of two alternative models of the universe was a key influence on Sandage, who devoted his work as an experimental astronomer to conducting measurements that might decide between these two *world models*, as cosmological models were then referred to. For this, he singled out two parameters, namely, the *Hubble constant* and the *deceleration parameter*,[8] arguing that a precise measurement of these two numbers could tell us which model of the universe is the correct one.

However, the decisive confirmation of the Big Bang came not from Sandage's measurements, but by the serendipitous discovery of cosmic microwave background radiation (CMB). Astronomers interpreted CMB as a relic radiation left over from the early hot stage of the universe. The acceptance of the Big Bang as the "standard" model of cosmology was followed by the slow but steady confirmation of the existence of *dark matter* in the universe in the late

1970s. Combining the Big Bang with dark matter, cosmologists pictured the universe as having evolved from an early hot state, and explained the formation of galaxies by the gravitational interaction of the dark matter.

Even though the existence of dark matter was accepted by the astronomical community on the basis of galaxy measurements, no direct detection of it had been made. From the perspective of particle physics, understanding the nature of dark matter was of fundamental importance. One key question to be answered was: *How much dark matter is there in the universe?* Within the standard cosmological model, this question was formulated in terms of the density parameter Omega (Ω), which symbolized the ratio of the mass of dark matter in the universe, to the critical mass needed to stop the expansion of the universe by gravity. Measurement of Ω thus became a key issue for the particle physics community.

These two research traditions crossed paths in the mid-1980s. On the one hand, the Hubble-Sandage program of determining the Hubble constant was on the agenda of astronomers. On the other hand, physicists were interested in understanding the physics of dark matter and structure formation in the early universe. A precise measurement of Ω was the *sine qua non* for these latter purposes. Astronomers knew that, in principle, supernovae were candidate objects for measuring the Hubble constant. In the 1970s, physicists became aware that these objects could also be employed to measure Ω.

Starting in the mid-1980s, a group of physicists at Lawrence Berkeley National Laboratory (LBNL) began working on this measurement, with the aim of "weighing the universe". This work brought them into contact with astronomers, who initially criticized their project on various grounds. A decade later, an astronomy team was formed with the direct aim of competing with the Berkeley physicists. This interaction, and the later competition of these two research traditions culminated in the joint discovery of the acceleration of the expansion of the universe. After dwelling on the history of these two groups of scientists, I will examine how the work of the Berkeley group can be considered as a positive example of scientific imperialism.

2 Formation of the Supernova Cosmology Project and their approach to the problem

The particle physicists who embarked on the measurement of the cosmological parameter Ω were based in LBNL, and were closely affiliated with the University of California, Berkeley. In order to understand the work of the Berkeley group, which later became the Supernova Cosmology Project (SCP), we need to examine the experimental culture that governed the laboratory of which it was part.

2.1 Alvarez and the experimental culture at the LBNL

In his extensive history of 20th-century experimental high-energy physics, Peter Galison studied the work of Luis Alvarez and his role in turning particle physics into big science (Galison 1997).[9] The particular importance of Luis Alvarez's work for understanding the evidential culture of the SCP stems from the fact that he was the scientific mentor of Richard Muller, who was the central scientist who began the SCP. Muller, in turn, was the mentor of the two scientists who played key roles in the completion of the project, namely, Carl Pennypacker and Saul Perlmutter. Pennypacker studied with Muller as an undergraduate at Berkeley, and obtained his PhD at Harvard, before coming back to Berkeley as a post-doc. His general interest at the time was in instrumentation and building detectors.[10] Perlmutter obtained his undergraduate degree at Harvard, and completed his PhD in physics under Muller at Berkeley.[11] Muller's mentorship played a significant role in the scientific styles of both these physicists, and, as we will see in the next section, Muller's style was heavily influenced by his doctoral advisor, Alvarez. In what follows, I aim to give a sketch of the evidential culture of the SCP, focusing on the careers of these scientists. However, first I need to say a few words about Alvarez's scientific style.[12]

There were two main sources of Alvarez's scientific style, both stemming from his work in World War II: "radar and the atomic bomb" (Galison 1997, 343). Both shaped Alvarez's laboratory style deeply. First, "[f]or Alvarez, as for many of his contemporaries on the Manhattan and Radar projects, it was the war that had taught them how to run a large scientific engineering project" (Galison 1997, 350). This influence of wartime activity on scientific practice was seen in many different aspects of LBNL research. One example that Galison provides concerns leadership culture: "in structuring his staff Alvarez put Don Gow in a role that was [according to Alvarez] 'not common in physics laboratories, but … well known in military organizations' – he made him chief of staff" (Galison 1997, 350). This leadership culture proved to be an essential aspect of the particle physics groups, and the SCP in particular. Second, the radar research that Alvarez conducted during the war years had a strong influence on his experimentation style. After suggesting that Alvarez had been "'indoctrinated' with 'radar philosophy',"[13] Galison explains:

> The radar model (which Alvarez cited over and over) was above all a *system*: transmitters, receivers, switching mechanisms, and antiaircraft guns, for example, all had to be designed, built, and tested together. In just this way he believed that both institutionally and conceptually, the … problems of big bubble chambers had to be solved of a piece. And data analysis, like the other components, had to be integrated into the system as well.
>
> (Galison 1997, 373)

The SCP exemplified Alvarez's "radar model" of system-building. The group designed and built their own camera, and produced their own software. Moreover, beginning with the earliest stages of the project, they created a supernovae search protocol that aimed to integrate human activity within the automatized computer search program,[14] in addition to a similar data analysis protocol. Concerning the transformation that experimental particle physics underwent under Alvarez's leadership, Galison writes: "Berkeley had become a powerful factory of physics" (Galison 1997, 369). It would not be an exaggeration to say that what SCP did was to apply this "factory" approach to a research problem in cosmology.[15] To clarify how this approach led to an imperialistic scientific endeavor, I now want to turn to Muller and his founding of the project.

2.2 Early pioneers of the SCP: Richard Muller and his students

The Supernova Cosmology Project succeeded in measuring the acceleration of the expansion of the universe under the leadership of Saul Perlmutter, at the end of the 1990s.[16] However, Perlmutter was not the founder of the group, and in fact, the history of the SCP goes back to the early 1980s, when Perlmutter was still a graduate student at Berkeley. Before the advent of the "Perlmutter era," Richard Muller was the first principal investigator of the project, with Pennypacker as his post-doc at the Berkeley Lab.

I visited the LBNL in January 2013, with the aim of interviewing the LBNL scientists who were members of the SCP during the research that led to the discovery of dark energy. Among other things, I was interested in probing the distinct attitudes of particle physicists vis-à-vis astronomers towards research practices such as instrumentation, data collection and analysis, and evidential reasoning in general. Even though my main aim in these interviews was not geared specifically towards understanding scientific imperialism, my interviews with Muller and Pennypacker contain several exchanges that may shed important light on this question. In what follows, I plan to probe the inception and the early work of the SCP, mostly relying on the oral-history interviews that I conducted with these two scientists, hoping to highlight some crucial points concerning imperialistic breaches in the context of experimental science, in addition to clarifying the research styles of these two scientists.

Richard Muller did his PhD work under the supervision of Luis Alvarez at the University of California, Berkeley, obtaining his degree in 1969. Perhaps unsurprisingly, his scientific career exemplified many aspects of the research style of his mentor. He began as an experimental particle physicist, which he was keen on emphasizing during our interview, and then did work on experimental cosmology, paleontology, geophysics, astrophysics, and climate science. Apart from his role in the formation of the SCP, he was also the initiator and the first principal investigator of another major research endeavor: the cosmic microwave experiments at Berkeley. Muller eventually quit this project, which was later completed by George Smoot, who earned the Nobel Prize in Physics in 2006 for the discovery of the anisotropy in cosmic microwave radiation. In our interview, I asked Muller what led him to quit the cosmic microwave project. In his response, he explained to me that it was when it became clear that continuing the project required "put[ting] a satellite into space" (Muller 2014). This meant working closely with NASA (the National Aeronautics and Space Administration), and for Muller, this would be "a miserable life," given the heavy bureaucracy involved (Muller 2014). This point provides a significant clue for understanding Muller's work. His scientific interest lay mainly in conceiving a particularly interesting and promising experimental project, and then designing and constructing the required instruments, rather than following it through the intricate bureaucracies of research which, in his opinion, leave little freedom for the scientist.

On his personal website, Muller explains how he initiated the CMB work as follows:

> Astrophysics was hot, and I wanted to create a new experimental project in that field. I read several books, including Steve Weinberg's new book on General Relativity, and Jim Peeble[s]'s book, *Physical Cosmology*.[17] I finally concluded that my project would be to detect the expected cosmic microwave anisotropy.
>
> (Muller 2007)

As one sees from this autobiographical excerpt, Muller had already identified cosmology as a fertile research area for experimentation. Hence, after leaving the CMB project, he decided to move on to another project in experimental cosmology, which he saw as a new field within a discipline that was in dire need of empirical support. On the question of how he decided to embark on the supernova project, Muller further elaborated:

> I was interested in cosmology, I was knowledgeable with the issue of the Big Bang. *And I thought one of the most important things I could measure* was the deceleration of the universe[18] … And I was aware of using supernovas to do this and it struck me as a very exciting way to do it. Brand new!
>
> (Muller 2014)

It was in one particular talk, given by the Stanford theoretical physicist, Bill Wagoner, that Muller learned the basics of the measurement that could be carried out. In an e-mail exchange, I asked Muller whether he went to talks at Stanford regularly, or he specifically chose to attend Wagoner's talk. He explained to me that it was not a chance event, and that he was consciously searching for a project in cosmology. This was particularly because at the time, "experimental cosmology was a new field of physics," within which, Muller was convinced, there were "very important projects that could be done" (Muller e-mail communication, 2016). As he further explained:

> *I've always felt that my role in physics has been to explore the cracks between the disciplines, since that is often the place where great discoveries are waiting to be made.* I don't go to many talks at Stanford, but I went to Wagoner's because I had heard about his ideas, and it sounded like there might be a project there.
>
> (Muller e-mail communication, 2016, my emphasis)

When Muller mentioned to Alvarez that he was planning to measure the deceleration of the universe using supernovae, Alvarez informed him that Stirling Colgate, a theorist at the Los Alamos National Laboratory, had already attempted this measurement without getting any results. Alvarez said to Muller that he "should do Stirling Colgate's experiment," adding that "Stirling Colgate … had built an automated telescope for detecting supernovas" (Muller 2014). This information was critical for Muller for two reasons. First, as Colgate was no longer pursuing the project, there was no competition, and he shared all his knowledge with Muller. Second, Colgate's method suggested to Muller that he could carry out this project without using "big telescopes" (Muller 2014). Colgate's main idea was to build an automated telescope that would search for supernovae, though he ran into difficulties in creating the search software. As Muller was not trained in astronomy, he was not well versed in the details of observational astronomy and he "didn't know how to use big telescopes" (Muller 2014). Colgate's idea gave him a way out: by automating the project, he could avoid the problem altogether by reducing the problem of observation to the problem of writing the automation software. At this stage, Muller hired Carl Pennypacker, his first post-doc. Immediately afterwards a full-time computer programmer was hired, and two graduate students also joined the team. Initially, the group worked at the MIRA telescope at Monterey, but later on switched to the Leuschner Observatory telescope, which was then operated by the Astronomy Department of the University of California, Berkeley. The astronomy department "was happy to help," as the telescope was not frequently used due to its "bad location" (Muller 2014). Furthermore, it was already automated, so the group simply had to "come up with a list of possible galaxies that [they] would look at" (Muller 2014). However, the astronomers were suspicious of the team's efforts: "*people made fun of us, because we didn't know anything about supernovas, as if that's important*" (Muller 2014). Muller was forthcoming about his lack of detailed knowledge on supernovae, for he did not think this was a serious concern. For Muller, this research did not require him "to understand how supernovas worked," because "supernova was a tool" to be used to "measure the slow-down of the universe from the Big Bang" (Muller 2014). Yet, this attitude was not welcomed in the astronomy community, who "were kind of annoyed that [he] was not that interested in the fundamentals of the supernova" (Muller 2014).

This form of confrontation that Muller experienced with the "astronomers" continued to exist until the very end of the SCP research. It is difficult to say whether he was aware of the full implications of it at the time, but Muller was certainly attempting something entirely new for the astronomy community. As he himself said, "[n]obody had really done this before" (Muller 2014). Muller's team was aiming to collect data on astronomical objects not with the intention of learning about the objects themselves, but only to use them as "tools". This notion of a "tool" is not simply a metaphor, for the knowledge that Muller was seeking actually depended on making the supernova data *useful* in the calculation of the deceleration parameter. Even though many astronomers doubted that supernovae are standard candles, Muller thought that this was not an issue, as he believed that "to the extent that they are not, we'll find some way to calibrate them, there'll be some signature that they have" (Muller 2014). When astronomers asked Muller how he planned to calibrate them, he quoted a line from his mentor, Alvarez: "I don't know but I will figure it out" (Muller 2014). According to Muller, astronomers did not like this answer; they did not "want you to go ahead with a project before you figure out every last bit of detail" (Muller 2014). In contrast with this attitude, Muller was confident that even if at the beginning of the project the calibration issue was not solved, a solution would eventually appear. For Muller,

> when you are doing experimental physics nobody's ever done before, there is no way you can solve every problem ahead of time. You have to know what kind of problems you can solve, what kind of problems you can't solve.
>
> (Muller 2014)

This notion of attempting a measurement that had not been done before was Muller's main line of defense, and justification of the epistemic and disciplinary transgression in which he was engaged. From Muller's perspective, astronomers were defending an epistemic territory that was too important to be left to their jurisdiction. It was to be explored, and if the astronomers did not engage in this sort of exploration before, it was not their place to criticize an outsider who saw this need and acted on it. And in this sort of exploratory work, it is necessarily the case that not everything can be known beforehand. As he put it, "you can never solve all the problems ahead of time; if you did, then someone else would have done it. And you don't know which problems are gonna be easy and which ones are gonna be hard" (Muller 2014).

Muller was experienced in this sort of confrontation, for he went through a similar situation in his CMB research as well. During his work on the microwave measurements, Muller needed a design that would keep two receivers at uniform temperature. Even though he did not have any specific background in the subject, he "knew [he] could learn how heat flow worked" (Muller 2014). His particular design was criticized, and he was accused of not knowing "anything about cosmology" (Muller 2014). This was especially a problem when he applied for funding:

> *It seems they don't wanna give money to anybody who didn't get their PhD thesis in the field.* Then you're OK. But we're always outsiders. *Everything innovative, you always come as an outsider.*
>
> (Muller 2014)

For an experimentalist such as Muller, the challenge was identifying the most important measurement to make, and developing the necessary instrumentation that would work best. Developing the data analysis tools, or acquiring the needed theoretical knowledge, was secondary to this. The experimentalist does not seem to be bound by the traditional dividing lines between disciplines. When I asked him whether his early work involved any astronomy or astrophysics, his answer was:

> MULLER: There was no astrophysics in my PhD, it was in particle physics, elementary particles, that's what my PhD was in. Soon as I got out, I was developing instrumentation, I was doing medical instrumentation. My first real project, first great project was an astrophysics project based on microwaves. I had no professional experience on microwaves whatsoever ... just didn't look like that hard a field to master. Cosmology? Never done ... published a paper in cosmology, OK? *But it was the most important thing to do, it was an exciting field.* And likewise for the supernovas. It was a cosmology thing, I published then in cosmology, but of course I knew nothing about supernovas but that's ... come on, it's not that hard to learn about supernovas [...]
>
> And then there are certain things you can answer, like "how you gonna calibrate them?," we'll figure out a way, believe me, look I've done this a dozen times, it's just not that hard a problem. The people who solved it of course think it was their genius, and they deserve a Nobel Prize for it. (Muller 2014)

Hence, Muller criticized the astronomers who, from his perspective, presented a straightforward aspect of experimental science as a special kind of success story.

After the formation period of the SCP ended and Saul Perlmutter became the principal investigator, several astronomers joined the group. These were mostly from a younger generation of astronomers, who were well versed in statistical analysis techniques that the group now needed. Muller mentioned this fact during our conversation before I brought it up, crediting Perlmutter for bringing in the astronomers. When I next inquired how astronomers joined the group, and what the relationship between the physicists and the astronomers in the group was like, he first volunteered his definition of astrophysics as "*astronomy done by physicists.*" Continuing in this vein, Muller explained:

> MULLER: Astrophysics was when people in the physics departments decided they were gonna do it and pretty soon the graduates from the physics department would be hired by the astronomy department and astronomy became astrophysics. But really the astronomers used to measure ... get photographic plates, and they would sometimes have a spectrometer that was built for them by somebody else. And that was astronomy when ... that was the astronomy department as I remember it. Astronomers could do great things but they didn't build new instruments. But when the physicists got in, they always wanted to build something new. And eventually the astronomy departments switched over to become astrophysics. But what was your question?

GURALP: How did the astronomers join? Why did Saul bring them?
MULLER: The original reason was that we had to get time on the telescopes.
(Muller 2014)

From Muller's perspective, the joining of astronomers was not strictly epistemic but was (at least, partially) related to the access that was needed by the group to collect data in big telescopes. He also acknowledged how astronomers were critical of their group's work. One such astronomer was Robert Kirshner, who is one of the prominent supernova experts in the world:

MULLER: And Kirshner ... I mean yeah he was the expert on supernovas. He was the world's expert on supernovas. And so he could identify all the problems we should have to solve along the way, and if we didn't have solutions, he'd say "cut their funding," which he did. He recommended cutting the funding. And he never would have gotten involved in the discovery, if he had been successful cutting our funding.
(Muller 2014)

Without the institutional support of the LBNL, a project such as SCP could not have been possible. When the SCP did not show satisfactory progress, an external committee was appointed to review the group's work, chaired by Kirshner. His critical review of the SCP led to a funding crisis for the group, which was eventually resolved within the LBNL.

At the end of the interview, I asked Muller about his general views on scientific knowledge, and in particular, confirmation of experimental results. To my surprise, he likened confirmation theory to *accounting*, and argued that the key distinguishing factor of valuable research was its *exploratory* character. Continuing with the analogy of scientist as *explorer*, he made some significant remarks that, with their anthropological undertones, bear directly on the issue of scientific imperialism:

MULLER: So I see science as more akin to exploration. If you imagine the explorer ... and I told you the story about the early days of the supernova project. Put this in ... as an explorer. Here I am an explorer.
GURALP: Yes.
MULLER: And what am I doing? I am crossing the sea, to get to some island.
GURALP: Yeah.
MULLER: Well it's a supernova island. Oh, what are you going to do if you ran out of water? I don't know, I'll find some somewhere. What are you going to do if your ship has leaks? I'll find some wood. Well, what kind of trees are there in this new land? I don't know! I'll find something that will work! How are you gonna keep your crew happy? Well, leave that to me ... and you get over there, are you welcomed? *No, the natives throw spears at you. That's all the astronomers.*

> [Laughter.] *The natives attacked us. So you're exploring this land and you're doing it in a new way.* And you don't get support of the natives. And you're running out of resources. Twice our project got canceled. What does it take to have a successful project? You had to be so resilient, that even when the head of the division tells you your project is canceled, you keep at it.
>
> (Muller 2014)

In his metaphoric illustration of the disciplinary breach of the physicists in terms of *natives versus explorers*, Muller gave a vivid description of physics as a *colonial* science.

Pennypacker's experience with the astronomers was not different. In our interview, he first suggested the notion of *disruptive innovation* to characterize the work of the Berkeley group in the field of cosmology: "I think about a model, you see innovation is very hard to do, and I think *disruptive innovation* is kind of what happened here in some ways. We were outsiders, and there are all these fabulous, game-changing, extremely appropriate technologies coming along" (Pennypacker 2015). For Pennypacker, *innovation* was key to research, a point Muller insisted on as well. Through *innovation*, Muller and Pennypacker believed, they could enter new fields and succeed in producing significant results. This led to a vivid *insiders versus outsiders* distinction in Pennypacker's mind. That they were not being welcomed by the "supernova community" was clear to him:

> In some ways the other group were clearly deep insiders … Kirshner was widely regarded as Mr. Supernova. I eagerly sought his counsel, wanted to talk to him at a conference in fact, concerning our preliminary plans, as physicists, to do the supernova search. *He just, he was clearly annoyed by me, blew me away, and more or less insulted me.*
>
> (Pennypacker 2015)

As confident physicists equipped with new technology, they were perceived by the astronomers as invading a territory without understanding its subtleties.

By 1992, Saul Perlmutter took over the leadership of the SCP, and recruited international particle physicists as collaborators, in addition to newly hired astronomers. He eventually solved the calibration problem with what came to be known as the "stretch method," which carried the mark of a particle physicist's formalistic approach. By specifying a typical supernova to be the "standard" one, he parameterized any given supernova light-curve by measuring how much it needs to be "stretched" along the time axis. Although relatively simple and unappealing from an astronomer's perspective, this method worked well for the calculations, and the level of precision needed. For Perlmutter, as a physicist, the proof of the pudding was in the eating.

The resistance between the astronomers and the particle physicists continued to exist within the group as well, regarding both data collection and analysis procedures. Due to space limitations of this chapter, I cannot go into this later period of the SCP. However, the above testimonies of the two pioneering scientists of the Berkeley group give a fairly complete picture of the physicists' perspective.

3 Astronomers' resistance and the formation of the High-*z*

The Berkeley search that eventually culminated in the SCP began in 1980. During this time, the Berkeley physicists came into contact with the astronomical community, and the reception was hostile. However, the astronomers did not immediately form their own group to compete with the intruders, and the High-*z* Supernova Research Team (High-*z*) was officially formed only in 1994.[19] There are several structural reasons for this fact, and here I would like to discuss them briefly, to the extent that this will be important in my discussion of scientific imperialism in experimental contexts.

In section one, above, I referred to the two research traditions, and contrasted the search for dark matter of the particle physicists, to the Hubble-Sandage program in astronomy, whose fundamental goal was to measure the Hubble constant (H_0). Due to the immense influence of Sandage on the whole field of astronomy, and the infamous *Hubble Wars* debate between him and Gerald de Vaucouleurs on the true value of H_0, the astronomy community became convinced that without a satisfactory measurement of this constant, our understanding of the distance scale of the universe could never be precise.[20] The two members of the High-*z* team who were awarded the Nobel Prize for the discovery of the accelerating universe, namely Adam Riess and Brian Schmidt, both wrote PhD dissertations under Robert Kirshner, on using supernovae as distance indicators to measure the Hubble constant.

In my interview with Robert Kirshner, I asked him how the High-*z* team came into being, and what his views were on the work of the "other" team. After explaining the background history of the Hubble constant measurements, and his work with Schmidt and Riess, he continued:

KIRSHNER: So at that point, *we kinda had the tools, and* … Brian who had finished [his PhD] and who was a post-doc, went down to visit at Cerro-Tololo,[21] and he came back and said, you know, *"we could …" because the other guys had been measuring* …

GURALP: So you knew about [the other team's work]?

KIRSHNER: Oh, I knew a lot about it because I was on the visiting committee, or whatever it was called, for the Center for Particle Astrophysics at Berkeley …Whenever they came to give … the annual report, or whatever it was, I'd go to the committee meeting, hear what they had to say. So I was very well informed about what they were doing […] They asked me several times to join what they were doing. I really did not want to.

GURALP: Why?

KIRSHNER: *I didn't think they were competent guys*, even though … you know, they're pretty good.

(Kirshner 2014)

There are two important points that this exchange evidences. The first is that what prompted Schmidt to attempt measuring distant supernovae was that by the early 1990s, the astronomical community developed "tools" to tackle this measurement, such as search software, statistical methods, and CCD (charged-couple device) cameras available at big telescopes. In the case of the SCP, these were all produced within the group. In addition to this, the Harvard astronomers' close collaboration with the CTIO group meant that they would have access to their data, in addition to the telescope. Second, this exchange also confirms Muller's views concerning how astronomers viewed the particle physicists: Kirshner did not think the particle physicists were competent to do this measurement. As he further explained:

> KIRSHNER: The other guys were not from the astronomical world [...] They just ... they didn't even wanna know about the interstellar dust, this stuff. They were not so interested in sub-dividing types of supernovae [...] We knew a lot more of the astronomy, about what dust did, types of supernovae, and stuff. And they weren't ... they didn't pay that much attention. They had focused on the technique for finding the supernovae in big digital arrays, which was very commendable. And they built their own software from scratch. Well it turned out there is a whole set of astronomical software out there.
>
> (Kirshner 2014)

The last sentence in the above excerpt connects to the point made above. For almost a decade, astronomers watched the particle physicists' work on supernovae measurements, and lacking the institutional and technological means, they could only resent the situation. However, by the mid-1990s, the astronomical community caught up. By that time, both open software and large CCD cameras became available to the community, to be employed for specific purposes. The High-*z* was officially formed in 1994, with the joint initiative of Nick Suntzeff from CTIO, and Brian Schmidt. The group was a collaboration between astronomers based in CTIO, Harvard, and the European Southern Observatory (ESO).[22] Building on their earlier work, and their general expertise in fields such as supernovae search and photometry, they were able to publish their acceleration result at the end of 1998.

In contrast to SCP, the High-*z* team employed two complementary methods of data analysis, based on two different methods of calibration, to ensure the validity of their conclusions. One of these methods was developed by the CTIO astronomers during the Calan/Tololo survey, and the other was Adam Riess' dissertation work at Harvard. As their data set was smaller compared to that of the SCP, the High-*z* argued that the consistency of their two different fitting methods (SCP used only one), made their result as reliable as that of their competitors.

4 Evidential cultures and scientific imperialism

In this section, I aim to expand Mäki's notion of imperialism of scope by bringing his treatment of scientific imperialism into dialogue with Collins' notion of *evidential culture* (Collins 2004, 2014). The general framework that Mäki proposes consists of three main elements, all of which can be identified within the case at hand. However, the framework is laid out mainly in the context of theoretical science, and in order to employ it to cover cases of *experimentation* as well, some modifications are needed.[23] These modifications can be effected, as I aim to show, following Collins' work, who builds on an early study by Pinch (1985). In this section, I will first introduce the concepts of *evidential significance* and *evidential culture*, following Collins and Pinch. I will then pass on to a discussion of Mäki's tripartite model and then discuss how it can be applied to the case at hand, once one introduces the concept of evidential significance to this model.

4.1 Collins and Pinch on evidential cultures

Collins proposes the concept of *evidential culture* in his study of the history of gravitational wave measurements (Collins 2004, 2014). This concept builds on an earlier work by Pinch (1985). The key concept that Pinch employs, and which is later generalized by Collins, is that of *evidential significance*.

Pinch begins his analysis by emphasizing that observations in science are undertaken for specific purposes such as testing a theory, or reproducing a previous result in order to confirm it. In other words, observations are always conducted within specific evidential *contexts*. According to Pinch, "[t]he evidential context can include such things as a body of knowledge, a theory, law or hypothesis, a classificatory scheme or one class of objects, a set of observations or one observation" (Pinch 1985, 10). On this basis, "[t]he relationship between the observational report and the evidential context defines the evidential significance of the observation" (Pinch 1985, 10). In other words, how significant an observation is, depends on the context within which it is interpreted. As Pinch observes, "*observational reports may take on significance in a variety of different evidential contexts*" (Pinch 1985, 10). That is to say, scientists can attribute different levels of significance to their experimental results by placing their observations in distinct evidential contexts. Pinch elucidates this idea with the example of the solar neutrino measurements. When a solar neutrino experiment produced negative results (that is, no neutrinos were detected), there were two evidential contexts in which it was interpreted:

> One context ... was a test of the pp-chain (and the significance of the results was that they indicated that the expected neutrinos were not present). However, the results were also put in the context of a test of another cycle of nuclear reactions that can occur in stars – the CNO-cycle. The significance of the results in this context was that they definitely ruled out the CNO-cycle in the Sun.
>
> (Pinch 1985, 10)

This point that Pinch makes is that the same observation can assume different levels of significance within different evidential contexts. This is readily applicable to the case of dark energy. For the astronomers, the evidential context of their measurement of supernova brightnesses was the determination of the Hubble constant, and the extra-galactic distance scale. However, for the physicists, the evidential context was finding out the amount of dark matter in the universe.

Pinch introduces the notion of "degree of externality" to *specify* the significance of an observation within an evidential context (Pinch 1985, 9). Scientists can report an observation at different levels of abstraction, or *externality*. To use a hypothetical example that Pinch provides, at the end of an experiment, the same result can be reported with "progressively increased externality" as follows:

a Splodges on a graph were observed.
b Ar^{37} Atoms were observed.
c Solar neutrinos were observed. (Pinch 1985, 9)

According to Pinch, as one moves from (a) to (c), the externality of the observational reports increases in the sense that elements external to direct sense experience are included in these reports. These elements may include the role of detectors, statistical analysis, and theoretical interpretations. As one reaches a higher level of externality, as in (c), the evidential context becomes more specific, and this makes the report more significant. Thus, as Collins explains, the "directly" observable splodges:

> can be said to be evidence of the existence of radioactive argon, solar neutrinos, events taking place in the sun, or whatever. This is an increasing scale of *evidential significance*. The higher the evidential significance, that is, the longer the chain of inference, *the more important the findings*.
>
> (Collins 2004, 402, my emphasis)

Before reporting an experimental result, scientific teams carefully deliberate on the level of externality at which they will state their findings. This deliberation needs to take into consideration several factors, such as the strength of the statistical analysis, and confidence in the absence of error. In the case of dark energy, the following possible observation statements were debated:

i High-redshift supernovae look dimmer than expected (in a decelerating universe).
ii High-redshift supernovae dimming is kinematical (ruling-out of light extinction by dust).
iii The expansion of the universe is observed to be accelerating (ruling-out of a local gravitational void).
iv A non-zero cosmological constant is discovered (assuming a standard model).

Similar to the example above, the significance of the result increases from (i) to (iv). I will discuss how the two teams reported on their measurements, and how these decisions were made, in the context of my discussion of scientific imperialism below.

Collins generalizes the notion of *evidential significance* into that of an *evidential culture*, by adding two extra dimensions, namely, *evidential collectivism* and *evidential threshold. Evidential collectivism* refers to how the scientists, or scientific teams, manage the process of confirmation of a result. Evidential collectivists "believe that it is the job of the scientific collective – the whole network of other individuals and laboratories – to join the process of assessment of results from an early stage" (Collins 2004, 398). As opposed to this, *evidential individualists* "believe nothing should be published until the result is certain; the development and debate that takes place prior to the public 'announcement' should be conducted behind closed doors" (Collins 2014, 201). Finally, *evidential threshold* denotes the statistical limit that is required for a result to be accepted as a *discovery*. A well-known threshold in particle physics demands that for an experiment to report a *discovery*, it should be at least a 5 sigma result.

All these elements of evidential culture played a role in the formation of both supernova search teams, as well as the announcement, and the acceptance of the results.[24] First, the notion of evidential significance was the impetus for the beginning of the Berkeley search. As we have seen above, Muller's decision to embark on the supernova research was directly related to the significance that he attributed to the measurement. As he says, "I thought one of the most important things I could measure was the deceleration of the universe". As at this point the existence of dark energy was unknown, for Muller measuring the deceleration meant measuring the amount of dark matter in the universe, i.e., "weighing the universe". Muller did not possess the astronomer's knowledge on supernovae. Yet, the supernova models constructed by physicists (such as Wagoner) convinced him that they could be treated as identical objects. This convinced him not only that this measurement could be done, but also that it would be possible to make the transition from (i) to (iii) above. For astronomers, the possibility of the argument from (i) to (iii) required ruling out dimming of supernova light due to dust, and various other errors. If these could not be ruled out, the significance of the measurement would be much less, for even (ii) could not be convincingly established. The significance of the measurement that the Berkeley group was attempting was a matter of contention between the physicists and astronomers from the very beginning.

The contrast of evidential collectivism versus individualism created a rift between these two evidential cultures as well. The SCP exemplified several aspects of evidential individualism quite strongly. I will mention one example concerning how they treated their "data points". One aspect of individualism of a particle physics group is that the data are kept confidential until they are fully analyzed by the scientific team, and a publication is submitted. This is not only to prevent "outsiders" from gaining credit by analyzing the data and claiming a discovery. Rather, as the team knows the work of the detector well, they have an "experiential" knowledge of which parts of data that might otherwise be interpreted as a new particle, actually are due to a glitch in the equipment that only a person who is familiar with the hardware could identify. In a similar fashion, the SCP wanted to keep their data confidential, before making it part of an official publication. On several occasions, however, they needed outside help to confirm an object as a supernova, and this brought them in close contact with the astronomers. In an oral history interview, Brian Schmidt describes one occasion as follows:

SCHMIDT: They [the SCP group] called Bob Kirshner up and had him observe an object [...] When, Bob went to it with the MMT [Multiple Mirror Telescope in Arizona], and I believe Adam Riess was out there, as was Peter Challis, he saw something, took a spectrum. I got the fax in the morning. Pete had already looked at it. I played around with it as well. We were all convinced it was a Type 1A Supernova a few days after maximum, at a redshift of 0.42. That was the first distant object. We were trying to get them to put it into the IAU, the International Astronomical Union circulars. I was working very hard with Bob to convince them. *They were just refusing.*

PAVLISH: They did not believe you?

SCHMIDT: No, no. They believed us. It was just not the way they did things. They were not astronomers.

(Pavlish 2007, my emphasis)

In the words of Kirshner, "they believed that they owned the object," and that this was a "cultural mismatch" between the astronomers and the particle physicists.[25] Whereas in astronomy one is encouraged to announce a discovery as soon as possible, so that others can also observe it (*high collectivism*), particle physicists are evidential individualists, in the sense that announcements come only after a full confirmation within the group is made.

4.2 Mäki on scientific imperialism

In a series of publications, Mäki studied scientific imperialism, and the possibility of giving a conceptual account of it. Recognizing that scientific imperialism is a "complex and fluid phenomenon," and thus "no single compact definition [of it] can be given," he proceeded via introducing several distinctions to circumvent this difficulty (Mäki 2013, 333). In this section, I will first provide a quick summary of his account, and argue that it is based on a *theory-dominated* view of scientific knowledge. Next, based on my case study, I will propose a modification to this account, so that it is applicable in the empirical context as well. I will finish by arguing that the discovery of dark energy is a positive case of scientific imperialism, for it led to a *robust* confirmation of the result by two competing research teams.

A key distinction that Mäki makes is the one between *domain-only imperialism (DOI) versus disciplinary imperialism (DI)*. In DOI, one discipline extends its *scope* into a domain that is traditionally considered as belonging to another discipline. The mechanism of this extension is that of *explanatory unification*. As Mäki writes in the context of economics imperialism, "[e]conomics expansionism is a matter of a *persistent pursuit to increase the degree of unification provided by an economic theory* by way of applying it to new types of phenomena" (Mäki 2009, 359). Here, it is important to note that DOI simply consists of scope extension: "What is involved in domain-only imperialism is just the … imperialism of scope" (Mäki and Marchionni 2011, 653). The full definition of *imperialism of scope* is given as follows:

> **Imperialism of scope**. An expansionist discipline seeks to explain phenomena that belong to the perceived domain of another discipline. This is the pursuit of explanatory unification that is disrespectful for disciplinary boundaries.
>
> (Mäki 2013, 334)

DOI may trigger DI if the expanding discipline exports, or imposes, its *styles* of research on to the host discipline:

> If a given style of doing research is deemed scientifically superior, what was *initially an application of a theory to a new domain of phenomena* can turn into a widespread commitment to this allegedly superior style, also within the target discipline.
>
> (Mäki and Marchionni 2011, 658, my emphasis)

Mäki calls this *imperialism of style*. DOI may also lead to another form of scientific imperialism, namely, *imperialism of standing*: if "a high-status discipline starts *explaining phenomena* that used to belong to the domain of a lower-status discipline, this may gradually have consequences for the customary practices and even survival of the lower-status discipline" (ibid., 657, my emphasis). Here, we see that both forms of DI (style and standing) contain the DOI (scope) in their core, for they require an initial act of explanation. Thus, to the extent that the defining mechanism of DOI is scope-extension through explanatory unification, which is essentially a *theoretical* exercise, Mäki's account can be seen to be a theory-dominated one. Below, I will argue that by modifying the imperialism of scope to include experimental contexts as well, one can apply it to the case of dark energy.

During the discovery of dark energy, all aspects of scientific imperialism that Mäki mentions – namely, *scope, style*, and *standing* – played salient roles. However, in the context of experimental practice, explanatory unification did not seem to constitute a major motivation. Neither in their publications, nor during the interviews that I conducted with them, did the originators of the Berkeley group mention the notion of extending the explanatory power of a cosmological theory. On the other hand, the notion of *evidential significance* appears to constitute the mechanism of their imperialistic breach, and it can be seen as an experimental counterpart to explanatory unification. As I explained above, the evidential significance of a given experiment is related to its having the feasibility to provide epistemically specific results. In other words, if an experiment with a high degree of externality is feasible, it would have evidential significance. As Muller and Pennypacker pointed out in the interviews, the Berkeley scientists operated within the culture of "attempting the impossible". Muller's characterization of this was quite straightforward. Anything that can be done, he argued, would already have been done. So groundbreaking research should "attempt the impossible". Still, there must be some indication for the "imperialist" experimenter that the technology available at the time could be sufficient to accomplish the task. Hence, the proposed experiment must lie within the limits of feasibility. We can then state the *evidential significance* condition for an imperialistic breach to occur as follows:

> **Evidential Significance Condition (EC):** The experiment should have high evidential significance, namely, it must have the *feasibility* to yield important results. This means that the experiment should be capable of engendering a long inferential chain of significance within the material and epistemic constraints that the imperialistic science is bounded by.

Even though the attempted experiment lies at the *limits of feasibility* for the host discipline (and hence is quite difficult, and risky[26] to conduct), it would yield large epistemic benefits if successfully executed. One benefit can be that of providing evidence for a particularly fundamental theory of the science in question; another can be the refutation of an empirical claim to which most practitioners are committed, or the measurement of a quantity that could put constraints on a family of favored models. An experimental practice becomes expansionist when the measurement in question is *perceived to be* feasible by the invading discipline. This expansionism does not seem to be motivated by any form of empirical unification, but rather that the measurement in question is worth doing, i.e., it is *evidentially significant.* We can thus modify the imperialism of scope in the experimental context as follows:

> **Imperialism of Scope (Experimental):** An expansionist experimental practice presupposes the evidential significance criterion, i.e., it aims towards important measurements that are feasible. This is the pursuit of *evidential significance* which is disrespectful of disciplinary boundaries.

The particle physicists' embarkation on and completion of the supernova project for a measurement of Ω constitutes an instance of imperialism of scope. They pursued an experimental project outside the area in which they were trained and were acculturated as scientists. This was motivated by a belief that this was an important measurement to make and that their technical and institutional capacities made it feasible.

Nevertheless, this imperialistic breach led to a positive outcome. After a period of resistance to the particle physicists' invasion of their experimental territory, astronomers formed their own group to attempt to conduct the same measurement, relying mostly on what I termed their *communitarian* resources. These included open software for search algorithms for supernovae, and recently available CCD cameras on large telescopes. As we have seen in the Schmidt and Riess interviews, even though the astronomy group started much later than the SCP, they could keep up with them by relying on the epistemic resources of the astronomy community.

The ability of the astronomers to form their own group, apply their own calibration scheme, and obtain an independent measurement result, led to a higher confidence in the accelerating universe. For the cosmology community in general, that two competing experiments obtained the same result increased its *robustness*. This played a considerable role in the quick acceptance of the result.

Finally, when we compare the main articles of the two groups in which the result of the accelerating universe was announced, we see that they employed similar strategies to achieve high evidential significance (Riess et al. 1998; Perlmutter et al. 1999). Both teams identified several sources of error, such as data selection bias, or effects of dust, and provided statistical arguments backed up by physical reasons why these possible systematic effects do not pull the significance of the result down. In this way, they aimed to convince their readers that the supernova research, which was conceived from the very beginning as a high-significance area, was capable of delivering on its promises.

5 Conclusion

The discovery of dark energy constitutes one of the most important developments in recent cosmology. The history of this discovery contains several valuable lessons for understanding interdisciplinary relations in modern science, and scientific imperialism in particular. In this chapter, I singled out two. First, I showed that scientific imperialism can occur in experimental contexts as well as theoretical, and that any account of scientific imperialism should therefore include experimental science. On this basis, I have offered an extension of Mäki's account, by modifying his notion of *imperialism of scope* so that it is also applicable in experimental contexts. Second, I argued that the case of the accelerating universe exemplifies a positive case of imperialism. The imperialistic breach of the particle physicists prompted astronomers to form their own team for finding and employing distant supernovae for cosmological purposes. Relying on the local resources of the astronomical community, the High-z was able to compete successfully with the SCP, and this competition led to a *robust* confirmation of the existence of dark energy in our universe. This gives us reason to opt for a neutral definition of imperialism, at least when it occurs in science.

Notes

1 In his comprehensive history of 20th-century physics, *Quantum Generations*, Helge Kragh writes that "there is a sense in which physics can be said to be the most fundamental and general of all the sciences," adding that this "'imperialist' point of view is far from new, but it is only in this century that it has been substantiated and has become more than an article of faith and self-congratulation" (Kragh 2002, 447). For a critical take on the imperialistic tendencies of physics from a philosophy of science perspective, see Cartwright (1999).

2 A central aim of this volume is to make headway on this very question of definition. See the editors' Introduction for a detailed discussion of the issues involved.

3 See Dupré (1995, 2001).

4 I should note that Mäki's usage of "scientific imperialism" as a *descriptive* term is contested by Steven Clarke and Adrian Walsh, who propose to treat it as a *normative* term, referring to a disciplinary overreach that should be strictly opposed (Clarke and Walsh 2013). Due to space limitations, I will not address this issue any further in this chapter.

5 Hubble worked with the 100-inch Hooker telescope at the Mt Wilson Observatory, which was the largest telescope in the world from its completion in 1917 until 1948. In 1948, the 200-inch Hale telescope on the Palomar Mountain in California was built, which was the largest optical telescope from 1948 until the mid-1970s. Hubble worked on this new instrument until he passed away in 1953.

6 Although a full treatment of the socio-epistemic consequences of this fact is beyond the limits of this chapter, a few words must be said in order to clarify its importance for my purposes. The career of a modern observational astronomer depends on obtaining observation time on large telescopes. As these are expensive machines to build, telescopes capable of cutting-edge research have always been few in number. On the other hand, and in contradistinction to the case of modern particle physics, early career astronomers are encouraged, if not required, to publish single-author papers to establish their autonomy as researchers. These two factors together resulted in astronomical research being conducted within a close-knit community, where access to telescopes, or research programs, require a community membership, and abiding by its norms.

7 For a classical statement of this model, see Bondi and Gold (1948).

8 "Deceleration parameter" (q_0) is a quantity that measures how the expansion of the universe evolves with time. Before the discovery of dark energy, q_0 was taken to be a positive number, as the gravitational attraction of matter would have slowed down the expansion of the universe.

9 Alvarez, who is considered to be one of the greatest experimentalists in the 20th century, was awarded the Nobel Prize in Physics in 1968.

10 Personal communication.

11 Perlmutter was Muller's doctoral student at Berkeley physics department during the years 1981–1986.

12 I use the term "scientific style" in a very general sense to refer to all kinds of research practices, such as organization of collaborations, technical and analytical preferences, as well as goals of inquiry.

13 Galison 1997, 341.

14 Within the LBNL, there were debates on how the *human* versus *machine* components of the measurement process should be organized. According to Alvarez's "interactionist" picture, even though the "[m]achines would *aid* the human," it was nonetheless necessary to accept that "human beings, by virtue of their peculiar capacities, had to remain central to the processing …" (Galison 1997, 371). The SCP adopted this interactionist approach early on, and abided by it throughout its research program.

15 Note that the notion of "factory" in the context of SCP has become an actor's category. A project that was started by the SCP group right after the 1999 announcement of the acceleration of the universe, with the aim of studying low-redshift supernovae, was called the *Nearby Supernova Factory*.
16 Perlmutter received the Nobel Prize in 2011 for this work, sharing it with two members of the competing High-z team of astronomers.
17 Peebles 1971.
18 All emphases in the oral-history transcriptions are mine.
19 In the acronym High-z, the letter "z" symbolizes the quantity known as the *redshift*. The measurement of redshift enables scientists to infer the velocity of distant astronomical objects.
20 The Hubble Wars was a debate on the value of the Hubble constant. For an in-depth analysis of the history and the resolution of this debate, see Guralp (2016).
21 By "Cerro Tololo," Kirshner refers to the Cerro Tololo Inter-American Observatory in Chile (CTIO). There were several astronomers at the CTIO who worked on supernovae, and the Harvard group collaborated with them on observations and data sharing. Schmidt visited CTIO several times, beginning in 1991.
22 In 1994, Brian Schmidt's post-doc at Harvard ended, and he took up another post-doc position at The Australian National University. The scientists joining from ESO were Bruno Leibundgut, a former post-doc of Bob Kirshner, and Jason Spyromilio.
23 Mäki does not explicitly say that his account is only geared for theoretical contexts. I will clarify this claim in section 4.2.
24 Here I will focus on mainly evidential significance, and evidential collectivism, as the question of threshold was not a contentious issue in this episode.
25 Oral history interview with the author.
26 "Riskiness" relates to the career of the scientist, and not to the physical risks of experimental science.

References

Bondi, Hermann and Thomas Gold. 1948. "The Steady-State Theory of the Expanding Universe". *Monthly Notices of the Royal Astronomical Society* 108: 252–270.

Cartwright, Nancy. 1999. *The Dappled World: A Study of the Boundaries of Science*. Cambridge: Cambridge University Press.

Clarke, Steve and Adrian Walsh. 2013. "Imperialism, Progress, Developmental Teleology, and Interdisciplinary Unification". *International Studies in the Philosophy of Science* 27(3): 341–351.

Collins, Harry. 2004. *Gravity's Shadow: The Search for Gravitational Waves*. Chicago, IL: University of Chicago Press.

Collins, Harry. 2014. *Gravity's Ghost and Big Dog: Scientific Discovery and Social Analysis in the Twenty-First Century*. Chicago, IL: University of Chicago Press.

Dupré, John. 1995. "Against Scientific Imperialism". In M. Forbes, D. Hull and R.M. Burian (eds). *PSA 1994: Proceedings of the 1994 Biennial Meeting of the Philosophy of Science Association* 2. East Lansing, MI: Philos Sci Association, pp. 374–381.

Dupré, John. 2001. *Human Nature and the Limits of Science*. Oxford: Oxford University Press.

Galison, Peter. 1997. *Image and Logic: A Material Culture of Microphysics*. Chicago, IL: University of Chicago Press.

Guralp, Genco. 2016. "Calibrating the Universe: Beginning and End of the Hubble Wars". In O. Schlaudt and L. Huber (eds). *Standardization in Measurement: Philosophical, Historical and Sociological Issues*. London and New York: Routledge, pp. 125–138.

Kirshner, Robert. 2014. Oral-history interview with the author. January 9.
Kragh, Helge. 2002. *Quantum Generations: A History of Physics in the Twentieth Century*. Princeton, NJ :Princeton University Press.
Mäki, Uskali. 2009. "Economics Imperialism: Concept and Constraints". *Philosophy of the Social Sciences* 39(3): 351–380.
Mäki, Uskali. 2013. "Scientific Imperialism: Difficulties in Definition, Identification, and Assessment". *International Studies in the Philosophy of Science* 27(3): 325–339.
Mäki, Uskali and Caterina Marchionni. 2011. "Is Geographical Economics Imperializing Economic Geography?" *Journal of Economic Geography* 11(4): 645–665.
Muller, Richard. 2007. "Pre-history of Berkeley's Role in COBE – the Cosmic Background Explorer Satellite". http://muller.lbl.gov/COBE-early_history/preCOBEhistory.html (accessed March 2016).
Muller, Richard. 2014. Oral-history interview with the author. January 22.
Pavlish, Ursula. 2007. Oral-history interview with Brian Schmidt. July 24, 25. Niels Bohr Library and Archives, American Institute of Physics, College Park, MD. www.aip.org/history-programs/niels-bohr-library/oral-histories/33746-1 (accessed April 2016).
Peebles, Phillip J.E. 1971. *Physical Cosmology*. Princeton, NJ: Princeton University Press.
Pennypacker, Carlton. 2015. Oral-history interview with the author. May 6.
Perlmutter, Saul et al. 1999. "Measurements of Ω and Λ from 42 High-redshift Supernovae". *The Astrophysical Journal* 517(2): 565–586.
Pinch, Trevor. 1985. "Towards an Analysis of Scientific Observation: The Externality and Evidential Significance of Observational Reports in Physics". *Social Studies of Science* 15(1): 3–36.
Riess, Adam G. et al. 1998. "Observational Evidence from Supernovae for an Accelerating Universe and a Cosmological Constant". *The Astronomical Journal* 116(3): 1009–1038.

Index

Note: Page numbers in *italic* indicate a figure and page numbers in **bold** indicate a table on the corresponding page.